82 Advances in Polymer Science

Polymer Physics

With Contributions by
F. Boué, Y. Hori, M. Kamachi,
H. Kashiwabara, Y. Osada, D. Rigby,
R.-J. Roe, M. Sakaguchi, S. Shimada

With 197 Figures and 21 Tables

Springer-Verlag
Berlin Heidelberg GmbH

ISBN 978-3-662-15176-1   ISBN 978-3-540-47817-1 (eBook)
DOI 10.1007/978-3-540-47817-1

Library of Congress Catalog Card Number 61-642

© Springer-Verlag Berlin Heidelberg 1987
Originally published by Springer-Verlag Berlin Heidelberg New York in 1987
Softcover reprint of the hardcover 1st edition 1987

2154/3020-543210

Editors

Table of Contents

Conversion of Chemical Into Mechanical Energy by Synthetic Polymers (Chemomechanical Systems)

Yoshihito Osada
Department of Chemistry, Ibaraki University, Mito 310, Japan

The isothermal conversion of chemical energy into mechanical work underlies the motility of living organisms. Chemomechanical systems based on synthetic polymers are the only artificial systems able to convert chemical energy directly into mechanical work. They may have potential uses where power supply is limited (e.g. under water, in space). They are considered as "transducers" or "receptors", whereby semimicroscopic deformation or strain plays an essential role for displaying functions. Several mechano-chemical systems have been investigated during the past three decades. This article describes the principles, fundamental behaviors and potential applications of these systems. Recently developed electro-activated chemomechanical systems using water-swollen polymeric gels will also be described.

Advances in Polymer Science 82
© Springer-Verlag Berlin Heidelberg 1987

1 Introduction

A variety of systems converting energy from one form to another has been developed up to now; energy conversion from thermal to mechanical (for example internal combustion engine), from electrical to mechanical (motor), from photo to electrical (solar battery), and vice versa. Some examples are shown in Table 1. Similar, but usually more efficient, energy conversion systems can be found in many biological systems.

Transformation from chemical into mechanical energy is the only system which has not yet been developed and utilized practically. The isothermal conversion of chemical energy into mechanical work underlies the motility of all living organisms, and can easily be seen for instance in muscle, flagella and ciliary movement. All these biological systems are characterized by an extremely high efficiency of energy conversion. An animal's muscle converts more than 50 % of the energy from food into useful work, whereas the best modern steam-turbin has an efficiency of about 40 % and the common gasoline engine one of 20–30 %. The high efficiency of the biological systems is largely due to direct conversion of chemical energy, without unnecessary intermediate passes producing heat.

Do we have any possibility to build conversion systems capable of directly transforming chemical energy into mechanical work? The answer is "yes".

The concept of systems transforming chemical energy directly into mechanical work has been investigated for many years. Such systems would be much more efficient in terms of fuel consumption than conventional thermal engines which first convert chemical energy into heat or electricity and then the latter into mechanical work. A one-step conversion is also expected to produce less waste products and work noiselessly.

J. H. van't Hoff already showed the possibility of conversion of a certain amount of chemical "free energy" into mechanical work. He also suggested a system of osmotic cells with pistons which converts free energy directly to work. Of course van't Hoff's theoretical osmotic engine had no similarity with the processes taking place in living muscles, where the motility is generated not by a difference of osmotic pressure, but by a pertinent assembly system of macromolecular compounds (proteins), which contracts and expands as the consequence of a chemical reaction.

In 1948 W. Kuhn [1], J. W. Breitenbach [2], and A. Katchalsky [3] independently found that water-swollen macromolecules can convert chemical energy directly into mechanical work under isothermal conditions. The principle of reversible contraction and dilatation is based on the reversible ionization of suitable groups (for example carboxylic acid groups) of a polyacid by alternating addition of alkali and acid, whereby the former produces an electrostatic repulsion of ions along the macromolecular chain and causes an expansion of the coiled polymer. A. Katchalsky enlarged the molecular extensions and contractions from the microscopic scale to macroscopic systems by fabricating a three-dimensional polymer network in the form of fibers or films [4]. These macromolecules were not analogous to biological ones of natural muscle, nor is their reactivity based on the high energy compounds of living organisms. However, Katchalsky developed useful models for the physico-chemical investigation of the principle, indicating the possibility of converting chemical energy directly into mechanical work or, conversely, of transforming mechanical into potential chemical energy. Following a proposal by Engelhardt [5], A. Katchalsky denoted these

Table 1. Energy Transformations in Artificial System

	Mechanical	Thermal	Electromagnetic	Photo	Chemical
Mechanical		Friction	Generator Piezoelectric	Triboluminescence	Electroluminescence Gaseous discharge
Thermal	Internal Combustion Engine		Seebeck effect Thermoionic emission Pyroelectric	Temperature radiation	Endothermic chemical reaction
Electromagnetic	Motor	Joule's heat		Reverse chemomechanical	Electrochemical reaction
Photo	Light pressure	Radiant heat	Photovoltaic effect		Photochemical reaction
Chemical	Chemomechanical [a] Muscle [a] Flagella [a] Protoplasmic Streaming	Exothermic Chemical Reaction	Dry cell Battery Fuel cell	Chemoluminescence	

[a] Denotes analogous functions observed in biological systems

transformations as "mechanochemical reactions" or "mechanochemical systems"[1] and in 1972 appeared the "Journal of Mechanochemistry and Cell Motility" (1972 to 1976, Gordon & Breach Science Publishers), which lasted until A. Katchalsky's accidental death.

Chemomechanical systems are the only artificial systems at present which can convert chemical energy directly into mechanical work. They may have potential uses under water and in space where any power supply is limited or difficult to obtain. This type of compounds can also be considered as *"metamorphic-active"* polymers, meaning that the chemomechanical reaction may be used not only to generate mechanical energy on a macroscopic level, but also to transform informations as a "signal" or "receptor", whereby semimicroscopic deformation or strain play an essential role for displaying functions such as those of a switch or a sensor. It should be noted that this kind of "semimicroscopic" chemomechanical work can also be seen in many biological systems. For example, enzymes can change their conformation and accumulate strain when they form an enzyme-substrate complex ("activated state"). This mechanical energy of deformation is partially liberated as chemical energy and utilized to promote the enzyme reaction. Hemoglobin also changes the conformational state if one of the four globulin molecules binds with an oxygen molecule. This induces a submicroscopic (about 7 Å) displacement of the globulin chain which is transferred to the adjacent globulin as a "signal" and provides for displaying an "allosteric effect". Thus, the chemomechanical process can be considered as a rather common phenomenon dominating the dynamics and functions of biological systems.

The term "chemomechanical system" refers to thermodynamic systems capable of transforming chemical energy directly into mechanical work or conversely of transforming mechanical into chemical potential energy [6]. These transformations are carried out in an "engine" capable of performing cyclically, reverting after each cycle to its initial state. At the completion of a cycle a transfer of reactive substances from one chemical potential to another has taken place, with a corresponding performance of mechanical work. The reactant falls from a high chemical potential to a lower one and the engine liberates mechanical energy. However, opposite cycles may be envisaged in which the application of mechanical energy may rise the chemical potential of a substance and permit its transfer from a reservoir of a lower chemical potential to a reservoir of a higher potential (reverse-chemomechanical system). Therefore, it is necessary to introduce at least one feedback system to realize a self-regulating chemomechanical system in which the synthetic macromolecules involved are sensitive to changes in the environment (such as pH, ion species, ionic strength, temperature, and chemical groups reacting) and undergo subsequent conformational changes. In other words, the system must contain macromolecules possessing chemically active groups or functions sensitive to conformational changes at either the local or total macromolecular level. Thermodynamic equations treating chemomechanical cycles [7], as well as functions such as "availability" and "maximum work" [8] for contractile systems have been discussed in the literature.

During the last three decades a large number of simple mechanomechanical systems

---

[1] We would like to prefer the term "chemomechanical reaction", "chemomechanical system" rather than "mechanochemical reaction", "mechanochemical system" in order to difine more precisely and to avoid confusion with other terminology.

have been investigated. They include polyelectrolyte fibers and membranes which expand and contract upon changing their degree of ionization; ion-exchange fibers which change their dimension upon exchanging a monovalent counter ion for a divalent; partly crystalline polymers which undergo melting of the crystallites with concurrent contraction; some fibrillar proteins, such as collagen and keratin, which perform chemomechanically upon interaction with strong salt solutions or "hydrogen-bond breakers"; polymer membranes which associate cooperatively with "complementary" polymers. Recently a novel chemomechanical system consisting of polyelectrolyte gels contracting reversibly under electric stimulus was reported.

The chemomechanical systems may be classified according to the principle of reaction:

1) hydrogen-ion transfer (pH muscle),
2) ion exchange or chelation,
3) redox reaction (redox muscle),
4) steric isomerization,
5) phase transition or order-disorder transition,
6) polymer-polymer association or aggregation,
7) electrokinetic processes,

Each of these systems will be described and discussed in detail in the following chapters.

2 Principles of Conversion of Chemical Into Mechanical Energy by Crosslinked Polymer Network Systems

2.1 Chemomechanical Systems Based on Hydrogen-Ion Transfer (pH-Muscle)

The shape of ionizable polymeric chains depends on the degree of ionization. Changing the state of the macromolecular chains, for example, by ionizing the carboxylic groups of a polyacid, produces an electrostatic repulsion along the chain and causes an expansion of the originally coiled molecule. The principle of a chemomechanical reaction of this type is based on reversible ionization. A. Katchalsky prepared a three-dimensional network of poly(acrylic acid) (PAA) or poly(methacrylic acid) (PMAA), in form of a membrane or filament, which swells in water [4]. On alternating additions of acid and alkali, the membrane (or filament) contracted and expanded reversibly, and while doing this lifted up and down a considerable load (Fig. 1). For example, a certain filament, to which a piece of weight 5000 times the weight of 1 cm of the dry filament was attached, changed its length reversibly by up to 30%. Chemical energy is thus transformed directly into mechanical work. Similar chemomechanical behavior was observed by using various kinds of polyelectrolyte gels such as polyvinyl phosphorate obtained by the phosphorylation of polyvinyl alcohol (PVA) [9], or PAA laminated with PVA [10, 11].

A thin film could lift a fairly heavy weight and operate reversibly for many cycles. The principle described made possible the construction of a continuously working chemomechanical system. The operation was rather sluggish and the efficiency low, but it demonstrated the possibility of building a novel type of actuators made of water-swollen soft materials. Later, Tatara [12] developed a chemomechanical piston

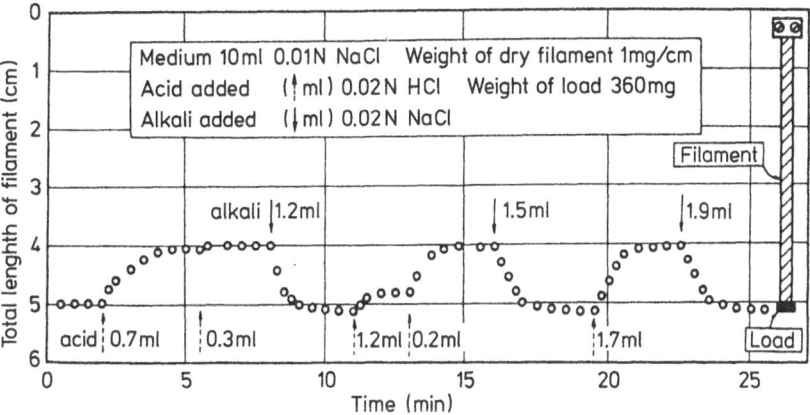

Fig. 1. Reversible lifting and lowering of a load by a filament of crosslinked poly(acrylic acid) on adding acid and alkali alternately to the surrounding medium [4]

and punch in which spherical ion-exchanging particles were packed. Using a chemo-mechanical plunch, a rubber sheet with a thickness of 1 mm could be pierced within 60 s by alternating addition of 2 N NaOH and 2 N HCl. In similar manner, an auto-mobile with a weight of 800 kg has been lifted by 5 mm, when 840 ml of the resin was packed on a piston and used as a jack. It was mentioned that the pressure of the resin in the swollen state was high enough to use the piston as a "rock" or "brake".

In order to discuss the swelling equilibrium, the elasticity was expressed in terms of a free energy having only a C term, as it is usually applied to common elastic materials [13-15]. Later Flory [16] and Tatara [17] derived an equation for the expansion and contraction of water-swollen polymer network assuming that the chemomechanical polymer films undergoing contraction and dilation under a load behave similarly to elastic rubbers under moderate deformations:

$$[f] = fv_2^{1/3}/(\alpha - \alpha^{-2}) = (2c_1/L_{i0}) [1 + (c_2/c_1) v_2^{4/3-m}\alpha^{-1}]$$

Here [f] is the reduced force as defined by the first equality, f is the force, v_2 the volume fraction of polymer, α the elongation ratio relative to the length in the isotropically swollen state, L_{i0} the length of the dry undeformed sample, and c_1, c_2, and m are arbitrary constants.

An electrically activated pH muscle system in which weak polyelectrolytes are sensitive to pH changes was also reported [18]. The pH changes are produced through electrodialysis of a solution compartment containing the muscle membrane. A unique method of electrical control was developed in this work, which made use of cationic and anionic exchange membranes on either side of the weak acid or weak base "muscle" membrane (Fig. 2). Initially the two compartments contain an equal concentration of a mixture of NaCl and HCl. The muscle membrane is in its un-dissociated and, therefore, less swollen state. Passing an electric current through the system (see Fig. 2) causes a decrease of the electrolyte concentration in the central compartment in which the muscle membrane is located, due to the difference of permselectivity of the two ion exchange membranes. (The same principle operates in the desalination of water by electrodialysis.) Since the H^+ ion mobility is higher, its

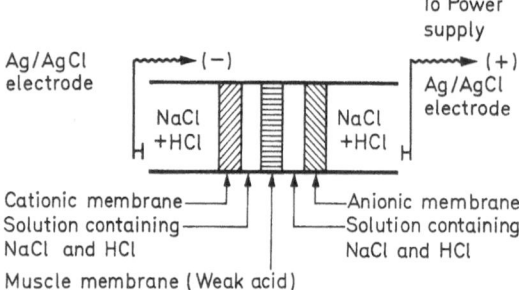

Fig. 2. Scheme of an electrically-activated pH muscle [18]

concentration decreases faster. When the H^+ ion concentration falls below the value of the muscle membrane dissociation constant, the weak acid groups ionize causing increased swelling of the muscle membrane with concomitant increase of its dimensions. The reverse effect was achieved by starting with a solution in which the H^+ ion concentration is less than the weak acid dissociation constant and passing electric current in the opposite direction. However, only about a 1 % total dimensional change was observed using this membrane. Therefore, the force at constant membrane length was recorded continuously and changes of the state of dissociation of the membrane polymer were reflected in changes of the membrane tension.

More recently, Suzuki developed chemomechanical system using composite films consisting of PVA-PAA and PVA-PAA-poly(allylamine). The composite films thus obtained were much stronger with high tenacity and exhibited a rapid responces from 4 to 40 s by alternative changing of solvent between ethanol and 0.01 N NaOH. (M. Suzuki, Biorheology 23, 274 (1986)).

2.2 Chemomechanical Systems Based on Change of Counter-Ions or by Chelation

Phenomena of dimensional changes of ion exchange resins are widely known [19, 20]. These changes are particularly pronounced if monovalent ions are exchanged with polyvalent ions. Since these changes are reversible, ion exchange has been sometimes utilized in chemomechanical systems [21].

A. Katchalsky and his school carried out measurements with fibers immersed in different equilibrium solutions in the apparatus shown in Fig. 3 [21]. A bundle containing about 3,000 fibers was suspended in à bath with which it maintained an exchange equilibrium. Because of the very large surface area, equilibrium was attained rapidly. The fibers swelled in sodium hydroxide to 11.5 times the dry volume, while the degree of swelling in barium hydroxide was practically the same as in water, i.e. 5 times the dry volume. The stresses applied to the wet fibers ranged from 0.50 to 25 kg/cm^2 (cross section referred to dry fibers). The dependence of the force-elongation curves on the degree of neutralization with sodium hydroxide was investigated.

N. Hojo and H. Shirai prepared water-swollen PVA films and observed contraction of the film by chelate-formation with copper (II)-ammine complex [22]. They also carried out partial (3.9%) phosphorylation of a PVA film which served as a model for chemomechanical reaction by chelating a variety of mono-, di-, and tri-valent metal

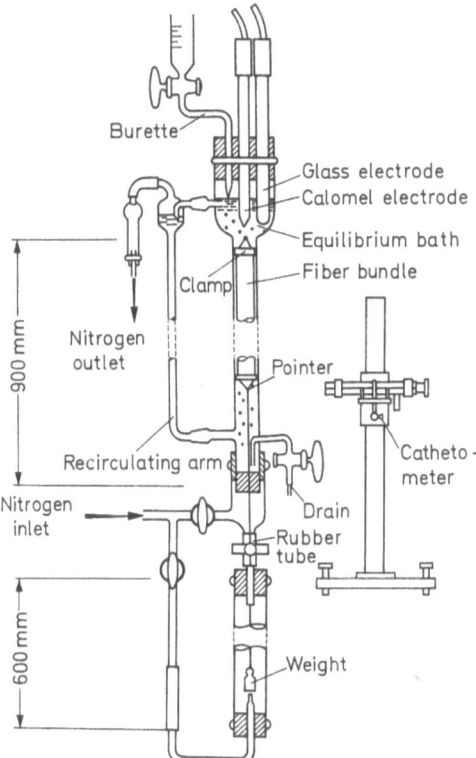

900 mm

600 mm

Burette
Glass electrode
Calomel electrode
Equilibrium bath
Fiber bundle
Clamp
Nitrogen outlet
Pointer
Recirculating arm
Catheto-meter
Nitrogen inlet
Drain
Rubber tube
Weight

Fig. 3. Apparatus for mesuring the force-elongation relation for fibers immersed in an equilibrium bath [21]

ions. In general, the contraction of the film was sensitively affected both by the ionic radius, and the electronegativity. The degree of contraction decreased in the series [23]:

$$Cu(II) > Cd(II) > Co(II) > Fe(II) > Ca(II) > Al(III) > Zn(II) > Ba(II) > K^+.$$

2.3 Chemomechanical System Based on Redox Reaction

The construction of a system in which the changes in length of the membrane are produced by an alternating chemical action of reducing and oxdizing agents is of interest. W. Kuhn studied a system consisting of a copolymer from vinyl alcohol and allylalloxan [24]. This copolymer was obtained by first copolymerizing vinyl acetate with n-allylbarbituric acid, then oxidizing the barbituric acid to alloxan, followed by alcoholysis of the vinyl acetate groups to vinyl alcohol. The obtained copolymer contained about 20 mol % of allylbarbituric acid.

The alloxan and dialuric acid groups which arise from the oxidation and reduction processes and which constitute a redox system, were fixed to the gel. The reduction of an aqueous solution of alloxan to dialuric acid with the use of hydrogen gas was possible only in the presence of platinum. The alloxan present in the copolymer could not leave the gel and, therefore, the soluble redox system 2-methyl-naphthoquinone/ 2-methyl-naphthohydroquinone (vitamin K component) was added to the solution. The soluble redox system transferred the redox action to the fixed system in the interior of the gel.

The strips showed a completely reversible change in length: on reduction, the original length was reduced by about 20%; on oxidation the strips dilated to the original length.

2.4 Chemomechanical Systems Based on Steric Photo-Isomerization

Spirobenzopyrane, azobenzene, stilbene, indigo, and thioindigo contain groups which perform cis $\rightleftharpoons$ trans isomerization under UV irradiation. Therefore, by introducing these compounds into macromolecules one can obtain photo-responsible polymers. Ueno and Osa [27] synthesized poly(benzyl asparagate) containing azobenzene in a side group, and found that the polypeptide chain with left-stranded helical structure transforms to that of right-stranded upon photo irradiation.

Aromatic polyimides containing an azobenzene in the chain backbone (4,4-di-aminoazobenzenepyromellitamide) showed reversible photo and thermal contraction. The effect was associated with trans $\rightleftharpoons$ cis isomerization [28]. However, the length changes was not large (0.23% contraction after 300 min irradiation with UV at 200 °C) and had a half-life of contraction of about 40 min.

The photocontractile behavior of crosslinked poly(ethyl methacrylate) containing spirobenzopyrane derivatives was observed by G. Smets [29, 30]. On irradiation of stretched samples at constant temperature, 2.3% of shrinking took place, while the original length was recovered in the dark. The cycle was repeatable and reproducible. The magnitude of the contraction depended on parameters such as stress applied to the sample, temperature, degree of crosslinking, and concentration of the photochromic moiety in the film. The dark length recovery took place within 2 min, which corresponded to less than 5% decoloration of the photochromic moiety incorporated. However the shrinking and dark recovery cycles could be repeated several times without apparent fatigue by alternate irradiation and dark periods, each as short as 2 min [31] (Fig. 4).

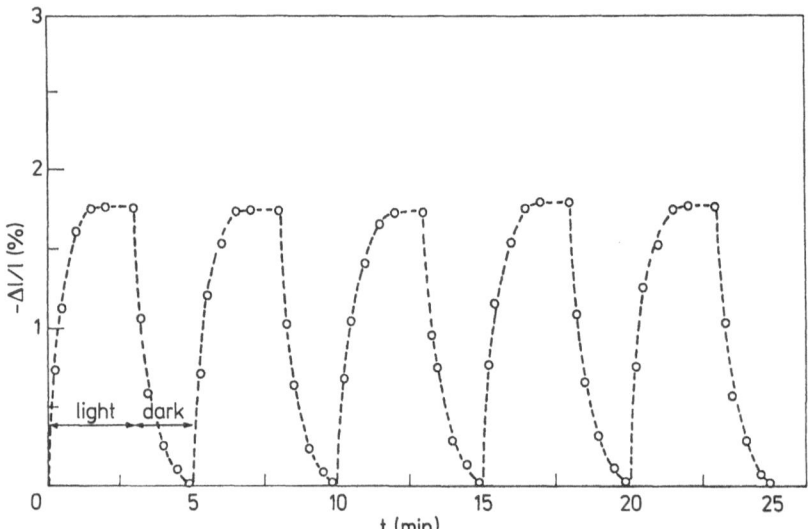

Fig. 4. Contraction-dilation cycle of photochromic crosslinked poly(ethyl acrylate) [31]. $l_0 = 3.97$ cm; width = 5 mm; load = 115 g; thickness = 0.48 mm

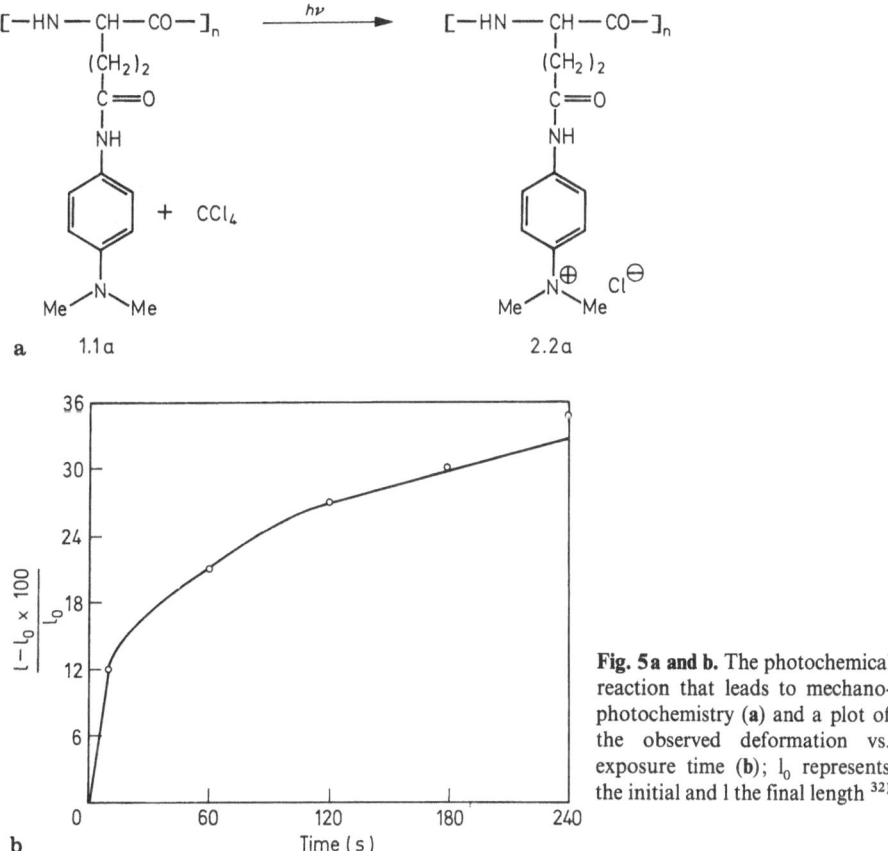

Fig. 5a and b. The photochemical reaction that leads to mechano-photochemistry (**a**) and a plot of the observed deformation vs. exposure time (**b**); l_0 represents the initial and l the final length [32]

A. Aviram synthesized poly(γ-glutamic acid) membranes containing N,N-dimethyl amino groups [32]. This polymer undergoes photoionization in the presence of carbon tetrabromide or carbon tetrachloride upon UV irradiation, as shown in Fig. 5. The main chain expands due to electrostatic repulsion which causes the chemomechanical deformations. As shown in Fig. 5, these films displayed a maximum of 35% deformation in length, corresponding to a 146% volume change. It was mentioned that photochemomechanical behavior of this type has a potential application for relief imaging if the photoirradiation is carried out through masks.

Attempts of photo-dilatation and contraction of polyamide and polyimide containing stilbene in the aromatic chain backbone were made in water. It was found that polyimide-poly(iminoisophthaloylimino-1,3-phenylenevinylene-1,4-phenylene), if stretched under constant tension, relaxed on irradiating with UV light, and recovered to initial stress when the irradiation was stopped [33]. This type of polyimide is stable in the dark in the trans conformation which is more stretched, and transforms to the cis form which is the more coiled conformation under irradiation. Therefore, the relaxation by photoirradiation should be associated with an increase in polarity of the polymer film.

Shinohara and Sakurai synthesized polymers containing photochromic pendant groups such as azobenzene or spiropyrane and observed the photo-induced change in hydrophobicity on the surface of the polymer film [34]. For example, azoaromatic polymers increased the water wettability from cos θ = 0 to cos θ = 0.2 (θ = contact angle) under UV irradiation and recovered to the original value by irradiating with visible light or by heating (Fig. 6). The observed reversible change in wettability is associated with a significant increase of the dipole moment across the azo bonds. Similar photo-induced swellability increases were observed for polymer films containing spiropyran. This can be explained by ion cleavage of the nonpolar spiropyran into the melocyanine form [35, 36].

Ishihara and his group prepared photoresponsible polymeric adsorbents by coating an insoluble support with poly(p-phenylazoacryl-anilide) [37]. This polymer should change the wettability on irradiating with UV light and accordingly should change the adsorption behavior of protein. In fact, a decrease of the adsorption of proteins was observed on irradiation. The result was presumed to be due to the increased polarity or decreased hydrophobicity of the adsorbent, which both would weaken the hydrophobic interaction of proteins with the adsorbent. The binding isotherm for various proteins in the photo stationary state was calculated and compared with that in the dark. The behavior observed has a potential application to affinity chromatography, and some attempts to separate mixtures of pigment and antibiotics by photoirradiation were made recently by the same authors [38].

Azobenzene derivatives can be used as photoantennas because of their large geometrical changes and good reproducibility, and several groups have attempted the photocontrol of chemical and physical functions of membranes [39], emulsions [40], polymers [4, 42] and cyclodextrins [43].

Shinkai et al. have reported that a photoresponsible crown ether compound which combines within the same molecule both a crown ether and a photoresponsible chromophore changes its conformation in response to photoirradiation, which results in a change of the complexation ability with alkali metal ions [44, 45]. For example, an azobis(benzo-15-crown-5) which has two crown ethers and an azo linkage as a photoantenna was synthesized. The photoisomerized cis-form exhibited greater binding ability for large alkali metal cations than the trans-isomer, due to the formation of intramolecular sandwich-type 1:2 cation/crown complexes (Fig. 7a) [46]. On the base of this phenomenon the authors showed that ion extraction and ion transport through the membrane can be controlled by light. The same authors immobilized molecules

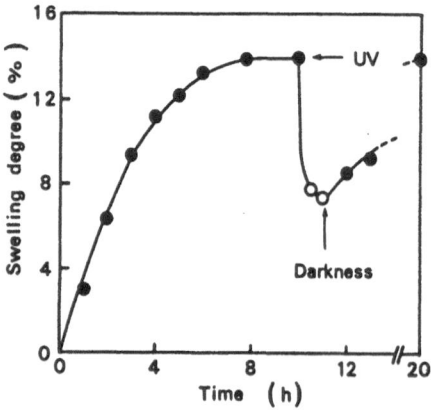

Fig. 6. Photoinduced change in the degree of swelling of an azoaromatic polymer membrane in water at 25 °C [34]. Unit mole fraction of the azobenzene moiety in the polymer: 0.024; (●) in the dark, (○) under UV irradiation

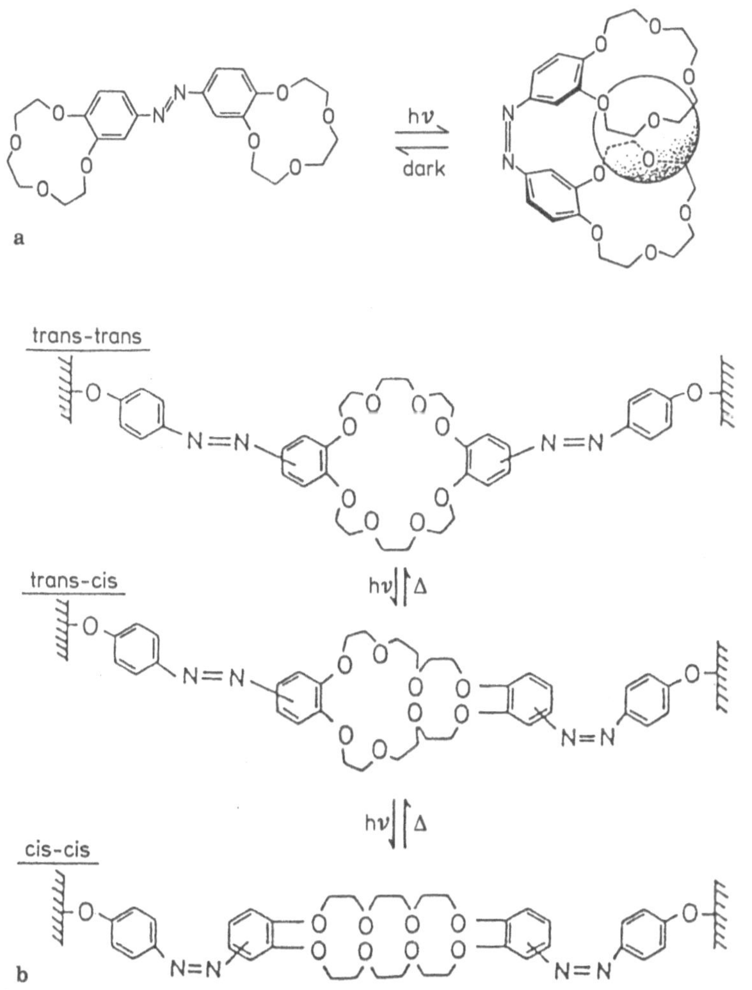

Fig. 7a and b. Examples of photo-responsible crown ethers synthesized by Shinkai [46]

containing both a crown ether and an azobenzene on a crosslinked polymer support (Fig. 7b) [47]. Since the distance between the 4- and 4'-positions of trans-azobenzene is 0.90 nm, whereas that of the cis-form is 0.55 nm, the photoinduced trans-cis isomerization should induce a contraction of the chromophores. This, in turn, will be compensated by an elongation of the crown ether, changing its complexation ability. Figure 8 shows the influence of photoirradiation on the binding of cesium p-nitrobenzoate. It is seen that photoirradiation induces a gradual increase in the concentration of cesium p-nitrobenzoate, indicating that the trans-to-cis- isomerization of azobenzene stretches the shape of the crown ether resulting in reduced complexation ability.

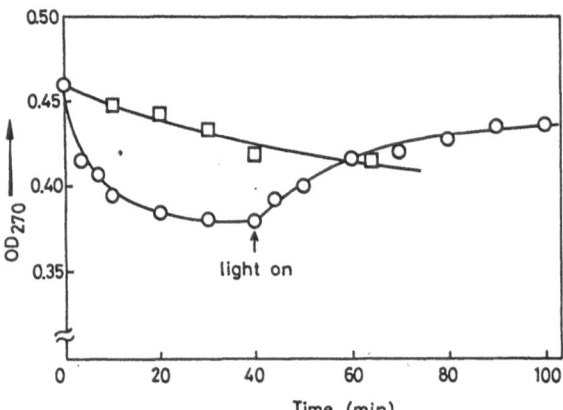

Fig. 8. Influence of photoirradiation on the binding of cesium p-nitrobenzoate to polystyrene beads. The reaction solution (7.7 ml of DMF) contains 100 mg (0.0145 mmol of crown ether) of polystyrene beads and 0.0145 mmol of cesium p-nitrobenzoate: (○) 0 → 40 min in the dark and then photoirradiated; (□) photoirradiated for 1 h in the absence of cesium p-nitrobenzoate and then mixed (time 0) with cesium p-nitrobenzoate in the dark [47]

2.5 Chemomechanical Systems Based on Phase Transition or Order-Disorder Transition

H. Yuki and his coworkers [48] synthesized three-dimensional polypeptide networks composed of l-glutamic acid and dl-cystine residues, and their reversible contractility upon changes in pH and ionic strength was studied. A foil made of this polypeptide contracted in its isoelectric region and extended in both acidic and basic media.

A. Katchalsky constructed chemomechanical engines working on chemical melting and crystallization of crosslinked collagen fibers by treating the fibers with solutions of salts such as LiBr, KSCN, or urea [49]. The engine exhibited rapid, reversible and reproducible contraction. At zero force the fiber contracted to half its length; if a force was applied the fiber was able to lift a weight several thousand times its own. Figure 9 shows the principle underlying the construction of such an engine. A closed loop of fiber is wound around pulleys as indicated in the figure. If pulley A is dipped into a strong salt solution (such as 8–10 M aqueous LiBr), the collagen undergoes contraction in the salt solution and exerts equal forces on the rims of pulleys C and D. Since the radius of C is larger than that of D, a net rotary moment acts on the compound pulley C–D and causes a counter-clockwise rotation. The rotation brings a new part of the collagen belt into the salt solution, which reproduces the same effect. For a steady state operation of the engine the contraction due to the chemical interaction of callagen with salt has to be counterbalanced by the reverse process. This occurs at pulley B which dips into water or a dilute salt solution, washing out the salt and relaxing the collagen. While the engine is working, salt is transferred from the concentrated solution to the dilute one.

Figure 10 shows photographs of a family of experimental engines [49]. The upper two pairs of wheels in Fig. 10a were added to facilitate the entrance of the collagen loop into the water bath. M. S. Sussman and A. Katchalsky built a modified chemo-

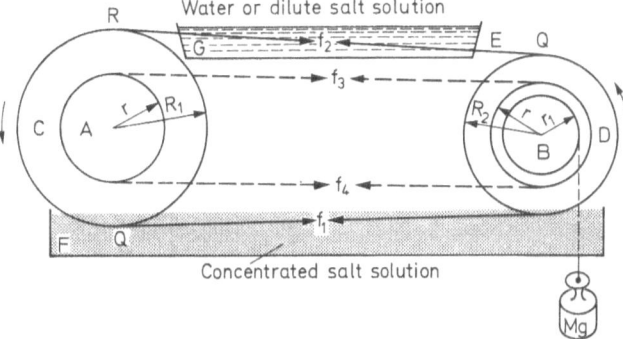

Fig. 9. Schematic illustration of a chemomechanical engine according to A. Katchalsky [49]

mechanical turbin (Fig. 10b) that rotates smoothly [50]. This engine is actuated by a collagen fiber in the form of a loop, a portion of which is wrapped helically around a pair of contoured spindles that are canted with respect to each other. The spindles are partly immersed in a bath of salt solution. The remaining portion of the fiber passes through a tank of fresh water below the salt solution bath. The salt solution causes the collagen to shrink, developing tensile force and rotating the tapered spindles. After spiraling through the brine, the fiber is pulled from the solution by a belt-driven roller and lowered into the water bath. The elimination of salt from the collagen into the water recoveres the material to its original length.

Fig. 10a. "Chemomechanical Engine" [49], **b)** modified "Chemomechanical Turbine" [50]

The fiber shrinked to about 60% of its normal length when it was transferred from the bath of fresh water to 8 M LiBr solution. It could be stretched to its original length by exerting on its ends a tension of 52 kg/cm² of cross sectional area.

Figure 11 shows the tension-length dependence of collagen fiber in LiBr. The cycle begins at point A, proceeds through points B and C returns to A. At point A, the fiber is at its normal length and in zero tension. In this state the fiber is carried to the top of the spindle. The material is then carried into the salt solution (point B). Tension increases as the fiber absorbs salt. Tension is relieved as the shrinking material spirals through the brine and exerts torque on the tapered part of the rotating spindles (from point B to point C). The cycle is closed by washing the salt off the collagen, which

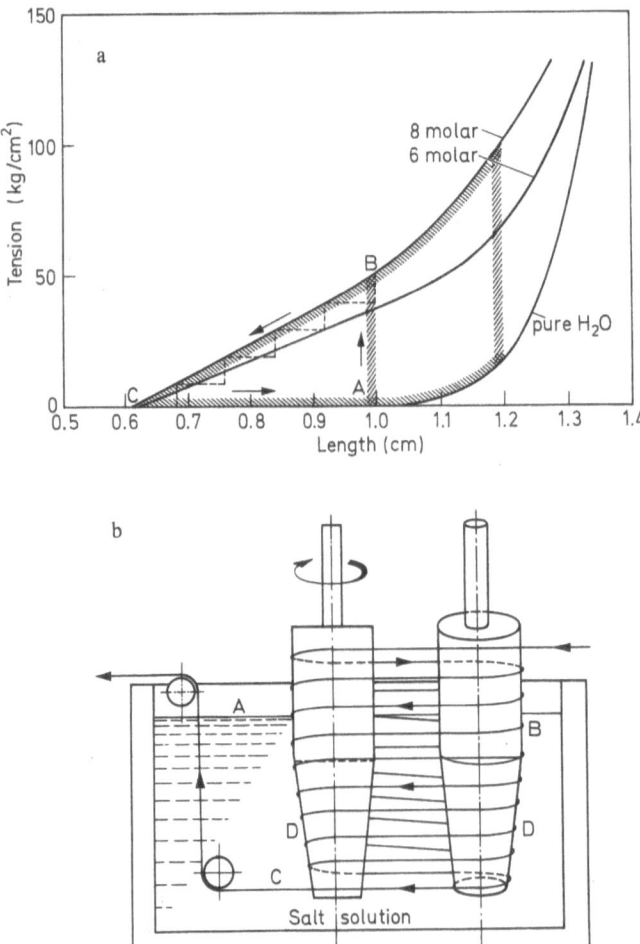

Fig. 11a. Tension-length dependence of callogen fiber in LiBr; **b.** contractile fiber-actuated turbine in side view [50]. Fiber tension builds up on cylindrical spindle sections between (A) and (B) at constant specific length. Stepwise contraction occurs between (B) and (C) as the fiber helix descends the conical portions of the spindles (D)

restores the fiber to its normal length (from point C to point A). The machine delivered about 30 milliwatts at its output shaft, which corresponds to a mechanical efficiency of about 40%.

2.6 Conversion of Mechanical Into Chemical Energy (Reverse Chemomechanical System)

It should be possible that mechanical stretching or compression of polyelectrolyte fibers provide a pH change of the embedding fluid. This is the reverse situation of the chemomechanical contraction of polymer fibers provoked by a pH change of the fluid (reverse chemomechanical system).

W. Kuhn [51] demonstrated this phenomenon for the first time, using crosslaminated fibers consisting of crosslinked poly(vinyl alcohol) and poly(acrylic acid). After obtaining the equilibrium pH value chracteristic for the unloaded state of the laminated fiber, a stretching of the fiber was undertaken with a mechanical force of up to 10 kg/cm². Figure 12 illustrates the change in pH of the embedding fluid as the cross-laminated fiber was stretched. With increased loading weight or, in other words, increased stretching of the fibers, the H^+ ion concentration of the embedding fluid incrased (see curves CC^1C^2 and $C E^1E^2E^3E^4$). The reverse was observed when the fiber was unloaded, i.e. the H^+ ion concentration of the outer fluid decreased with the removal of the weight. Thus, a pH shift of the embedding fluid as a result of the stretching of the cross-laminated fiber was observed. In a similar manner, the pH value of the embedding fluid of a nucleohistone fiber decreased from 4.8 to 4.7 by

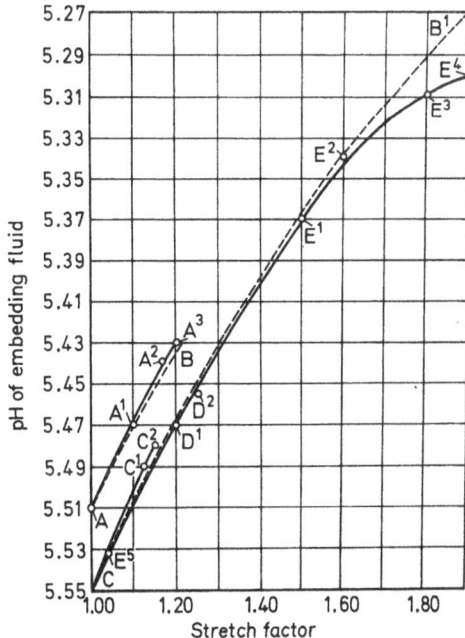

Fig. 12. The pH-stretch curve of the cross-laminated system [51]. The equilibrium pH value for the unstretched systems was 5.51 in the case of one system (experimental points marked $AA^1A^2A^3$) and 5.55 for the three other systems. The broken curves (AB and CB′) represent theoretical curves (see [51])

stretching the fiber to 4 times its original length [52]. A shift of the isoelectric point of protein fibers (keratin) was also observed [53].

All these examples demonstrate that mechanical work can be transferred to chemical potential energy. Examples of the conversion of free energy change of association between a solvated polymer or drug and a membrane by stretching and contracting of the membrane will be shown later (Fig. 17, 18).

Frenkel investigated the influence of external stress on the "critical point" at which the hydrochemical and chemical contraction was induced [54]. As indicated by Frenkel earlier, proteins with helical structure contract to a relatively symmetrical random-coil structure upon breaking of the cooperative system of hydrogen bonding by temperature or chemical environmental changes. Therefore, the existance of an external force should increase the "critical" transition temperature at a "critical" concentration of breakers of hydrogen bonding, and should eventually diminish the contraction of the protein fibers under a certain load. This kind of influence of load on the order-disorder transition of crystalline polymers and proteins had been theoretically proposed by Frenkel [55], Gee [56], and Flory [57], and is the equivalent to the famous Clapeyron-Clausius equation describing the boiling point-vapor pressure relationship. Fig. 13a shows changes of the length of collagen fibers (from rat tail tendons) by application of different external loads. The transition temperature increases and the magnitude of contraction diminishes when the stress is increased. As predicted theoretically, a certain critical load under which no contraction occurs was obtained by plotting stress vs maximum relative contraction, and extrapolating to contraction zero. A value of 165 ± 10 kg/cm^2 was found, and this value was the same when the contraction was induced by chemical environmental change (Fig. 13b).

These results indicate that the free energy of phase transition from ordered to disordered structure is changed by the application of external stress and this can be considered as a good example of the conversion of mechanical to chemical potential energy (reverse chemomechanical system).

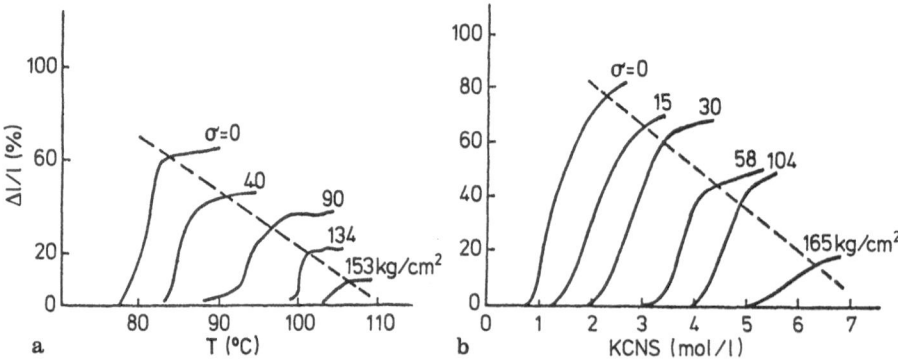

Fig. 13a and b. Contraction of collagen fibers under different fixed loads [54], σ; **a)** dependence on temperature, **b)** dependence on KCNS concentration; (σ in kg/cm^2)

3 Mechanical Systems Based on Polymer Association

3.1 Isotonic and Isomeric Contraction

The study of the interaction between two "complementary" macromolecules and the aggregation of the resulting complexes to a supermolecular structure in solution is of special interest, because many biological phenomena such as enzymatic processes, supermolecular assemblies in virus shells, and muscle contraction depend on specific protein-protein interactions. Studies on synthetic macromolecules may serve as models of such phenomena.

Macromolecules undergo drastic conformational changes as a result of mutual interaction in solution, and sometimes even insoluble complexes are formed. Complex formation has been obtained in many systems of uncharged or ionized synthetic polymers [58-77]. If such drastic changes of shape of macromolecular chains can be translated from the microscopic level to a macroscopic scale, a conversion of chemical free energy of complexation into mechanical work should be realized.

However, complexation of this type often proceeds irreversibly [58-60]. Hence the difficulty in constructing such energy conversion systems is to find suitable reversible systems from synthetic polymers.

Previously, V. A. Kabanov and I. M. Papisov have found that poly(ethylene glycol) (PEG) of molecular weight (MW) 2000 undergoes reversible complexation-dissociation with poly(methacrylic acid) (PMAA) in an aqueous medium by a change of temperature [59,61].

PMAA   PEG   Complex

Figure 14 shows the temperature dependence of the viscosity of an aqueous solution of PMAA (curve 1) and equimolar PMAA-PEG mixtures containing equal concentrations of the repeating units of the two components (curves 2 and 3) [78]. A pronounced viscosity drop in the temperature range 25–45 °C in curve 2 demonstrates complexation between PEG with a molecular weight of 2000 and PMAA. The viscosity recovers completely on lowering the temperature. On the other hand, a solution mixture prepared from the same PMAA but with PEG having a molecular weight of 20,000 exhibits low viscosities over the entire temperature range (curve 3), indicating the formation of a stable complex. The use of PEG with a molecular weight lower than 1000 did not result in any viscosity change. The endothermic complexation, favored by a rise in temperature, is due to hydrophobic interactions between the α-methyl groups of PMAA and the ethylene backbone of PEG [61].

We have examined this thermoreversible complexation, as a means of transforming chemical energy into mechanical work. A crosslinked PMAA membrane under a load of 490 mg was suspended over water maintained at 10 °C. An aqueous solution of PEG was then added to the embedding fluid after constant length had been attained. Figure 15a shows the contraction of the PMAA membrane observed on addition of PEG and a subsequent temperature change (isotonic contraction). It is seen that the membrane in solutions of PEG contracts sharply with rising temperature, especially in the region of 20–30 °C. The dimensional changes are reversible; more than 40% of these changes occur over a temperature range as small as 10 °C. The work done per contraction by one gram of contractile substance is 5×10^{-3} cal. Thus, a 4.7 mg dry membrane to which a load 100 times its weight was attached underwent a reversible contraction-dilation by over 70% in the temperature range of 10–40 °C.

The internal stress of the membranes was measured at constant length for the corresponding system (isometric contraction) [81], and it was found that the stress developed

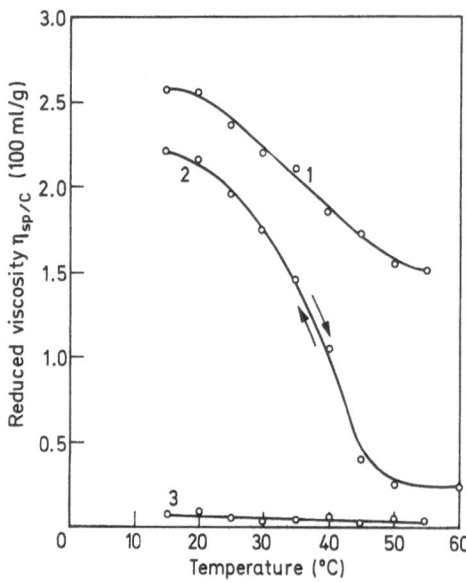

Fig.–14. Temperature dependences of reduced viscosities of PMAA and PMAA-PEG complexes in water [78]: (1) PMAA; (2) PMAA-PEG, MW of PEG: 2,000; (3) PMAA-PEG, MW of PEG: 20,000. [PEG]/[PMAA] = 1.0 (per repeating unit); PMAA = 0.05 g/100 ml

in the PMAA membrane was 4–6 kg/cm² which is almost the value found in natural muscles.

The temperature at which a specified contraction is obtained can be shifted by changing the concentration of PEG. This concentration effect is shown in Figure 15 b. The data indicate clearly the shift of the temperature for a specified contraction, although the recovery of the length of the membrane is incomplete in the lower concentration range of PEG.

As mentioned above, the contraction with increasing temperature is related to hydrophobic interactions between the α-methyl groups in the PMAA and the ethylene in the PEG backbone. Therefore, if the hydrophobic interaction could be chemically

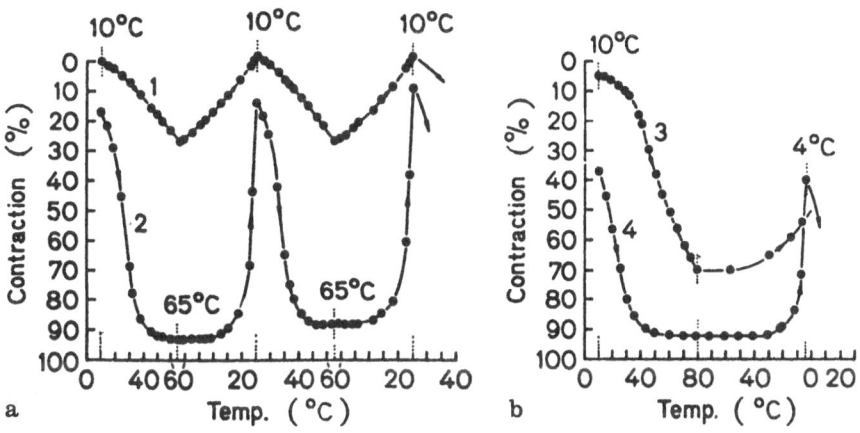

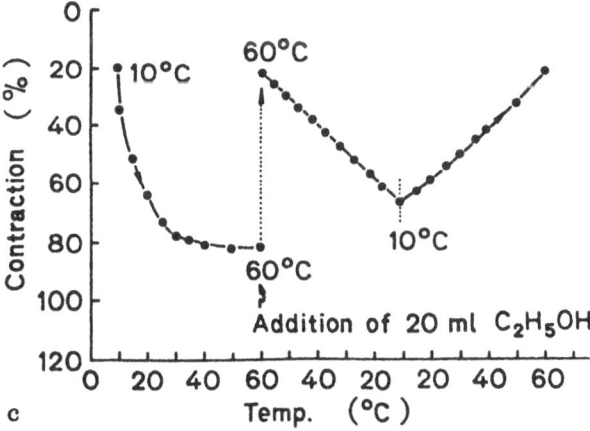

Fig. 15a–c. Temperature dependence of the contraction (in % of the dry membrane) of PMAA membranes [79], dry membranes: 10 mm wide, 23 mm long, 4.7 mg weight, loaded with 490 mg. Embedding fluids: **a)** curve 1: 70 ml water, curve 2: 70 ml of 0.066 % PEG in water; **b)** curve 3: 70 ml of 0.022 % PEG in water, curve 4: 70 ml of 0.11 % PEG in water; **c)** 70 ml of 0.066 % PEG in ethanol-water solution. PEG molecular · wt. 2000

weakened, the membrane should dilate with increasing temperature, since under these conditions the complexation is promoted mainly by hydrogen bonding which becomes weaker with increasing temperature. Figure 15c shows that this in fact happens, if 20 ml of ethyl alcohol are added to the embedding solution of PEG (which corresponds to 23 % alcohol-water solution) in the course of the cycle. Thus, the temperature coefficient of the membrane dimension can be reversed by changing the property of the embedding solution.

Figure 16 shows the profiles of isothermal contraction of PMAA membranes observed on the addition of a variety of complementary polymers [82]. It is clear that PEG with a molecular weight of 2000 produces the most rapid and pronounced contraction, which tends to saturate after a certain period of time. PEG's of molecular weight higher than 2000, such as 7500, 20,000, and 83,000, also show considerable contraction over a long period of time, but the velocity of contraction is rather low. On the other hand, PEG's with molecular weight of 1000 and 600 cause rapid but small contraction. In other words, PEG with high molecular weight contracts slowly but continuously and PEG with low molecular weight contracts rapidly and easily attains equilibrium.

The described effects of the chain length of PEG on the profiles of the membrane contraction can be associated with the rate of penetration of PEG molecules into the membrane. Another interesting feature is the fact that PEG is the only macromolecule for which a pronounced contraction of the PMAA membranes has been observed. As shown in Fig. 16, neither poly(N-vinyl pyrrolidone) (PVPdn), nor poly(vinyl alcohol) (PVA), or poly(methyl vinyl ether) (PMVE) exhibit appreciable contractions, despite the fact that all these polymers induce pronouced shrinkage of dissolved

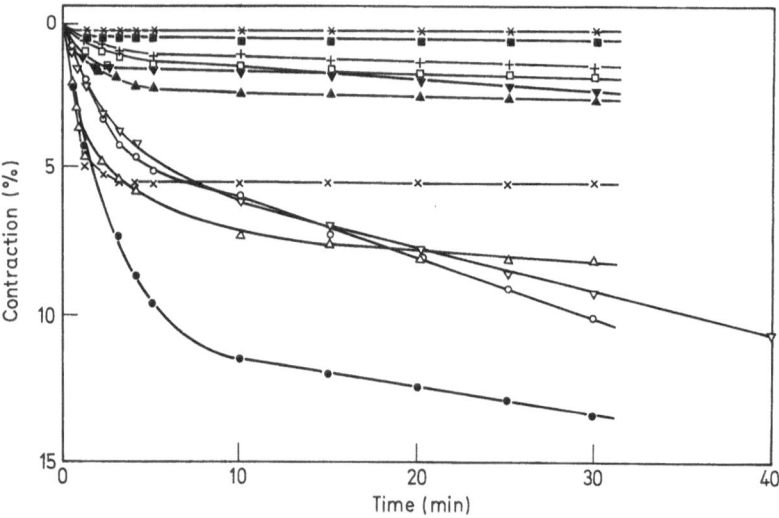

Fig. 16. Time dependences of the PMAA membrane contraction [82]. Membrane: 10 mm wide, 20 mm long, 3 mg weight, loaded with 500 mg, prepared from PMAA of MW $= 10^6$. Polymer added: 1.5 $\times 10^{-2}$ mol/l; temperature, 40 °C. Molecular weight of polymers: (□) PEG, 600; (×) PEG, 1000; (●) PEG, 2000; (△) PEG, 7500; (○) PEG, 20,000; (△) PEG, 83,000; (▲) PVPdn, 10,000; (○) PVPdn, 40,000; (■) PVPdn, 360,000; (○) PMVE, 64,000; (□) PVA, 22,000

PMAA in water [82, 83]. It is possible that the contraction of the PMAA membranes is not only related to the stability constant of the complexes, and to the extent of viscosity drop in solution, but also to a certain structural complementarity between the dissolved polymer and the membrane. If this is true, the contraction of the PMAA membranes should be discussed in terms of size and geometry of the polymers in solution and the chemical structure, both of which should influence the rate of penetration (penetration mechanism). The modest contraction observed with PVPdn, PVA, and PMVE seems to support this assumption. The penetration mechanism of the membrane contraction was proved experimentally and will be described in more detail in the paragraph of "chemical valves" (Section 3.4.1).

Figure 17a shows the profile of isothermal contraction of a PMAA membrane observed on addition of poly(ethylene glycol)-9-dianthroate (PEG-ant); this is PEG with two anthracene moieties incorporated as chromophores at both chain ends. Figure 17b shows the change in concentration of PEG-ant in the embedding fluid. It is seen that the addition of PEG-ant produces an effective contraction of the membrane which saturates after ca. 10 h. The decrease in the PEG-ant concentration in the embedding solution indicates that an effective adsorption of PEG-ant on the membrane occurs parallel with the contraction. Stretching of the membrane by applying a load at [A] (Fig. 17a) in order to recover the initial length of the membrane causes an additional adsorption of PEG-ant, and the amount of PEG-ant adsorbed attains nearly 90 % of total PEG-ant. On the contrary, a shrinking of the membrane by removing the load at [B] results in a desorption of PEG-ant from the membrane, and the amount of PEG-ant adsorbed decreases from 90 to 50 %.

Thus, the adsorption-desorption equilibrium of PEG-ant on the PMAA membrane can be closely associated with the mechanical conditions of the membrane. This fact indicates the possibility of controlling the chemical equilibrium by applying a mechani-

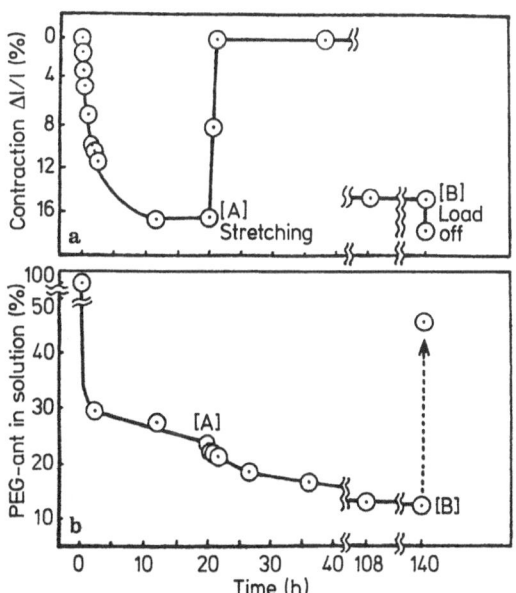

Fig. 17a and b. Profiles of contraction of PMAA membranes (**a**) and concentration change of PEG-ant (**b**) [84]. PMAA membrane: 65 mm × 10 mm × 28 μm, 14.4 mg (dry) weight, load: 500 mg, PEG-ant: 12 mM (referred to repeating unit of PEG); temperature: 25 °C. (PEG-ant = poly-(ethylene glycol)-9-dianthroate.)

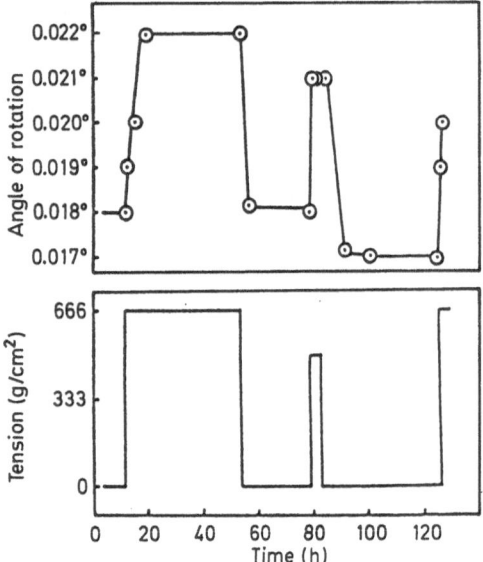

Fig. 18. Reversible adsorption and desorption of pilocarpine by PMAA membrane in strained and relaxed state [85]; ⊙ indicates the initial value of the angle of rotation (initial pilocarpine concentration: 1 mM/l)

cal energy to the system (reverse chemomechanical reaction). Similar adsorption-desorption behavior was observed under isometric condition: adsorption of 62% of the total PEG-ant developed 1.85 kg/cm² of stress, and the complete relaxation of the membrane released 70% of the PEG-ant adsorbed. Figure 18 is another example of equilibrium control by mechanical energy. It shows the change of concentration of the drug pilocarpine (cholinergic, ophthalmic) produced by stretching and relaxing of a crosslinked PMAA membrane [85]. It is seen that stretching of the membrane by loading about 600 g/cm² stimulates the release of pilocarpine, and the response is rather quick and reproducible. (The angle of rotation produced by the optically active pilocarpine is used as a measure of the concentration of the latter.)

3.2 Formation of Higher-Ordered Structures by Polymer-Membrane Association

If a PMAA membrane is immersed in a PEG-ant solution and allowed to stand at constant temperature, a spontaneous adsorption of PEG-ant is observed giving large crystals with an interesting morphology on the surface of the membrane. Microcrystals, with needle-like shape and 0.5–1.0 μm long, were initially observed under an optical microscope with crossed nicols; the microcrystals tended to aggregate within a few hours (Fig. 19a). The crystal growth was rather rapid and after 5–10 h massive crystal aggregates were formed as shown in Fig. 19b. Numerous needle-like crystals formed in the initial stages seem to gather side by side into larger plate-like crystals, or combine head to head into a sort of fibrous structure. Increasing the concentration of PEG-ant and the aging time resulted in more dense aggregates and more rapid growth. The interaction between PEG-ant and PMAA on the surface of the membrane evidently plays an essential role in the crystallization, particularly in the nucleation

Fig. 19a–c. SEM photographs of PEG-ant crystals formed on the surface of a PMAA membrane (**a, b**) and the electron diffraction pattern (**c**) [84]

process, since the PMAA membrane is the only membrane that does adsorb PEG-ant from the solution and build these crystallites. Neither crystal growth nor PEG-ant adsorption were observed on membranes of poly(acrylic acid) (PAA), polyethylene or poly(ethylene terephthalate).

We have reported that the polymer complex between PEG and PMAA is much more stable than that formed from PEG and PAA (for example, the stability constants of PEG(MW 3000)-PMAA and PEG-PAA are 250 and 9 l/mol, respectively, at 30 °C [86]; cf. Table 2, Sect. 3.4). It was also found that the fluorescence intensity of PEG-ant increases more than 10 times by the addition of PMAA due to the complex formation, in water [87]. In contrast, no fluorescence increase was observed by the addition of PAA. Since an increase in fluorescence intensity can be associated with "binding" or "fixing" of anthracene chromophores in the hydrophobic domains of the complexes, the cooperative complexation of the macromolecular chains is adequately considered to promote the nucleation of the crystal growth.

It should be noted that PEG (with no anthracene end groups) does also adsorb on the PMAA membrane and form complex-aggregates. However, they are of undefined shape and size with no higher-ordered structure and are quite different from those obtained with PEG-ant. Thus, the combination of strong intermacromolecular hydrogen bonding between PEG chains and the membrane, and the mutual stacking of anthracene groups is probably responsible for this interesting crystalline organization.

In order to investigate the molecular orientation and structure in the platelike crystals formed in the later stages of the organization, a selected-area electron diffraction study was made using a microcrystal 1.7 μm long and 0.5 μm wide [84]. Shape spots (Fig. 19c) clearly indicate that the microcrystal is a single crystal (rings are due to evaporated gold used for calibration purposes). Needle-like crystals smaller than 0.1 μm wide formed in the earlier stage showed a similar diffraction pattern although the intensity was weaker. The lattice distances of the single-crystal were calculated using (100) and (010). The diffractions corresponding to the reflections from the ((hko)) planes of the unit cell show that the molecular chains are aligned perpendicular to the surface of the crystallite. From the results obtained it was concluded that the PEG-ant single-crystal grown on the surface of the PMAA membrane has a unit cell of orthorhombic symmetry with spacings a = 0.6732 nm and b = 0.6177 nm [84].

Table 2. Stability Constant and Degree of linkage of the complexes at various temperatures

Complexation system	mol wt. of PEG or PVPdn	Stability constant K (liter/mole) (degree of linkage θ)					
		10 °C	20 °C	30 °C	40 °C	50 °C	60 °C
PMAA-PEG	200	2 (0)	0 (0)	0 (0)	4 (0.02)	10 (0.05)	5 (0.03)
	1000	3 (0)	0 (0)	0 (0)	5 (0.02)	14 (0.07)	16 (0.08)
	2000	15 (0.08)	16 (0.08)	62 (0.22)	280 (0.47)	520 (0.58)	440 (0.54)
	3000	10 (0.05)	68 (0.23)	250 (0.45)	470 (0.55)	790 (0.63)	1130 (0.68)
	7500	760 (0.62)	2600 (0.77)	4700 (0.83)	6200 (0.85)	6830 (0.85)	7200 (0.86)
	20,000	1450 (0.71)	2860 (0.78)	4200 (0.82)	5000 (0.83)	5000 (0.83)	4580 (0.82)
PMAA-PEG in ethanol-water (ethanol %)	(10)	22 (0.11)	55 (0.20)	162 (0.37)	198 (0.41)	171 (0.38)	131 (0.34)
	(17)	328 (0.49)	268 (0.46)	207 (0.41)	118 (0.32)	61 (0.22)	49 (0.19)
	(23) 2000	nd	208 (0.41)	72 (0.24)	39 (0.16)	33 (0.14)	20 (0.10)
	(37)	0 (0)	0 (0)	0 (0)	0 (0)	0 (0)	0 (0)
	(50)	0 (0)	0 (0)	0 (0)	0 (0)	0 (0)	0 (0)
PAA-PEG	2000	1 (0)	10 (0.05)	8 (0.04)	3 (0.02)	0 (0)	0 (0)
	3000	2 (0)	7 (0.04)	9 (0.05)	8 (0.05)	1 (0)	0 (0)
	7500	13 (0.07)	18 (0.09)	24 (0.11)	20 (0.10)	13 (0.07)	3 (0.02)
	20,000	64 (0.22)	84 (0.27)	150 (0.36)	180 (0.39)	190 (0.40)	160 (0.37)
PMAA-PVPdn	10,000	950 (0.66)	920 (0.65)	1260 (0.69)	1850 (0.74)	3920 (0.81)	7560 (0.86)
	40,000	1460 (0.71)	1700 (0.73)	1780 (0.73)	2200 (0.76)	5010 (0.83)	11,900 (0.89)
	160,000	12,200 (0.90)	14,300 (0.89)	18,000 (0.90)	22,200 (0.91)	27,400 (0.92)	(0.92)
	360,000	15,500 (0.90)	18,700 (0.91)	24,700 (0.92)	33,500 (0.93)	43,500 (0.94)	48,600 (0.94)
PMAA-PVA	22,000	nd	nd	nd	130 (0.34)	nd	nd
PMAA-PMVE	64,000	nd	nd	nd	1840 (0.74)	nd	nd

PVA: poly(vinyl alcohol)
PMVE: poly(methyl vinyl ether)
PVPdn: poly(N-vinyl pyrrolidone)
nd: not determined

These values are different from the unit cells of PEG with a monoclinic (a = 0.805 nm, b = 1.304 nm) or triclinic (a = 0.471 nm, b = 0.444 nm) system, and of anthracene (a = 0.856 nm, b = 0.604 nm) with a monoclinic system.

3.3 Equilibrium Study of Polymer-Polymer Association

The procedure chosen for the calculation of the degree of linkage (θ) and the stability constant of the complex (K) for a polycarboxylic acid and an electron-donating polymer is indicated by the following equations for complexes with stoichiometric composition [86, 88]:

$$\theta = 1 - ([H^+]/[H^+]_0) \tag{1}$$

$$K = \theta/C_0(1 - \theta) \tag{2}$$

where C_0 is the normality of the polymeric acid, $[H^+]$, $[H^+]_0$ are the hydrogen ion concentrations in the presence and absence of the complementary macromolecule, respectively.

The results obtained from equations (1) and (2) by using pH measurements as a function of temperature are in full agreement with the data obtained from hydrodynamic measurements (Fig. 14) and with the chemomechanical behavior (Fig. 15).

The values of θ for various PMAA-PEG complexes as a function of temperature are summarized in Table 2 [86]. As predicted from the viscosity measurements, PEG with a molecular weight of 1000 or lower has very low θ values, whereas for PEG with molecular weights of 2000 or 3000 θ increases rapidly above a certain temperature. Complexes prepared from PEG with molecular weights of 7500 and 20,000 have large θ values nearly equal to 1.0 which are almost insensitive to temperature changes.

The complexation of PEG with PAA is quite different from that with PMAA. PEG with molecular weight of 7500 does not seem to form a complex at all; even PEG with a molecular weight of 20,000 has a relatively low θ value. This is also in agreement with hydrodynamic measurements [78, 86].

The stability constants for the complexation are also summarized in Table 2. It may be seen that the stability constant strongly depends on the species of macromolecule, chain length, and temperature. A dramatic chain-length dependence of the stability constant suggests that the complexation is proceeding by "cooperative interaction" between two macromolecules; that is, both constituents are required to have a "critical chain length" in order to from a complex [59, 61, 62, 72, 73, 76, 77, 89].

The temperature dependence of θ and K for PMAA-PEG in ethyl alcohol-water is also listed in Table 2 [86]. The data indicate that the addition of alcohol reverses the temperature dependence of complexation at alcohol concentrations greater than 17%. The change of sign of the temperature coefficient of the dimensional change in PMAA membranes observed when ethanol (23%) is added (Fig. 15) can be explained by the same reasoning. Table 2 also shows that complexation is completely inhibited at concentrations of ethanol greater than 37%.

The stability constant and its temperature dependence can be used to calculate the enthalpy and entropy changes for the complexation from d ln K/d(1/T) = $-\Delta H^0/R$

and $\Delta S^0 = -(\Delta G^0 - \Delta H^0)/T$, where $\Delta G^0 = -RT \ln K$ is the free energy change for the complexation.

In these systems both ΔH^0 and ΔS^0 have positive values, characteristic of hydrophobic interaction, except when ethyl alcohol is added and the sign of the parameters is reversed. Extremely large values of ΔH^0 and ΔS^0 were obtained at 30 and 40 °C in an aqueous solution in the system PMAA-PEG with molecular weights of PEG of 2000 and 3000, and they were related to the enthalpy and entropy of complex formation which is analogous to the melting transition of polynucleotide complexes with hydrogen bond bridges. Positive values of ΔS^0 in aqueous systems suggest a release of water during complexation. Conversely, in aqueous alcohol, in which ΔH^0 is negative, complexation is mainly due to hydrogen bonding.

Generally, the binding of small cations to a polyanion is anticooperative, and the apparent complexation constant decreases as more cations are bound and the charges are neutralized. Even small molecules with no electrostatic effect bind anticooperatively if more than one polymer site is occupied by each small molecule. For a simple noncooperative binding process the stability constant of complexation K is independent of θ. The presence of different types of noncooperative binding processes with different stability constants results in a decrease of K with increasing θ. As stated above, the complexation has a dramatic chain length dependence. This can be interpreted only in terms of a cooperative complexation of the polymers. A similar chain length dependence has been observed in Coulombic complexation systems between polycation oligomers and polyanions [73], in which case the standard free energy change is linear with the number of interacting sites n, and the stability constant K for complexation is expressed as

$$K = A \exp (Bn)$$

where A and B are constants depending on the chemical structure of the polyelectrolytes. The cooperativity of the PMAA-PEG system was discussed in terms of a one-dimensional Ising model [86].

3.4 Application of Chemomechanical Systems Based on Polymer-Polymer Association

3.4.1 "Chemical Valve"

If the chemomechanical contraction is developed isometrically, the contractile stress appearing in the membrane should expand the pores through which water and the macromolecular solute permeate [90]. This results in ultrafiltration membranes having chemomechanically expanding and contracting pores. A schematic illustration is given in Fig. 20 [90].

Figure 21 shows the effect of the chemomechanical contraction of a PMAA membrane on water permeability if small amounts of PEG with various molecular weight are subsequently added to the water [90]. It is seen that the PMAA membrane exhibits a marked increase in water permeation as soon as the membrane is contacted with the PEG solution. Once the PMAA membrane is treated with PEG, the membrane can

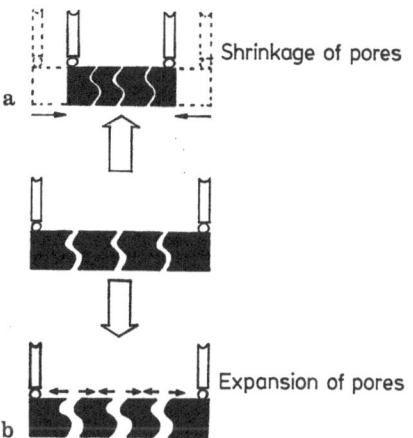

Fig. 20a and b. Schematic representation of chemomechanical contraction of a PMAA membrane by polymer-membrane complexation ("chemical valve"); **a)** isotonic contraction, **b)** isometric contraction [90]

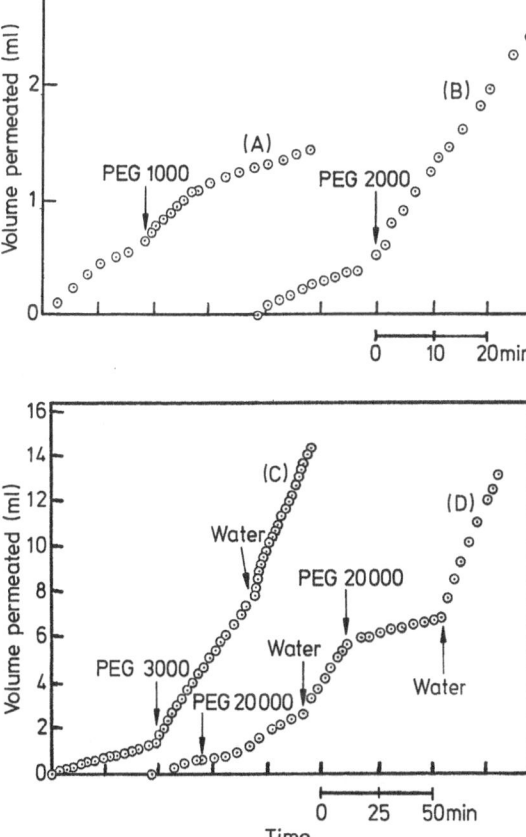

Fig. 21. Effect of chemomechanical contraction of a PMAA membrane on water permeation [91]. Mol wt. of PEG: (A); 1000, (B); 2000, (C); 3000, (D); 20,000. PMAA membrane: 3.4 cm² area, 30 μm thickness. PEG added: 5.8 × 10⁻² mol/l, transmembrane pressure: 0.2 kg/cm². Vertical arrows indicate the points at which water or PEG solution was added

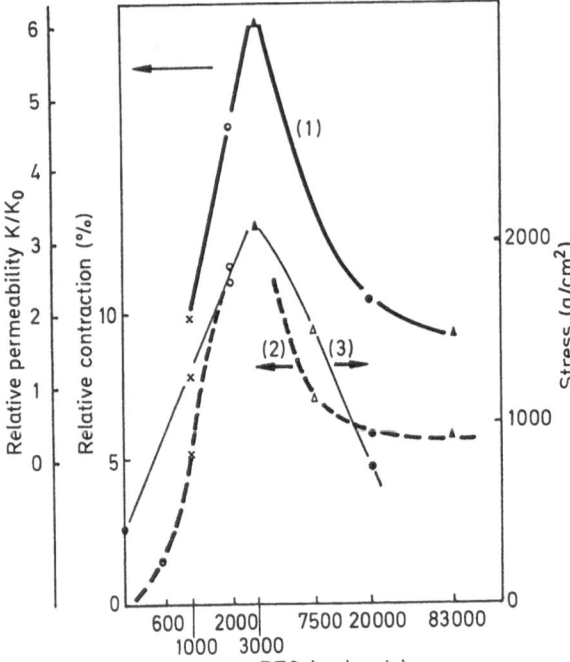

Fig. 22. Dependence of water permeability (1), relative contraction (2), and stress (3) of a PMAA membrane on mol wt. of PEG added [90]

maintain high permeability and water subsequently permeates the membrane with high velocity. If the membrane is rinsed with alkali solution (ca. pH 8) to dissociate the complex and remove the PEG, it returns to the initial low water permeability.

The increased water permeation in the presence of PEG is due to the expansion of pores, because the water permeability has the same dependence on the molecular weight of PEG (Fig. 22, curve 1) as the contraction and the stress generated under isometric contraction (curves (2) and (3)). The appearance of a maximum at a mol wt. of PEG of 3000 in the three curves in Fig. 22 indicates that the membrane-PEG complexation takes place not only on the surface, but also inside the swollen membrane due to the penetration of PEG molecules through the expanded pore channels (penetration mechanism) [89]. In addition, more than 98 % of the PEG added was found in the permeate, indicating that effective permeation of PEG molecules through the membrane occurred. It should be noted that other water soluble polymers capable of forming complexes with PMAA, for instance poly(N-vinylpyrrolidone), poly-(methylvinylether), and poly(vinyl alcohol), do neither contract the membrane nor increase the water permeation, and no polymers were found in the permeates [91].

Thus, the chemomechanical PEG treated PMAA membrane can behave as a "chemical valve", expanding and contracting the pore size. Flow tests with hemoglobin and albumin solutions were conducted on this chemical valve. It is seen from Fig. 23 that both protein solutions permeate the membrane without decreasing the high permeation velocity owing to the expansion of pores, whereas in the case of an untreated PMAA membrane, the proteins clog the membrane rapidly and hardly permeate. With the chemomechanical membrane the retention drops from 51 to 43 % for albumin and from 78 to 53 % for hemoglobin. The fluxes of water (F_0) and of protein

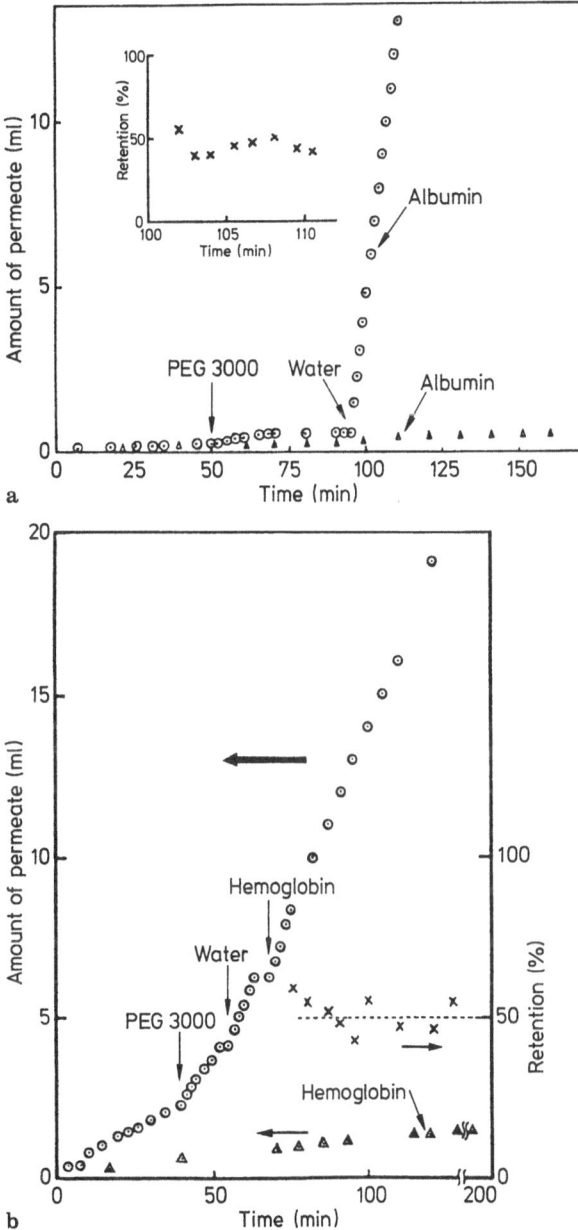

Fig. 23a and b. Effect of chemomechanical contraction of PEG treated PMAA membranes on albumin (a) and hemoglobin (b) permeation [90]. (a) PEG added: mol wt. = 3000, 5.8×10^{-2} mol/l solution, albumin (human) added: mol wt. = 67,000, $2.5 \times 10^{-2}\%$ solution, transmembrane pressure: 0.2 kg/cm^2, $\triangle$ denotes albumin permeation through an untreated PMAA membrane. Insert: percent of retention of albumin by the chemomechanical membrane. (b) PEG added: mol wt. = 3000, 5.8 $\times 10^{-2}$ mol/l solution, hemoglobin (from beef blood): mol wt. = 64,000, $2.5 \times 10^{-3}\%$ solution transmembrane pressure: 0.2 kg/cm^2, $\triangle$ denotes hemoglobin permeation through an untreated PMAA membrane

solution (F), the retention (R) and performance (V: the value of the velocity of permeation multiplied by % retention) are summarized in Table 3. It follows that the chemomechanical membrane can improve the performance of protein separation about 240 times for albumin and 55 times for hemoglobin compared with an untreated PMAA membrane. The membrane permeabilities of aqueous solutions of other proteins (amylase, invertase), sugars (glucose, raffinose), and their mixtures were also investigated for PMAA membranes treated and untreated with PEG [92]. In all cases the PEG-treated membranes allowed a much faster permeation of the macro- and micro-solutes as well as of water.

Table 3. Results of flow tests of proteins and sugars with PMAA membranes treated and untreated with PEG [92]

	Albumin		Hemoglobin		Amylase	
	Untreated	Treated	Untreated	Treated	Untreated	Treated
F_0 ($\times 10^{-5}$)[a]	4.7	1.9	4.7	27	46	38
F ($\times 10^{-5}$)[a]	0.51	57	0.32	150	45	106
F/F_0	0.11	31	0.07	5.2	0.97	2.8
R	0.51	0.43	0.78	0.53	0.035	0.035
V ($\times 10^{-2}$)	5.5	1300	5.2	290	3.4	9.8

	Amylase + Albumin				Raffinose	
	Untreated		Treated			
	Amylase	Albumin	Amylase	Albumin	Untreated	Treated
F_0 ($\times 10^{-5}$)[a]		127		13.2	395	39
F ($\times 10^{-5}$)[a]		4.2		2.9	15	59
F/F_0		0.03		0.22	0.04	1.52
R	0.06	0.38	0.06	1.00	0.02	0.02
V ($\times 10^{-2}$)	0.2	1.25	1.4	21.9	0.76	31

	Glucose + Raffinose			
	Untreated		Treated	
	Glucose	Raffinose	Glucose	Raffinose
F_0 ($\times 10^{-5}$)[a]		347		72.1
F ($\times 10^{-5}$)[a]		0.98		2.9
F/F_0		0.003		0.041
R	0.18	0.30	0.23	0.34
V ($\times 10^{-2}$)	0.52	0.85	9.3	14

[a] (cm³ s⁻¹ cm⁻²); F_0 for PEG-treated membranes was measured prior to treatment with PEG; F_0: flux of water, F: flux of solution, R: retention, V: performance (see text)

Ultrafiltration has never been widely used because of the unavailability of membranes combining high permeability with high permselectivity. The inherently high water permeability of the chemomechanical membrane, plus its controllable permselectivity to water-soluble solutes, can potentially lead to practical ultrafiltration processes.

Another type of chemomechanical membrane having "chemical valve" function has a composite structure. It consists of a porous substrate onto which methacrylic acid (MAA) has been graft-polymerized by the method of plasma-initiated polymerization [92-100]. The principle of the chemomechanical expansion and contraction of the pores in this type of membrane is based on the specific interaction between micro- and macromolecular solutes contained in the permeant and the polymer chains grafted onto the membrane, giving rise to a significant conformational change of the grafted polymer [101]. This interaction may be induced by changing the pH and ionic strength, or by the addition of metal ions or polymers capable of forming complexes with the grafted polymer. For example, at low pH the PMAA is mostly unionized, assuming a compact conformation, and therefore the composite membrane exhibits high permeability (Fig. 24a). In contrast, if the permeate is of high pH, PMAA is mostly ionized and assumes an extended conformation. The pores of the membrane are coverd effectively by PMAA chains and the apparent permeability is low (Fig. 24b). PMAA can also form chelate complexes with di- and trivalent metal ions (Fig. 24d), or from polymer complexes with complementary polymers like PEG, poly(N-vinyl-pyrrolidone (PVPdn) through cooperative hydrogen bonding (Fig. 24c). These specific interactions of PMAA will result in a drastic conformational shrinkage of the chains and lead to increased permeability.

The pH dependence of the water permeability of the graft membrane at constant ionic strength was investigated, and it was found that the water permeability increased about 7 times when the pH of the permeate changed from 12 to 2.

Chelation with di- and trivalent metal ions showed even more drastic permeability changes. The experiment was performed by alternating addition of Cu^{2+} and ETA (ethylenediamine-tetraacetic acid disodium salt), an effective polydentate ligand for Cu^{2+}. The result is shown in Fig. 25. It is seen that the alternating addition of Cu^{2+}

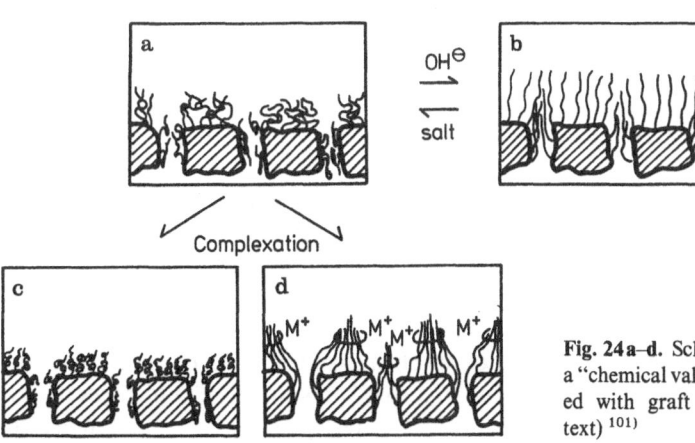

With polymers   With metals

Fig. 24a–d. Schematic illustration of a "chemical valve" membrane prepared with graft copolymer; (**a–d** see text) [101]

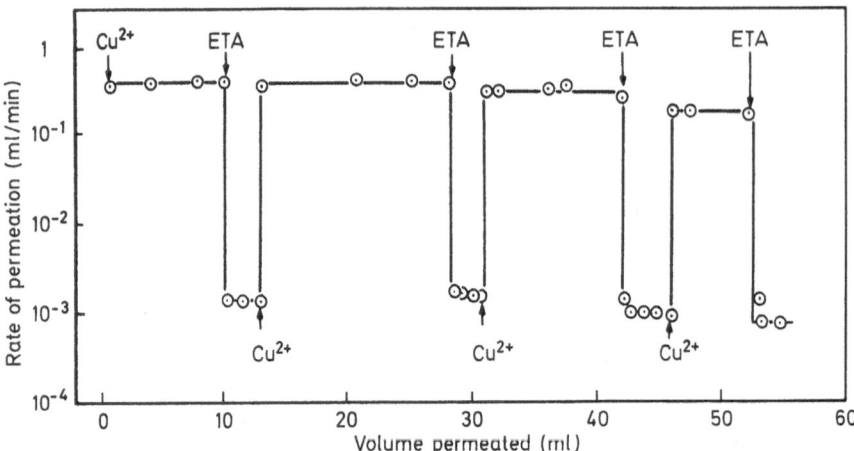

Fig. 25. Change in water permeability by alternating addition of Cu^{2+} and ETA [101]. $CuCl_2$: 20 ml of a solution of 5.0 mmol/l; ETA (ethylenediaminetetraacetic acid disodium salt): 20 ml of a solution of 10.0 mmol/l

and ETA brings about repeated changes in water permeability as large as by a facter 1000. Increased water permeability was also obtained by the addition of polymers capable of undergoing interaction with the PMAA chains to form polymer-polymer complexes [101].

3.4.2 Control of Slow Release and Diffusion Processes

Invertase is an enzyme with a mol. wt. of 27×10^4 which hydrolyzes saccharose to glucose and fructose. Due to its large molecular size, invertase cannot penetrate through a PMAA membrane treated with PEG. Figure 26 shows an experiment, in which an aqueous solution of invertase was separated from that of a substrate (saccharose) though an untreated PMAA membrane. The pore size of the membrane was allowed to change by varying the pH of the medium. The rate of saccharose hydrolysis depends strongly on the pore size of the membrane [91]. It is seen that the rate of reaction at pH 3.5 is much higher than that at pH 4.62. Since invertase exhibits the highest activity at pH 5.5, the obtained result is due to the enlargement of the pore radius of the PMAA membrane realized by isometric contraction. Thus, a mechanochemical membrane can change the apparent rate of enzyme reaction under isothermal condition.

Another type of reaction control realized by isotonic contraction of a PMAA membrane is shown in Fig. 27 [91]. In this case the invertase solution was placed in a glass tube, the lower part of which was loosely wrapped with the PMAA membrane. The contraction of the membrane, bringing about the shrinkage of the pore radius (isotonic contraction cf. Fig. 20a) interferes with the permeation of the substrate molecules from outside. Fig. 27 shows the effects of pH change and addition of PEG on the rate of hydrolysis. It is seen that the addition of a small amount of PEG impedes the hydrolysis completely, whereas a pH change has no effect on the rate. PEG molecules adsorbed on the PMAA membrane can be removed by rinsing with a dilute NaOH

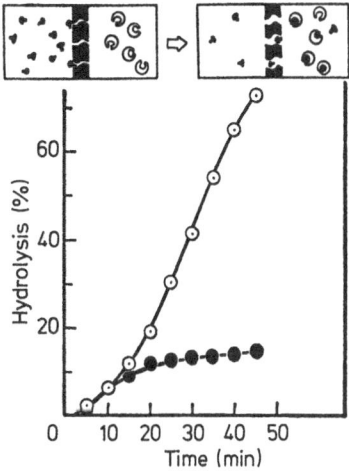

Fig. 26. Effect of the isometric contraction of a PMAA membrane on the rate of hydrolysis of saccharose [91]. Saccharose: 10 wt.%, Temp.: 25 °C, ●: pH 4.62, ⊙: pH 3.5

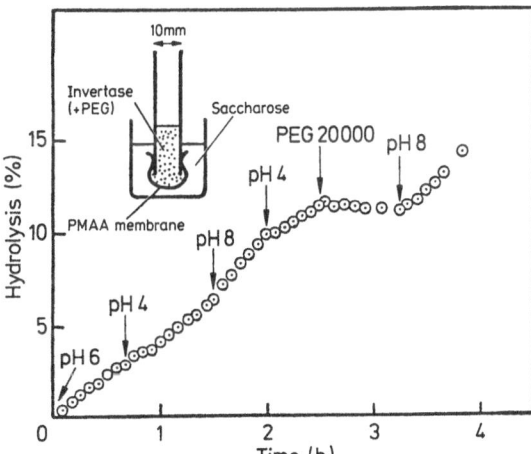

Fig. 27. Effect of isotonic contraction caused by pH change and PEG addition on the rate of hydrolysis of saccharose [91]. Invertase solution: 2 ml; Saccharose: 20 wt.% 10 ml; HCl: 0.1 mol/l; NaOH: 0.1 mol/l; PEG added: 5.8 × 10⁻² mol/l, Temp.: 25 °C

solution, after which hydrolysis starts again. The described contraction and expansion of the PMAA membrane can also be applied to control the release rate: the release of raffinose was completely stopped by the addition of a small amount of PEG to water in which PMAA gel containing raffinose was immersed. No effect was observed by the addition of PVPdn or PMVE (cf. Fig. 16).

4 Chemomechanical Systems Based on Electrokinetic Processes (Electro-Activated Chemomechanical Systems)

4.1 Principle and Behavior

The model of an electrically activated artificial muscle system which contracts and delates reversibly by an electric stimulus under isothermal condition was developed recently [102]. The electrical control makes use of crosslinked polyelectrolyte gels. A water-swollen polymer gel is inserted between a pair of electrodes which is connected to a DC source (Fig. 28). When the electricity is turned on, the polymer gel starts to shrink, releasing water droplets. The velocity of water release is high. For example, a water-swollen poly(2-acrylamido-2-methyl-1-propanesulfonic acid) (PAMPS) gel (12 mm wide, 19 mm long, 30 mm high, 5.6 g weight, 1 g dry PAMPS absorbs 2250 g water), reduced its weight by 70%, due to loss of water, when a 12 V/cm (0.7 mA) DC current was imposed for 20 min (Fig. 29). The minimum electric field to induce contraction of the gel was 1–2 V/cm, the rate of water release increased with increasing electric field. After immersing the gel in water, the dehydration could be repeated with the same velocity.

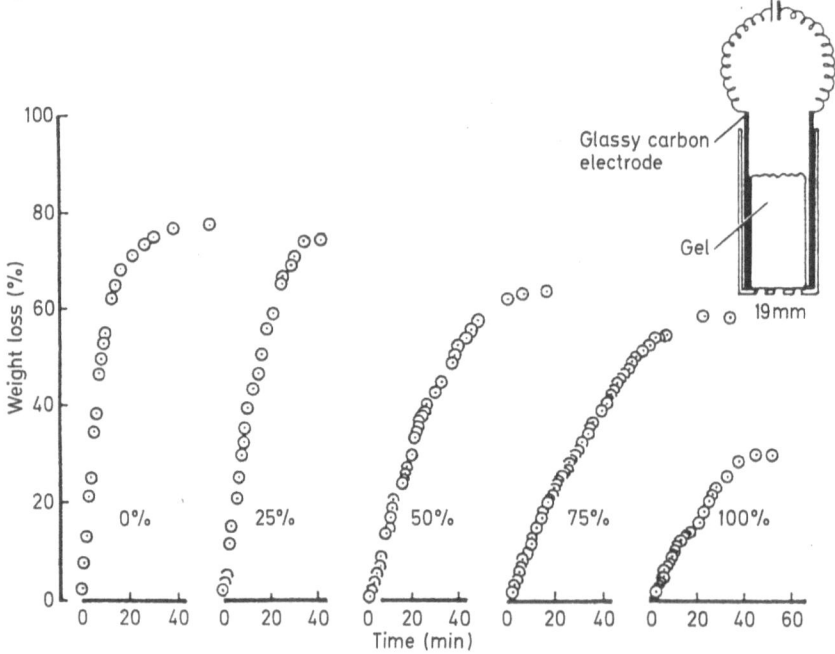

Fig. 28. Apparatus for the electro-activated chemomechanical system and time profile of weight change of the gel equilibrated in water-ethanol mixtures [102]. Gel: poly(2-acrylamido-2-methyl-1-propanesulfonic acid); electric field: 6.3 V/cm; the degree of swelling of the gel before the experiment was 2250 times in water and 1000 times in ethanol. Numbers denote volume % of ethanol

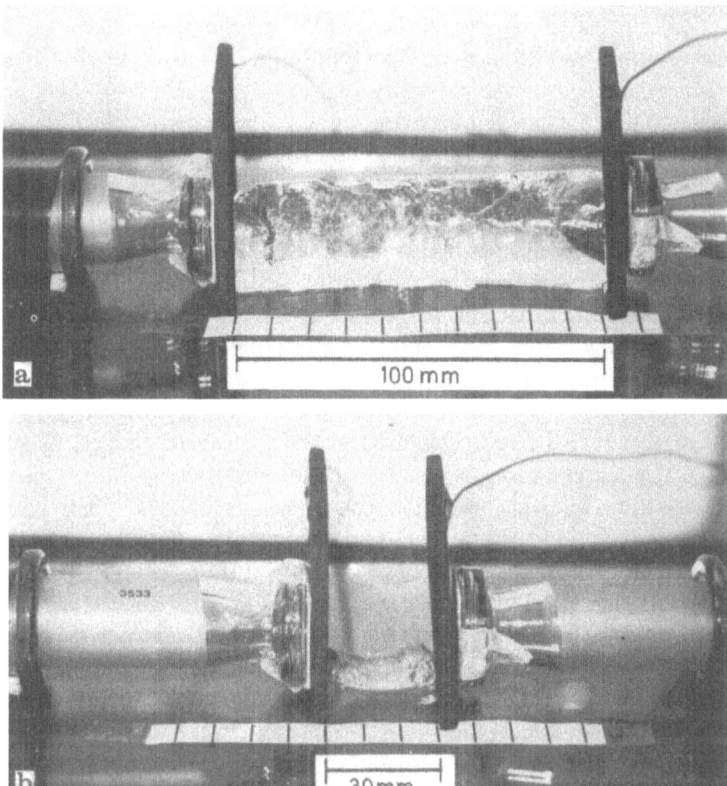

Fig. 29a and b. Photograph of a polyelectrolyte gel before (**a**) and after (**b**) imposing an electric field; gel: crosslinked poly(2-acrylamido-2-methyl-1-propanesulfonic acid); electric field: 15 V, 10 h

The presence of a salt such as NaCl in the gel increases the rate of dehydration, whereas organic solvents such as ethanol or acetone decrease both the rate and the extent of the contraction, as shown in Fig. 28. The rate of water release of the PAMPS gel in 50 and 100 % ethanol was 38 and 17 % of that in water, respectively. No contraction was observed if an AC in the range of 6 Hz to 5 kHz was applied at 20 V. Other gels prepared from synthetic polymers and copolymers containing ionizable groups, for example poly(methacrylic acid), partially-hydrolyzed poly(acrylamide), poly-(styrenesulfonic acid), quarternized poly(4-vinyl pyridine), poly(vinylbenzyl trimethyl ammonium chloride) exhibited also marked contraction. Gels from proteins such as gelatine and collagen, sugars such as alginic acid and its salts, ager-ager, and gum arabic also showed contraction by electric stimulus. In all these cases, a migration of the gel to the counter electrode was observed, i.e., the negatively charged polymer gel migrated to the anode and, consequently, the volume of the gel near the anode shrinked. Water droplets evolved at the cathode. Polymer gels containing no ionizable moieties such as poly(2-hydroxyethyl methacrylate) (HEMA), poly(acrylamide), poly(vinyl alcohol), and starch showed no contraction at all.

The phenomena of contraction and dehydration are apparently simple, but the mechanism is not fully clarified at present. Tanaka [103, 159] described that a phase transition occurs resulting in a sudden collapse of the gel when an electric field is applied across a partially hydrolyzed poly(acrylamide) gel placed in a 50 % mixture of acetone and water. The principle and behavior observed by us seem to be different from those observed by Tanaka since the contraction of our gel is not induced by a phase transition. We have found [105] that the gel shrinks most significantly and most rapidly in water and an increasing acetone component results in a gradual decrease of the contraction of partially hydrolyzed poly(acrylamide) gel. Increasing the electric field leads to a corresponding increase in contraction; no critical point of the electric field inducing a sudden substantial shrinkage was observed. The presence of an appropriate amount of NaCl increased the rate of contraction, whereas that of CaCl decreased it significantly. The rate of contraction was proportional to the electric current applied.

We tentatively suggest that the contraction occurs as a result of an electrostatic interaction between the charged macromolecules and the electrode, giving extensive desolvation (dehydration). For the polyanion gels it was found that contraction takes place only in the region of the anode and no contraction was observed in the cathode area. Release of water droplets was observed only at the edge of the gel near the cathode. The situation is reversed if a positively charged polymer gel, for example one of quarternized poly(4-vinylpyridine), is used, i.e., significant contraction occurs near the cathode and water evolves near the anode. These results suggest that negatively charged crosslinked macromolecule electrophoretically move to the anode. The positively charged counter ions of the polyelectrolyte and water molecules should travel to the cathode due to electrostatic forces and the electro-osmosis phenomenon, respectively. In fact, when the calcium salt of a crosslinked poly(arginic acid) gel was allowed to contract for 2 h and then dried, a well-alligned meridional diffraction pattern was observed at the end of the gel near the anode; at the same time a strong pattern due to the calcium salt was observed at the opposite end of the gel. The contraction of the gel could therefore be associated with the decrease in hydration power of macro-ions and microcounter ions due to the interaction with the electrodes. In the case of the polyanion gels, hydrated H_3O^+ ions travel to the cathode together with water molecules and are reduced at the electrode releasing hydrogen gas and water droplets from the gel:

$$2\,H_3O^+ + 2e^- \rightarrow H_2\uparrow + 2\,H_2O\downarrow \tag{1}$$

Near the anode the following reaction may take place:

$$4\,OH^- \rightarrow O^2\uparrow + 2\,H_2O + 2e^- \tag{2}$$

Polycations will be adsorbed to the anode through electrostatic forces; they lose the hydrating water molecules which again travel to the cathode together with H_2O according to equation (1). The detailed mechanism of the contraction requires further investigation.

4.2 Electro-Activated Chemomechanical Devices

Several potential applications for electrically activated chemomechanical gels may be envisaged [102].

A chemomechanical device capable of moving a load up and down automatically and repeatedly is the simplest application of electro-shrinkable gels. Thus, two pieces of water-swollen gel (10–20 g) of AMPS were placed on the two plates of a balance and a DC current (10 V) was applied to one of the pieces. The weight of the gel decreased, bringing the device out of blance. DC was now connected to the other gel and started shrinking it. Thus, the balance could oscillate many times (more than 100 times) until most of the water of the gel was consumed.

Figure 30 shows the schematic diagram of an electrically activated "chemical valve" membrane which reversibly expands and contracts the pore size by an electric stimulus. If the chemomechanical contraction is developed isometrically, i.e. keeping the membrane dimensions constant, the contractile stress appearing in the membrane expands the pore channels through which water may permeate. The principle is the same as described in Sect. 3.4.1. Figure 30 shows also the effect of a chemomechanical contraction of the PAMPS membrane on water permeation, if 6.5 V of DC is alternatingly imposed and shut off. The "chemical valve" membrane increases and decreases the water permeability repeatedly on electric stimulus. The increase is proportional to the DC current. The membrane might be used as a permselective membrane, continuously separating solute mixtures with different molecular size. This type of electrically-activated "chemical valve" membrane exhibits long term stability and has worked continuously for more than one month.

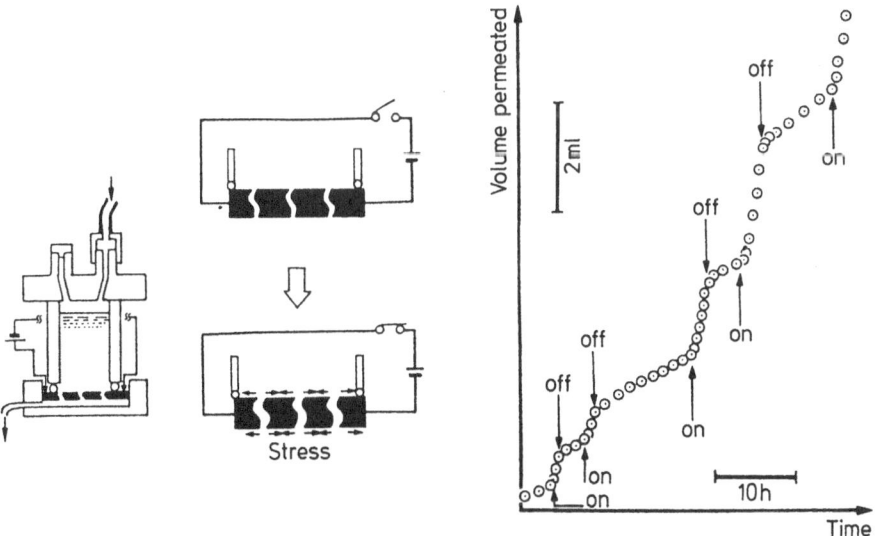

Fig. 30. Apparatus for the electro-activated "chemical valve" membrane and change of the water permeability by alternative "on" and "off" of an electric field [102]. Electric field: 2.6 V/cm. The membrane was prepared by polymerization of AMPS in the presence of a porous (average pore size 8 μm) poly(vinyl alcohol) sheet

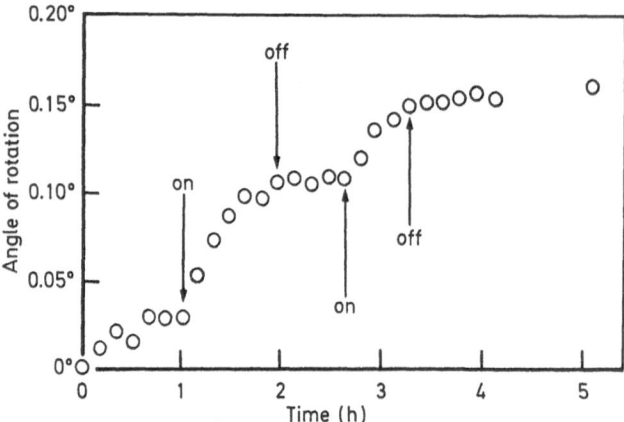

Fig. 31. Release profile of pilocarpine from PMAA gel by alternative "on" and "off" of an electric field [105]; electric field: 6 V/cm; Temp.: 25 °C

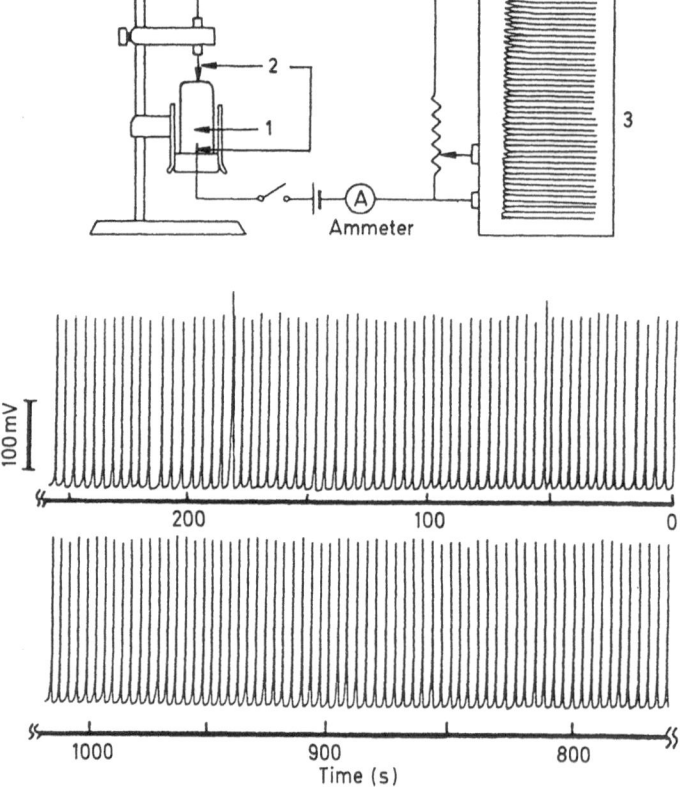

Fig. 32. Circuit for the display of dynamic oscillation and its characteristics [102]. 1) platinum wire electrode, 2) gel, 3) amplifier and recorder. Gel: copolymer of AMPS and HEMA, 12 mm diameter, 10 mm long; degree of swelling: 45 times, electric field: 1.4 V/cm

On the base of the same principle, the present author developed a "slow release control" system for drugs [105]. A piece ($10 \times 10 \times 10$ mm^3) of crosslinked PMAA gel containing pilocarpin, (a drug for glaucoma, cf. Sect. 3.1) was inserted between a pair of electrodes and immersed in water. Figure 31 shows the change of the concentration of pilocarpin in water as a function of time. It is seen that the concentration of pilocarpine changes stepwise indicating that the drug is flashed out by imposing the electric field. This kind of drug system has a potential application as a "missile drug", shooting the drug of a required amount to the required place at a required moment.

We have found that a metal/polymer gel/metal structure (gel sandwiched between two platinum-wire electrodes), is able to switch on and off reversibly between two stable states: one characterized by high impedance corresponding to the "off" state, and the other by low impedance corresponding to the "on" state. The "off" and "on" states seem to be realized by reversible shrinking and expanding of the gel or harmonized movement of polymer chains driven by electric stimulus. Figure 32 shows the scheme and result for this dynamic oscillation or switching device. The ratio of voltage at "on" and "off" states is higher than 10. The effect is reproducible, and the device could be switched 300 to 2000 times within 20 min. The frequency of switching depends on the electric field: higher field induces higher frequency.

However, we have not yet succeeded in determining the experimental conditions necessary to produce a certain type of oscillation. Switching and memory phenomena have been observed previously in films of certain metal oxides and in amorphous semiconductors, but to our knowledge, these phenomena have not been observed previously in water-containing organic materials.

The present author also constructed a polymeric switching device functioning like a "bimetal", from a laminated polymer gel/plastic structure [102] (Fig. 33). The

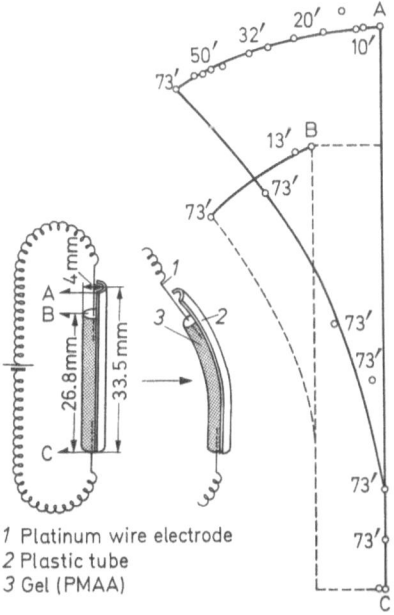

1 Platinum wire electrode
2 Plastic tube
3 Gel (PMAA)

Fig. 33. Schematic illustration of a polymer composite functioning like a "bimetal" and the profile of its "bending" [102]. Gel: poly(methacrylic acid), plastic: poly(vinyl chloride); electric field: 0.25 V/cm; numbers indicate time (min) elapsed after electric field was imposed

contractile force developed in the gel by an electric stimulus through a pair of platinum-wire electrodes bends the "nonshrinkable" poly(vinylchloride) tube. The rate and extent of bending are quite large. The composite device consisting of crosslinked poly-(methacrylic acid) (degree of swelling 3 times) and poly(vinyl chloride) tube, as illustrated in Fig. 33, bent as much as 24° after 30 min, and 43° after 70 min. It recovered the original linear form after dipping it into water.

Chemomechanical lifting behavior of the weight was investigated using mildly crosslinked polyelectrolyte gels under influence of direct current. It was fond that the rate of lifting of the weight attached to the bottom end of the gel and the power generated increased with an increase in the amount of weight [106]. As shown in Fig. 34, the rate of contraction initially decreased from 0.4 to 0.1 cm/h from 0 to 5.5 g of load. However, the lifting rate then increased with increasing amount of load. Thus, the gel being weighted with 22 g could lift at a rate 6 times faster than that with 5.5 g load. The efficiency of the work done by the gel with 22 g of load was 24 times larger than that done by the gel with 5.5 g. This anomalous behavior could be observed in variety of gels of weak polyacids and polybases such as: polyacrylic acid (PAA), calcium salt of alginic acid, and poly 4-vinylpyridine (PVP). Polymer-polymer complexes.

These anomalous lifting behaviours of the gels can be explained in terms of spontaneous ionization of ionizable groups which give rise an increased electric current by stretching (reverse chemomechanical reaction). It is well known that the dissociation of weak polyelectrolyte strongly influences its conformational state in solution, i.e., the increase in ionization results in more expanded conformation of polymer chain. Conversely, the stretching of the macromolecular chain by applying stress induces the additional ionization. This behavior of polyelectrolyte chain should be observed in polyelectrolyte networks as well, i.e., stretching of polymer network will induce an additional ionization of ionizable groups and consequently increase the rate of contraction of PMAA gel. In fact, we observed the ionization of carboxylic group, i.e., the decrease of pH in the gel by applying the stress.

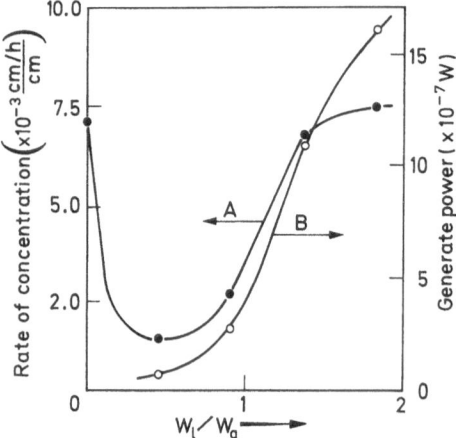

Fig. 34. Rate of contraction (A) and generate power (watt) (B) vs load applied (WI) per weight and length of PMAA gel sample (Wg) [106]. Conditions: sample dimensions: length — 60 mm, diameter — 17 mm, wet weight — 12 grams, Electric field: 50 V DC

This observed anomalous behavior may well provide the basis for an automatic or inherent control sensor which spontaneously adjusts the energy absorbed by a synthetic muscle system commensurate with that required to do the work, i.e., the heavier the load the more energy absorbed without external stimulus-similar to the mechanical devices with a sensor feedback control system. One can speculate about the similarity of biological muscle system wherein the force applied to accomplish a given work activity is in proportion to that required. Without this inherent control, the force applied to any work task, be it lifting a light object or a heavy one would be the same.

D. E. DeRossi [107] recently focused on contractile phenomena in polyelectrolyte networks activated by electric fields and investigated mechanism and fundamental contractile behaviors.

5 Future Aspects: A New Approach to Biomechanics and Artificial Muscles

Metal, ceramics, and plastics are three major industrial materials. They are strong and tough, but give us a feeling of "hard" and "cold". The materials described here are water-swollen gels feeling soft and exhibiting similar rheological behavior as natural muscle.

Chemomechanical systems converting chemical into mechanical energy may have a variety of applications. They have potential uses as antenna, switching device, thermostat, pH-sensitive electrode, and various kinds of sensors. They also may be particularly important to provide mechanical forces under water or in space where power supply is limited. Chemomechanical systems functioning on a small temperature gradient (an example of which was introduced in Sect. 3.1) are considered as a unique type of solid state heat engine which may function using exhaust gas, solar heat, or a temperature gradient of sea water as heat sources.

Systems which develop contractile force and change their dimensions by electric stimulus enable the construction of chemomechanical electronic devices transforming chemical energy into mechanical work, and may eventually lead to approaches to

Table 4. Potential applications of chemotechnical systems

Function	Application
Separation	Permselective membrane, chemical valve, affinity chromatography
Switch and sensoring	Thermostat, bimetal, pH control, functional electrode heat sensor, thermal actuator
Force	Jack, brake, lock, press, spring, wedge, rock and concrete breaker, self-construction and operation, connector, clip, clamp
Mechanical energy	Heat engine, chemical engine, toy
Biomechanical use	Artificial heart, artificial muscle, iris, clip for surgery, "missile" drug
Construction	Breathing wall, heat-absorbing material
Velocity control	Enzyme immobilization, slow release control

"biomechanics" based on water-swollen "soft" materials. From this point of view, the application of chemomechanical as the "actuator" for robots and other machine systems has particular interest. Robots made of these soft gels will behave more safely, more carefully, and more gently. They will not tear out the knob when they open a door, and they will shake hands softly instead of crushing the hand.

Combining a photovoltaic cell or solar battery with a polyelectrolyte gel one may construct a device the function of which resembles that of the iris diaphragm. The chemomechanical systems are also interesting as analogs to physiological muscle and biological systems exhibiting rhythmic phenomena, e.g., oscillation of nerve walls where the appearance of dissipative structures [107–109] is considered to play the predominant role. Finally, the chemomechanical systems using polyelectrolyte gels may serve as simple and suitable models which help to understand the mechanism of electrophysiological phenomena.

Examples of potential applications of chemomechanical systems are summarized in Table 4.

6 References

1. Kuhn, W.: Experimentia, *5*, 318 (1949)
2. Breitenbach, J. W., Karlinger, H.: Monatsh. Chem., *80*, 211 (1949)
3. Katchalsky, A., Oplatka, A.: Proc. Fourth Intern. Cong. Pheolgy, edit, by Lee, E. H., and Copley, A. L., Partl, 73 (Interscienc Publishers. 1965)
4. Kuhn, W., Hargitay, B., Katchalsky, A., Eisenberg, H.: Nature, *165*, 514 (1950)
5. Engelhardt, V. A.: Advances in Enzymology, *6*, 147, F. F. Nord Ed., Interscience, New York (1946)
6. "Size and Shape Changes of Contractile Polymers" Chap. 1 Ed., A. Wassermarn, Pergamon Press, New York (1960)
7. Katchalsky, A.: J. Polym. Sci., *7*, 393 (1952)
8. Sussman, M. V.: Nature, *256*, 195 (1975)
9. Katchalsky, A., Eisenberg, H.: Nature, *166*, 267 (1950)
10. Kuhn, W., Ranel, A., Walters, D. H., Ebner, G., and Kuhn, H. J.: Fortschr. Hochpolymer Forsch., *1*, 540 (1960)
11. Kuhn, W., Thurkauf, M.: Kolloid-Zeitschrift und Zeitschrift für Polymere, *184*, 114 (1964)
12. Tatara, Y.: Trans. Japan Soc. Mech. Engrs. (in Japanese), *76*, 1000 (1973)
13. Flory, P. J.: "Principles of Polymer Chemistry", Cornell University Press, Ithaca, New York, 1953 Chap. 13
14. Katchalsky, A., Kunze, O., and Kuhn, W.: J. Polym. Sci., *5*, 283 (1950)
15. Katchalsky, A., Michael, I.: J. Polym. Sci., *15*, 69 (1955)
16. Flory, P. J., Tatara, Y.: J. Polym. Sci., Polym. Phys. Ed., *13*, 683 (1975)
17. Tatara, Y.: J. Polym. Sci., Polym. Symp., *54*, 283 (1976)
18. Boyack, J., Enos, J., Lalone, A., Fragala, A.: "Development of an Electrically Activated Artificial Muscle System" (1970)
19. Mongar, J. L., Wasserman, A.: J. Chem. Sci., 500 (1952)
20. Gregor, H. P., Gutoff, E., Bregmann, J. T.: J. Colloid Sci., *6*, 245 (1951)
21. Katchalsky, A., Zwick, M.: J. Polym. Sci., *16*, 221 (1955)
22. Hojo, N., Shirai, H., Kameyama, Y., Nagasaki, T., Ichikawa, S.: Kogyo Kagaku Kaishi, *74*, 269 (1971)
23. Hojo, N., Shirai, H., Mori, T.: ibid., *74*, 273 (1971)
24. Kuhn, W.: Gazg. Chem. Ital., *92*, 951 (1962)
25. Kuhn, W.: Angew. Chem., *70*, 50 (1958)
26. Kuhn, W., Ramel, A., Walters, D. H.: ibid., *70*, *31*, Chimia (Zürich) *12*, 123 (1958)
27. Ueno, A., Anzai, J., and Osa, T.: J. Polym. Sci., Polym. Chem. Ed., *17*, 149 (1979)
28. Agolini, F., and Gay, F. P.: Macromolecules, *3*, 349 (1970)

29. Smets, G.: Polym. Symposium IUPAC-Congress in Bratislava, Chem., *39*, 225 (1974)
30. Smets, G., and De Bauwe, F.: Pure Appl. Chem., *39*, 225 (1974)
31. Smets, G.: J. Polym. Sci., Polym. Chem. Ed., *13*, 2223 (1975)
32. Aviram, A.: Macromolecules, *11*, 1275 (1978)
33. Osada, Y., Katsumura, E., Inoue, K.: Makromol. Chem., Rapid Commun., *2*, 411 (1981)
34. Negishi, N., Ishihara, K., Shinohara, I., Akaike, T., Okano, M., Kataoka, K., and Sakurai, Y.: Kobunshi Ronbunshu, *37*, 287 (1980)
35. Ishihara, K., Hamada, N., Kato, S., and Shinohara, I.: J. Polym. Sci., Polym. Chem. Ed., *22*, 881 (1984)
36. Ishihara, K., and Shinohara, I.: J. Polym. Sci., Polym. Lett. Ed., *22*, 515 (1984)
37. Negishi, N., Ishihara, K., Shinohara, I., Okano, T., Kataoka, K., and Sakurai, Y.: Makromol. Chem., Rapid Commun., *2*, 95 (1981)
38. Ishihara, K., Muramoto, N., Shinohara, I.: J. Appl. Polym. Sci., *29*, 211 (1984)
39. Kano, K., Tanaka, Y., Ogawa, T., Shimomura, M., Okahata, Y., and Kunitake, T.: Chem. Lett., 421 (1980)
40. Balasubramanian, D., Subramani, S., Kumar, S.: Nature (London), *254*, 252 (1975)
41. Chen, D. T., Morawetz, H.: Macromolecules, *9*, 463 (1976)
42. Irie, M., Menju, A., Hayashi, K.: Macromolecules, *12*, 1176 (1979)
43. Ueno, A., Yoshimura, H., Saka, R., Osa, T.: J. Amer. Chem. Soc., *101*, 2779 (1979)
44. Shinkai, S., Nakaji, T., Noshida, Y., Ogawa, T., and Manabe, O.: J. Amer. Chem. Soc., *102*, 5860 (1980)
45. Shinkai, S., Nakaji, T., Ogawa, T., Shigematsu, K., and Manabe, O.: J. Amer. Chem. Soc., *103*, 111 (1981)
46. Shinkai, S., Ogawa, T., Kusano, Y., Manabe, O., Kikukawa, K., Goto, T., and Matsuda, T.: J. Amer. Chem. Soc., *104*, 1960 (1982)
47. Shinkai, S., Kinda, H., and Manabe, O.: J. Amer. Chem. Soc., *104*, 2933 (1982)
48. Yuki, H., Sakakibara, S., Tani, T., and Tani, H.: Bull. Chem. Soc. Japan, *29*, 664 (1956)
49. Steinberg, I. Z., Oplatka, A., and Katchalsky, A.: Nature, *210*, 568 (1966)
50. Sussman, M. V., and Katchalsky, A.: Science, *167*, 45 (1970)
51. Kuhn, W.: Makromol. Chem., *35*, 2. Sonderband 200 (1960)
52. Pasinsky, A., Brokhina, V.: Doklady Acad. Nauk USSR, *86*, 1171 (1952)
53. Vorob'ev, V. I.: Doklady Acad. Nauk USSR, *137*, 972 (1961)
54. Kukhareva, L. V., Ginsburg, B. M., Vorov'ev, V. I., and Frenkel, S. Ya.: Macromolecules, *6*, 508 (1973)
55. Frenkel, S. Ya.: "Kinetic Theory of Liquids" Izd. Acad. Nauk USSR, Moscow, 1945
56. Gee, G.: Quart. Rev., Chem. Soc., *1*, 265 (1947)
57. Flory, P. J.: J. Amer. Chem. Soc., *78*, 5222 (1956)
58. Michels, A. S.: Encyclopedia of Polymer Technology, Wiley, New York, Vol. 16, (1968)
59. Kabanov, V. A.: IUPUC Macromol. Chem., *8*, 121 (1973)
60. Tsuchida, E., Osada, Y., and Sanada, K.: J. Polym. Sci. Polym. Chem. Ed., *10*, 3397 (1972)
61. Papisov, I. M., Antipina, A. D., Kabanov, V. A.: Dokl. Acad. Nauk USSR, *199*, 1364 (1971)
62. Tsuchida, E., and Osada, Y.: Kobunshi Kagaku, *30*, 515 (1973); Kobunshi, *22*, 384 (1973)
63. Morawetz, H., Hughes, W. L., Jr.: J. Phys. Chem., *56*, 64 (1952)
64. Chen, Haunn-Lin, Morawetz, H.: Macromolecules, *15*, 1445 (1982)
65. Ohkubo, T., Honjo, K., Enokida, A.: J. Chem. Soc. Faraday Trans. 1, *80*, 2087 (1984)
66. Schodt, K. P., Gelman, R. A.: J. Blackwell, Biopolymers, *15*, 1965 (1976)
67. Ohno, H., Nii, A., Tsuchida, E.: Makromol. Chem. *181*, 1227 (1980)
68. Baranovsky, V. Yu., Litmanovich, H. A., Papisov, I. M., Kabanov, V. A.: Europ. Polym. J., *17*, 969 (1981)
69. Izumrudov, V. A., Savitskii, A. P., Bakeev, K. N., Zezim, A. B., Kabanov, V. A.: Makromol. Chem., Rapid Commun., *5*, 709 (1984)
70. Margolin, A. C., Sherstyuk, S. F., Izumrudov, V. A., Zezin, A. B., Kabanov, V. A.: Europ. J. Biochem., *146*, 625 (1985)
71. Saito, M., Abe, K., Osada, Y., and Tsuchida, E.: J. Chem. Soc. Jpn., *5*, 977 (1974)
72. Tsuchida, E., Osada, Y., and Abe, K.: Makromol. Chem., *175*, 583 (1974)
73. Tsuchida, E., and Osada, Y.: Makromol. Chem., *175*, 593 (1974)

74. Anufrieva, E. V., Pautov, V. D., Papisov, I. M., Kabanov, V. A.: Dokl. Akad. Nauk SSSR, *232*, 1096 (1977)
75. Kabanov, V. A., Zezin, A. B.: Soviet Sci. Rev., Ser. B, Chem. Rev., *4*, 207 (1982), Ed. M. E. Volpin, Harwood Academic publishers GmbH
76. Zezin, A. B., Lutsenko, V. V. Rogachova, V. B., Aleksina, O. A., Kalushnaya, R. I., Kabanov, V. A., and Kargin, V. A.: Vysokomol. Soedin. Ser. A, *14*, 772 (1972)
77. Zezin, A. B., Lutsenko, V. V., Izumrudov, V. A., and Kavbanov, V. A.: Vysokomol. Soedin. Ser. A., *16*, 600 (1974)
78. Osada, Y., Sato, M.: J. Polym. Sci., Polym. Lett. Ed., *14*, 129 (1976)
79. Osada, Y., Saito, Y.: Makromol. Chem., *176*, 2761 (1975)
80. Osada, Y., Saito, Y.: Nippon Kagaku Kaishi, 171 (1976)
81. Osada, Y., Sato, M.: Polymer, *21*, 1057 (1980)
82. Osada, Y.: J. Polym. Sci., Polym. Lett. Ed., *18*, 281 (1980)
83. Fedofov, N. G., Vedeneev, V. I., Sarkisov, O. M.: Dokl. Akad. Nauk USSR, *208*, 401 (1984)
84. Osada, Y., Koike, M.: Chem. Lett., 209 (1984)
85. Koike, M., Osada, Y.: Polymer Preprints, Japan, *32*, 2805 (1983)
86. Osada, Y.: J. Polym. Sci., Polym. Chem. Ed., *17*, 3485 (1979)
87. Osada, Y., Koike, M., Katsumura, E.: Chem. Lett., 809 (1981)
88. Osada, Y., Sato, M.: Nippon Kagaku Kaishi, 175 (1976)
89. Osada, Y.: J. Polym. Sci., Polym. Chem. Ed., *15*, 255 (1977)
90. Osada, Y., Takeuchi, Y.: J. Polym. Sci., Polym. Lett. Ed., *19*, 303 (1981)
91. Osada, Y., Takeuchi, Y., Koike, M.: Nippon Kagaku Kaishi, 812 (1983)
92. Osada, Y., Takeuchi, Y.: Polym. J., *15*, 279 (1983)
93. Bell, A. T., Shen, M.: Eds. "Plasma Polymerization", American Chemical Society; Washington DC, 1979; ACS Symp. ser. No. 108
94. Osada, Y., Bell, A. T., Shen, M. J.: Polym. Sci., Polym. Lett. Ed. 16, 309 (1978)
95. Johnson, D., Osada, Y., Bell, A. T., Shen, M.: Macromolecules, *14*, 118 (1981)
96. Osada, Y., Takase, M., Iriyama, Y.: Polym. J., *15*, 81 (1983)
97. Osada, Y., Takase, M.: Nippon Kagaku Kaishi, 439 (1983)
98. Osada, Y., Iriyama, Y.: J. Amer. Chem. Soc., *104*, 2925 (1982)
99. Osada, Y., Iriyama, Y.: J. Phys. Chem., *88*, 5951–5956 (1984)
100. Osada, Y., Iriyama, Y.: Thin Solid Films, *118*, 197 (1984)
101. Osada, Y., Honda, K.: J. Membrane Sci., *27*, 327 (1986)
102. Osada, Y., Hasebe, M.: Chem. Lett., 1285 (1985)
103. Tanaka, T.: Scientific American, *244*, 110 (1981)
104. Tanaka, T., Nishio, I., Sun, S. T., and Nishio, S. V.: Science, *218*, 467 (1982)
105. Osada, Y., Hasebe, M.: Reprints of Ann. Meeting of J. Electrochem. Soc. Japan 12–14 (1984)
106. Osada, Y., Kishi, Y.: Macromolecules, submitted
107. DeRossi, D. E., Chiarell, P., Buzzigoli, G., and Domenici, C.: Trans. Am. Soc. Artif. Intern. Organs, *32*, 157 (1986)
108. Nicolis, G., and Prigogine, I.: "Self-Organization in Nonequilibrium System–From Dissipative Structures to Order through Fluctuations", John Wiley & Sons, Inc., New York, (1977)
109. Glansdorff, P., and Prigogine, I.: "Thermodynamic Theory of Structure, Stability and Fluctuations", Wiley-Interscience, London, (1971)
110. Prigogine, Ilya: "Time and Complexity in the Physical Sciences", W. H. Freeman and Company, San Francisco, (1980)

Editors: S. Olivé and G. Henrici-Olivé
Received May 27, 1986

Transient Relaxation Mechanisms in Elongated Melts and Rubbers Investigated by Small Angle Neutron Scattering

François Boué
Laboratoire "Léon Brillouin" Centre d'Energie Nucléaire de Saclay,
91191 Gif-sur-Yvette Cedex/France

The dynamics of polymer melt can be investigated by observing the return to isotropic state of a piece of melt after a sudden deformation, maintained during all the process of relaxation (a step-strain). This method of transient relaxation is applied to small angle neutron scattering measurements. The very large domain of time which is covered using the time temperature superposition allows to test the different models of dynamics, mainly the Rouse model (chains are assumed to behave as if they were free) and the tube-reptation model, successfully applied for the description of entangled polymer. The possible form factors for oriented polymers are also discussed for the melt as well as for rubbers. The labelling technique (deuteration) is very flexible and allows to observe chains of given molecular weight inside matrices of chains of another molecular weight.

Advances in Polymer Science 82
© Springer-Verlag Berlin Heidelberg 1987

1 Introduction

The material studied here is a particular solid: a polymer melt, i.e. a polymer in the bulk state above the glass transition temperature T_g and crystallisation temperature T_s. Because the chains are long and entangled, the maximum characteristic time is also very long. It is easy to reach considerably lower values, as in the experiment described here, in the lower time range: thus the material behaves macroscopically as a solid. In the upper time range, one approaches the maximum relaxation time and the material begins to behave macroscopically as a liquid. A few results will also be given on cross-linked melts (i.e. rubbers) which will never behave macroscopically as a liquid because of the permanent crosslinks. Conveniently for Small Angle Neutron Scattering (SANS) measurement the sample (polystyrene, PS) is quenched in the glassy state; PS does not crystallize, it just freezes in the melt state without any conformational change. Flows of solutions can be used in SANS for determining the dynamics of solutions (see end of section 2.1); however, data on this methods will not be discussed here. We basically consider the dynamics of a chain in a polymer melt.

Briefly, the experiment is the following: the sample is prepared in an isotropic state and contains labeled (deuterated) chains. It is then deformed as fast as possible and maintained under constant conditions (length, temperature). The small angle neutron scattering — due to the presence of the deuterated chains — is recorded for each value of the time t elapsed since the end of the stretching. This gives the form factor of the single chain as a function of t, denoted as $S_t(q)$. If the duration of the stretching is short compared to the time t, the deformation history is equivalent to a step strain. A typical variation of the stress with t is given in Fig. 1; the different regimes will be described later; the longest relaxation time is denoted as T_{ter} and was measured in some of the experiments. As the time t increases, the material returns to the isotropic state. Thus, $S_t(q)$ returns to the isotropic form factor for $t \gg T_{ter}$, i.e. the dynamics are observed via the return to equilibrium of a deformed system under the effects of the thermal fluctuations. This use of static SANS for dynamics appears as a *very powerful* tool, for several reasons:

(i) it is well suited to the melt as it can use nuclear labeling (deuteration) which does not cause demixtion, because the Flory factor χ between deuterated and hydrogenated

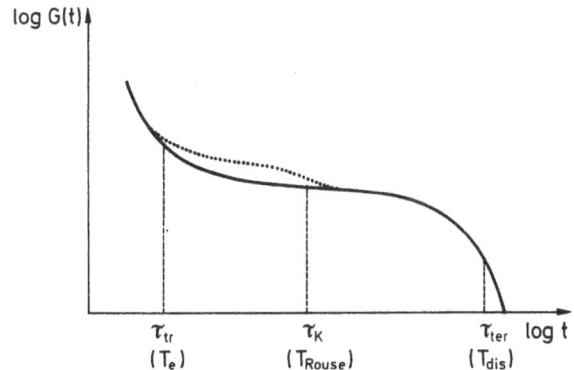

Fig. 1. Schematic variation of the modulus G(t) with time after a step deformation, for a small elongation ratio (solid line) and large one (dotted line)

species is very weak. The labeling allows also to observe chains of a given molecular weight diluted among chains with different molecular weights.

(ii) measurements can be precise or reasonably fast for a melt such as polystyrene because a large fraction of deuterated species can be used, which gives a large scattering signal [1], when χ can be neglected (see Ref. 62).

(iii) it covers a large part of the submolecular range of the scattering vector q for chains with molecular weights M which are relatively easily synthesized and handled and cover completely the range of scientific and industrial interest. Namely, q range from $5 \cdot 10^{-3} \text{ Å}^{-1}$ to $5 \cdot 10^{-1} \text{ Å}^{-1}$ and M from some thousands to $2 \cdot 10^{6}$.

(iv) it allows to study times t which are too long to be covered by the inelastic techniques observing the fluctuations of the system at equilibrium.

Other techniques used for dynamics do not combine these advantages. For example, light scattering fulfills only (iii) within a different range of q but not (i); X-ray scattering obeys (ii), (iii), and (iv) but again not (i). For optical techniques, (i) is fulfilled such that they can be applied to melts (see Sect. 15), but the advantage (iii) disappears. Inelastic neutron scattering combines (i), (i), and (iii) but excludes (iv).

The SANS method for dynamics has been developed while other techniques have also been much improved. This led to a renewal of the interest on the subject, also because at the same time new and stimulating theories were developed. The masterpiece of this theoretical work is the tube [2]-reptation [3] model: the main idea of reptation is to assume that, in the point of view of the dynamics of each chain, the obstacles formed by the other chains create a tube inside which the chain is constrained: it has then to *disengage* the tube by its two ends via a one-dimensionnal motion along the centre line of the tube; this gives a maximum relaxation time T_{dis} proportional to M^3 (de Gennes, 1971), and well separated from the shorter times of the relaxation spectrum. This 3rd power law is a great advantage of the theory, as the old Rouse model only predicts M^2. Specifications on the tube itself are not directly derived by the genuine theory; but as far as the effect of its fluctuations tends to a permanent constraint at long times, for large enough T_{dis}, i.e. long enough chains, the relevant process will be disengagement. This theory allows for a large crossover range of time and molecular weight, before reaching the pure disengagement regime. This idea has been precisely exploited in the Doi-Edwards model for melts [4]. The crossover is described by imposing a constant size D of the tube diameter, without knowing how this could arise: thus the model of dynamics is not selfconsistent. However, the fit of mainly rheological data is then possible and furnishes a value for D for each polymer. It is expected, as often in physics, that the crossover ranges are different depending on the physical quantity measured. The crossover in time is escaped in some experiments which measure the self diffusion coefficient; the time scales are then much larger than the maximum time of the chain. The mass crossover must still exist for the selfdiffusion but seems to lie at small values; the disengagement prediction $D \propto M^{-2}$ is verified in the range of accessible high masses. The situation is different for other physical quantities: for example, the maximum time for viscoelastic quantities is proportional to $M^{3.4}$, instead of M^3. This can be explained by a crossover effect [5] such that the test of the M^3 dependence would require extremely high masses ($> 10^7$ for polystyrene), which are difficult to synthesize as well as to handle. Quantitative specifications for the tube then become important, at least in commercial applications for which the molecular weight is not so large. These latter also involve very polydisperse samples for which the crossover regime can be very complicated. One interesting feature of the experiments reviewed here is the exploration of this eventual crossover regime in time, which can give new information on the still unknown motions involved at that stage. It will be seen that the limit model behaviour of an ideal free chain for the shortest time can be in agreement with measured data, while in the upper range the limit of pure disengagement is less obviously checked so far, and should at least lead to modifications of the simple specifications on the tube contained in the model of Doi-Edwards.

All cases, let us insist, will benefit from the huge advantage of observing the motions at different scales of length, or in other words, together the t dependence, via the q dependence.

Three parts will appear in the review. First the background of the experiment: technical remarks and the possible extensions or alternatives for using SANS in dynamics (Sect. 2), the background of the present interpretation of the data (Sect. 3), and a brief recall of the Doi-Edwards model. Second, a detailed description is given of the different characterizations of the (q, t) dependence, radius of gyration, asymptotic laws and comparison of the form factor with some calculated expressions; some recent results on crosslinked melts and mixtures of different molecular weights will be presented. The third part contains additional remarks on other more detailed approaches.

2 Technical and Theoretical Background

2.1 The Use of Elastic SANS for Polymer Dynamics

The different ways of investigating the polymer dynamics by elastic neutron scattering is focussed upon in this section, especially the case of the "step strain + quenching" method, as it is currently the only method used for uncrosslinked melts. A few other methods, mentioned in the text, are presently being developed (also see review by Oberthur [6]).

The Deformation History. Polymer dynamics can be investigated by SANS via special phenomena such as demixtion [7] observed by X-rays [8] or crystallization. A more direct way is simply to observe a sample mechanically displaced out of equilibrium. A classical approach used in other techniques is a *steady* deformation, characterised by a constant rate of deformation $s = 1/L \, dL/dt$, where L is a distance. The investigation in time is related to the dependence on s. Other procedures involve time-dependent deformation histories, which relate to the actual time: typical cases are a periodic deformation, e.g. oscillatory, and a stepstrain deformation. A time analysis is then needed. In the case of a periodic deformation, one can divide the period $2\pi/\nu$ into small intervals of phase ψ, $\psi + \Delta \psi$, within which $S_{\psi,\nu}(q)$ is measured. In the case of stepstrain there is only one time parameter, the time t elapsed after the stepstrain which will be divided into short intervals to measure $S_t(q)$.

The deformation can be stretching or shear. In the case of very liquid melts or solutions, a flow cell can be used (Couette flow or elongational cell), which is specially convenient for steady deformation.

Glass Transition: Time Magnification and Quenching. Polymers are glass-forming materials. If no crystallization occurs, this means that the spectrum of relaxation time depends on temperature via the difference $T - T_g$, where T_g is the glass transition temperature. For all kinds of glasses the dependence is similar: all the times of the spectrum are multiplied by the same factor $a_T = f(T - T_g)$ [12]. The closer to T_g, the larger is the variation of f; a_T can range over several decades (six is easily achieved for polymers). As glass transition freezes the dynamics, it allows a quenching of the relaxation process, which makes it convenient to observe the sample a long time, as it could be necessary with SANS.

Different Possible Time Analysis. At a given temperature the whole spectrum spreads for long chains also over several decades. Finally, a combination must be made between: the relevant physical time of the spectrum, the temperature, the use of quenching, and the counting time needed. There are three ways to obtain $S_t(q)$ after stepstrain deformation:

a) Direct acquisition in situ. One acquires the data for an interval (t, t + Δt), where Δt is the counting time needed for a full run but is small compared to t. Thus, the minimum t is proportional to the minimum counting time Δt_{min} needed for a reasonably large intensity recorded at each q.

b) Repeated acquisition in situ. Δt is taken shorter than Δt_{min} and the deformation is repeated several times (allowing for relaxation of the sample between two successive times). The minimum t is related to the minimum time of acquisition of the electronics of the multidetector and time channel system.

c) Quenching. One eliminates the limit of the counting time and does not need to relax the sample in situ. One sample may be used for each value of t, or the same reheated several times (both methods have been used). The duration of quenching and of eventual reheating must be short compared to t.

Exactly the same three methods may be used for any time-dependent deformation (e.g. also the oscillatory one). Quenching is easier for T not far from T_g. It can also be used to quench a steady deformation of a melt. However, it is not suitable for solutions nor crystallizing polymers. Direct acquisition in situ requires to observe a long time period within the spectrum of a long chain, or to work very close to T_g to magnify all times of the spectrum; this requires a precise temperature control, as a_g depends strongly on $T - T_g$. Repeated acquisition is suitable for times much shorter than the maximum time; this is especially convenient for rubbers for which this latter is infinite.

The Anisotropy of Scattering. In all experiments reported here the sample is oriented in order to have a scattering depending on the angle φ between the scattering vector q and the principle axis of the deformation (here generally the stretching axis). This supplies much information for all directions φ, but involves long counting times (see Sect. 12). It is possible to align the drawing axis of the sample with the beam axis, which gives an isotropic scattering, for example with an equibiaxial extension, or a uniaxial one if the sample is cut perpendicularly to the stretching axis $\vec{S}$ into narrow bands and consecutively piled vertically with their cross section perpendicular to the beam.

Temperature control. This is generally achieved via a fluid. For acquisition in situ, the temperature equipment must be transparent to neutrons; a gas (air) is then suitable for thermalising. If the acquisition is done after quenching, any system can be used; the use of a liquid (liquid salt or non-swelling oil (silicone)) has the advantage, compared to air, to allow fast reheating and to prevent the sample to flow under gravity during long time periods by matching the density of the liquid to the one of the melt. A complication arises when accuracy of the control is needed on a large length scale, as for uniaxial stretching which makes the samples very long. A detailed discussion of such experiments is given in Refs. [9] and [10].

Making an Unstrained Sample. If one uses a "solid" sample, i.e. neither a very liquid melt nor a solution, one has to ensure a perfectly unstrained state before the deformation. For melts this is related to the moulding, which should not induce too large strains and, mainly, to the annealing. An artefact from uncorrect annealing will appear at relaxation times of the order of the annealing time, such that studies at longer times need more careful annealing. The duration of the annealing will increase with the terminal, relaxation time i.e. considerably for molecular weights over 10^6 (typically one day at 180 °C). Under these conditions a new difficulty arises from the chemical degradation. First, the degradation time becomes of the order of the annealing time; second, large chains are much more sensitive to degradation than the smaller chains. This has so far limited our exploration of long times; progress has been made recently.

Currently Performed Experiments. Flow experiments with equipment similar to the Schlumberger elongational viscometer have been attempted, but encountered many difficulties; an elongational "cross cell" system (same flow as the Taylor four-roll-mill) is being developed [11]. Shear flow has been studied more extensively in a Couette viscometer (see Ref. [6] for review). A periodic oscillatory uniaxial deformation has been recently started on PDMS rubbers [6]. Up to now the range of available time

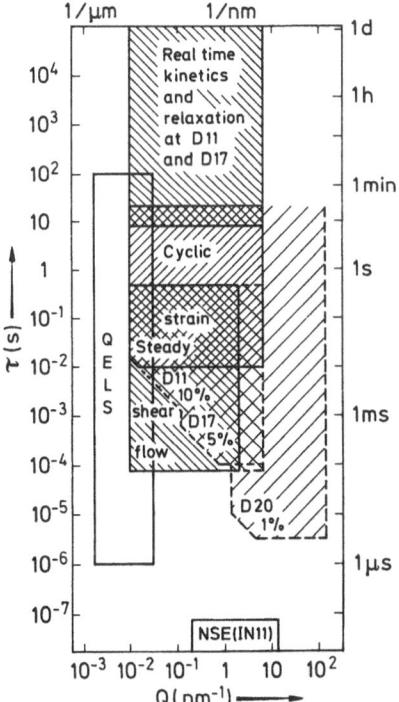

Fig. 2. Time (τ) and scattering vector (Q) range for various scattering experiments on dynamics: SANS ILL spectrometers D11 and D17: relaxation after stepstrain, cyclic experiments, steady couette shear. Elastic neutron scattering ILL spectrometer D20, also real time experiment. Neutron Spin Echo (NSE) ILL spectrometer, inelastic measurement and classical quasielastic light scattering (QELS) (from Ref. [6])

($> 10^{-2}$ sec) is large compared with the molecular times expected for this materials ($10^{-10}-10^{-5}$ s). The orders of magnitude of time and scattering vectors are given in Fig. 2 for these different experiments. The rest of the review is mainly devoted to stepstrain experiments involving quenching of uncross-linked, but also crosslinked, polystyrene. The whole set is reviewed in Sect. 4.

2.2 Theoretical Background of Melt Dynamics

For melts and concentrated solutions the current theory outlines the following: at short times, the Rouse model, in which the other chains only play the role of a viscous medium for individual chains, is accepted to fit all the data for polymer dynamics. Sufficiently long chains involve times longer than a threshold T* for which that model is clearly wrong: the chains are then called "entangled". The current idea is that the other chains now slow down the lateral motions of a particular chain above a typical distance much more strongly than its longitudinal motion which is still of the Rouse type. Then, for large enough chains only this latter motion is relevant; it is a "reptation" inside the obstacles due to the other chains imagined as a tube surrounding it. There is no theoretical explanation of this slowing down, i.e. of the typical size of the tube, as the collective dynamics are still unknown. Thus, the tube is not selfconsistent. The Rouse model uses Einstein's description of Brownian motion for each monomer and the entropic elasticity of the Gaussian chain (it fails, e.g., for distances smaller than the size, ϱ, of a set of a few monomer; this set is called the Rouse subchain). Reptation uses the same description. But, in addition, it needs an arbitrary description of the tube. Its typical diameter D can only be obtained from data on entangled material, in contrast to the friction coefficient ζ and the size ϱ which can be approached in monomeric or static experiments. There is an exception in semidilute solutions for which de Gennes proposed $D = \xi$ (correlation length), which exerts a certain concentration dependence (although some recent papers propose the existence of a second characteristic length in that range of concentration, see Adam and Delsanti). In concentrated solutions $\xi \propto c^2$, which gives the right dependence for the modulus.

For melts and concentrated solutions, Doi and Edwards presented a comprehensive model connecting the 3d Rouse model for the small scale with the tube model for the large scale (r > D). D can be taken as the radius of gyration of a Brownian subchain of molecular weight $M_e \propto D^2$. The Rouse model will then fail for $t > T^* = T_e = T_{Rouse}(M_e) \propto M_e^2$. Later on, the motion is uniquely a one-dimensional Rouse motion along the centre line of the tube; as predicted by de Gennes, a second and third process will appear. The second is an equilibration-fluctuation of the linear density along the central line of the tube, with the Rouse spectrum $\tau_p \propto (1/p^2)(M/M_e)^2$. In the third process the chain moves as a whole together with its centre of gravity along the tube centre line in a back-and-forth motion; each time one end approaches from the centre to the extremity of the tube it creates a new end part; the relaxation spectrum is then $\tau_p \propto (1/p^2)(M/M_e)^3$, p being an odd number only.

Doi and Edwards extended the model to the case of relaxation after deformation (in particular stepstrain). As already remarked by Daoudi, the length along the centre line at $t = T_e$ is then larger than the equilibrium length in the corresponding tube under isotropic conditions. Thus, the chain will retract inside the tube, which will shorten. Before and after this process Doi and Edwards describe the medium as a rubber with a number of crosslinks proportional to the length of the part of the tube which has not yet relaxed. This allows to calculate the stress following the classical theories of rubber elasticity. The shortening of the tube and the retraction of the chain then allow to explain the decrease in stress for high deformation observed at a certain $\tau_k \propto M^2$, with exactly the right variation of λ. The stress can also be calculated for the last process (disengagement). Only the centre part of the tube remains deformed; its length is:

$$L(t) = L_0 \mu(t)$$

with

$$\mu(t) = \Sigma_{p\,odd}(1/p^2) \exp(-p^2 t/T_{dis}) .$$

The number of active meshes is proportionnal to $\mu(t)$, which gives for the stress:

$$\sigma(t) = (\lambda^2 - 1/\lambda) \, G_N^0 \mu(t) .$$

One can then take the mesh size equal to M_e; this gives, via the rubber model, the plateau modulus:

$$G_N^0 = kT/M_e .$$

M_e from the plateau modulus agrees with M_e from the time threshold for failure of the Rouse model ($T_e \propto M_e^2$) [12], such that the reptation model is consistent.

It still fails for the dependence upon molecular weight, $M^{3.4}$, (experimental) instead of M^3 (theoretical). Doi suggested that the main disengagement process is combined with the linear density fluctuation, as the two characteristic times proportional to $(M/M_e)^2$ and $(M/M_e)^3$ are close if M/M_e is not too large. The 3.4 in $M^{3.4}$ is an apparent exponent and should disappear for very long chains (for which data are not yet available).

However, it is not obvious that, even apart from that effect, the crossover is as sharp between the Rouse and pure reptation, as suggested in the Introduction. One way of expressing this relationship is to modify the tube theory by varying the tube diameter: this is called tube renewal. There are several possibilities: a first proposal was the effect of disengaging of other chains [13]; the effect is important for a long chain M embedded in shorter chains P (the characteristic time being M^2P^3) but not for all chains of same length (M^5!). In case of deformed systems it was noted that in the contraction process the other chains also contract, increasing the tube diameter, which in return lowers the retraction effect. Selfconsistent calculations have been proposed [14]. Also the fluctuations of linear density of the chain allows a first disengagement (this being the main relaxation effect for star polymers [15]). The effect of the molecular weight of the matrix on the relaxation of a chain of given weight (Refs. [16] and [17]; also see Sect. 11) can be explained by these modifications of the genuine theory. Other theories proposing leaks in the tube [18] or selfconsistent power laws [19] are currently being considered.

Still other possibilities exist; e.g., a concept revealing a weakness of the Doi-Edwards model: the modelization of the melt by a temporary rubber leads to the use of the classical rubber theory: this is known to encounter considerable discrepancies for actual rubbers (crosslinked melts). A slightly different description of a rubber would lead to greater motion of the crosslinks, and consequently of more extensive rearrangement of the tube constraints in the melt (see Sect. 13).

3 Loss of Affineness and Time Dependence

Here we describe the qualitative variation of the form factor with the scattering vector q (of modulus $|q| = q$) and the duration of relaxation t. The covered time range must also be defined, which is usually achieved by referring to the viscoelastic behaviour. Furthermore, different possibilities of data interpretation are suggested.

3.1 Loss of Affineness

In Fig. 3 $q^2S(q)$ is plotted against q. The dot-dashed line represents the isotropic form factor of a Gaussian chain, the dotted lines represent the totally affine representation explained later. For any value of t, the curves have the same following characteristics: in the parallel direction they start from a very small value of $q^2S(q)$ and increase when q increases; for large q they are close to the isotropic form factor. In the perpendicular direction they start at small q from a small value, increase to a maximum, and then decrease back towards the isotropic form factor. Actually the two directions reflect the same basic feature: as q increases (i.e. as the length scale decreases), the form factor appears closer to the one of the isotropic chain. In the parallel direction the function is always increasing, but in perpendicular direction there is a maximum; this maximum is not present in a S(q) vs q plot, but comes from an artefact of the $q^2S(q)$ representation in the range $1/q \sim R_g$. R_g is the radius of gyration, i.e. the global size of the chain. For $q \ll 1/R_g$, the form factor is nearly a constant, as would be the Fourier transform of a delta function, because the chain appears as a small point; S(q) can be understood as the number n(q) of scattering points in a sphere of radius $1/q$. Thus, that number is equal to N and constant as long as $1/q$ is larger than the size of the chain, i.e. here $1/q \gg R_g$. Thus, $q^2S(q)$ increases in all directions for $qR_g < 1$; this regime corresponds to the increase of the perpendicular while this is not the case for the parallel direction in the available q range because $R_{g//}$ is too large: we have always $qR_{g//} \gtrsim 1$. The regime $qR_g > 1$ concerns the inside of the macromolecule. The number n(q) of scattering points thus starts to decrease. For an isotropic Gaussian chain, it decreases as $(1/q^2)$ (as $r \propto \sqrt{n}$), and $q^2S(q)$ is constant. For a deformed chain, n(q) is higher in the perpendicular direction, because the chain has been compressed, and smaller in parallel. When q increases, the curves return back to the isotropic curve; in parallel it produces an increase, in the perpendicular a decrease and consecutively a maximum appears the whole curve for the perpendicular direction. This return to isotropy at large q reflects the origin of the possibility of large recoverable

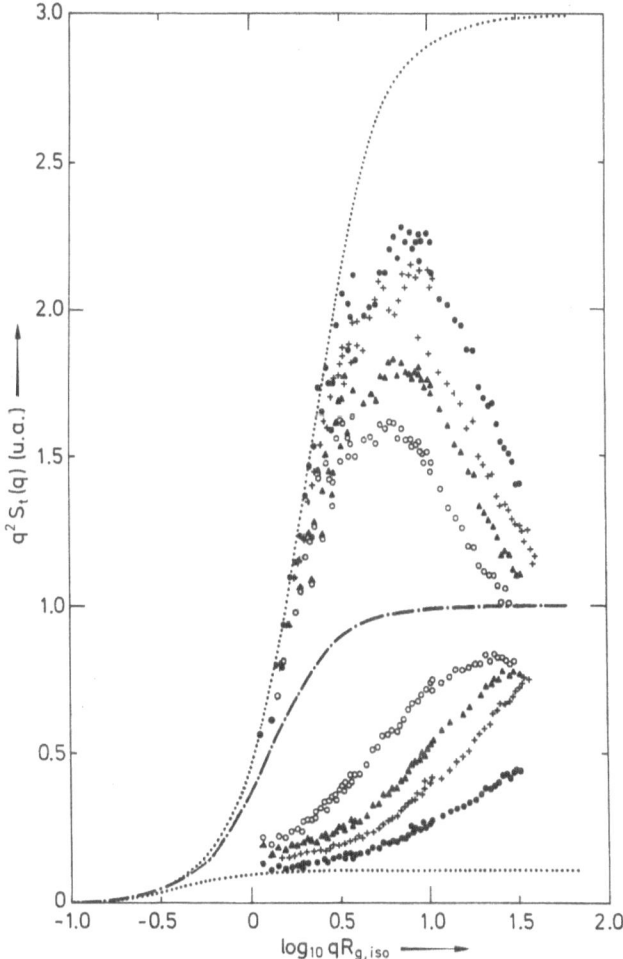

Fig. 3. Data for the single chain from factor $S_t(q)$ in $q^2S_t(q)$. Representation against $\log qR_{g\,iso}$ for the following sample of set I: ●, 71 ($t^{117} = 10$ s); +, 18 ($t^{117} = 60$ s); ▲, 49 (9000 s); and ○, 50 (130 000 s), Dashed dotted line is for the isotropic Gaussian conformation and dotted lines for the completely affine deformation of that latter in parallel and perpendicular direction

deformations in polymer materials: the system of chains causes a reduction in deformation from a macroscopic down to a quasinil deformation of the chemical bond. This can be called a loss of affineness in reference to the totally affine deformation, where it is assumed that the deformation remains the same down to the monomer (which is unrealistic). Two remarks on this concept of loss of affineness are necessary. First, when the deformation is totally recoverable, one expects a threshold in length above which the deformation is affine to the one imposed macroscopically. This corresponds to t/T_{max} sufficiently small. T_{max} is the maximum relaxation time of the system (for a network, $T_{max} = \infty$). At longer t, the deformation will relax even at the macroscopic level. But it is expected that the remaining deformation will still decrease when the scale decreases. Second, the scale of onset of the loss of affineness is not obviously the global size of the chains or smaller, but could rather be of a much *larger* scale. In that latter case the form factor is blind for distances larger than R_g, and and for $qR_g \ll 1$ always exhibits the deformation for any q. The unique way to explore larger distances is to increase the molecular weight of the chain.

Figure 3 also shows that, when the time increases, the form factor is less and less oriented, so that for $t = \infty$ the chain would be isotropic for any scale. We here consider the different ways of characterising the decreasing of orientation together with the decay of orientation with time.

3.2 Orders of Magnitude of Times. Difficulty of Estimation

To determine the required time range, it is necessary to first compare the classically obtained times from the viscoelastic behaviour (Fig. 1). The situation is complicated by the fact that these times depend on the temperature. The usual glass time temperature superposition (see Sect. 2) has been checked in the 50s for rheological data: the characteristic times τ_{tr}, T_{ter}, etc. depend on the temperature [12] T by the same factor $a(t) = 10(c_1(T - T_g)/c_2 + T - T_g)$. For the SANS data, a similar superposition has been proposed [20]: a form factor for $t = t_1$ at T_1 is compared to a form factor for $t_2 = (a_{T_1}/a_{T_2})t_1$ at T_2. A satisfying overlapping is possible by a correct choice of c_1, c_2 and T_g (see Fig. 4). A similar superposition was found in inelastic measurements for large q [21]. Unfortunately, an important uncertainty remains when choosing c_1, c_2 and T_g. For viscoelastic measurements the values differ [12]; one reason is that experiments using frequency measurements were more conveniently carried out at high temperatures for very liquid melts. Data from Ref. [12] and, more recent, from Ref. [22] have been compiled in Ref. [20] for:
— the Rouse time of large chains of molecular weight M. For this purpose we took the maximum time of a small chain (M′) inside the range of mass where the Rouse model appears valid, $T_{ter} \propto M'^2$, and multiplied by $(M/M')^2$. Among these Rouse times is the one for a chain of mass M_e, the mass between entanglements. We used an estimate of τ_{tr}, as classically admitted [12].
— the terminal time T_{ter} for large molecular weights. For this purpose we used different laws $T_{ter} \propto M^\alpha$ where α is between 3.0 and 3.8.

The obtained values are given in Table 1. These values will be used throughout this paper with a standard temperature of 117 °C; any time at any temperature can be given also in equivalent time at 117 °C, noted t^{117} (see Table I). Another source of error is the extrapolation of the molecular weight: the polydispersity is imprecisely described for some data, and its effect is still uncompletely known. It seems best to measure the characteristic viscoelastic times by measuring the stress in our stepstrain experiment, as was done for the data of set I[20] and II (see Table 1).

3.3 Evolution with Time and Parallel to the Viscoelastic Modulus G(t)

The data of set I, which are the most complete for the considered time range (representative data of Fig. 3), are now used for a qualitative description of the variation with time. In the initial range, the radius of gyration R_{gper} does not change; the relaxation

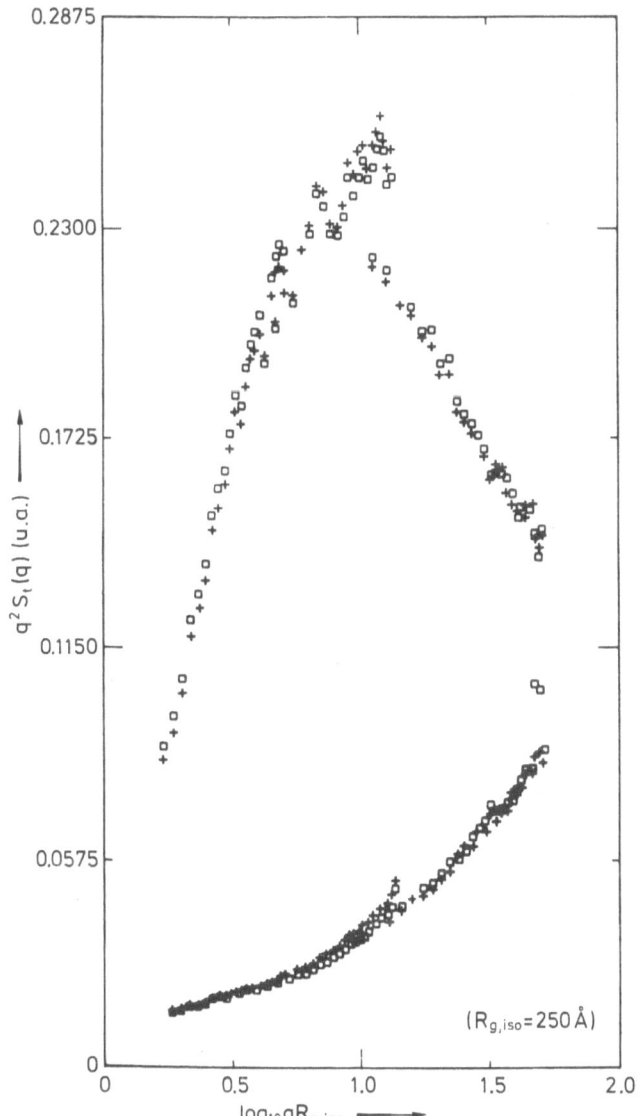

Fig. 4. Agreement with the time temperature superposition for two samples stretched and relaxed at different temperatures in order to have same reduced time $t^{117} = 60$ s. $+$, 18, T $= 117\ ^\circ$C; $\square$, 24, T $= 124\ ^\circ$C

occurs, in the perpendicular direction only at large q, and in the parallel direction for all values of q, but is much faster at large q. In the second regime, the radius of gyration starts to change; the relaxation occurs at any q; in the perpendicular, the height of the maximum decreases, and in the parallel direction, the curve increases for all values of q; however, the relaxation at large q is notably slower than before. This can be characterised in terms of the decay time τ as a function of q; for the initial time

Table 1.

Set	Masses	T_{ter}^{117} (M_{wD})	T	t	t^{117}	q range	Ref.
I Stretch-ed $\lambda = 3$	$\begin{cases} M_{wD} = 780\,000 \\ M_{nD} = 555\,000 \\ M_{wH} = 760\,000 \\ M_{nH} = 660\,000 \end{cases}$	$5 \cdot 10^5$–$5 \cdot 10^6$	113 117 122 128 134	30 s 30 s to 20 mn 30 s to 20 mn 1 mn to 30 mn 1 mn, 16 mn	6 s 30 s–1200 s 200 s–8 $\cdot 10^3$ s $2 \cdot 10^3$ s–$6 \cdot 10^4$ s 10^4 s, $1.3 \cdot 10^5$ s	(Å^{-1}) $7 \cdot 10^{-3}$, $2 \cdot 10^{-1}$.	[9] [10] [20]
II Stretch-ed $\lambda = 3$	$\begin{cases} M_{wD} = 95\,500 \\ M_{nD} = 81\,400 \\ M_{wH} = 117\,000 \\ M_{nH} = 105\,000 \end{cases}$	$5 \cdot 10^3$ s	113 117 122	10 s to 10 mn 30 s to 20 mn 30 s to 20 mn	2 s–120 s 30 s–1200 s 200 s–8 $\cdot 10^3$ s	idem	[9] [10]
III Stretch-ed $\lambda = 3$	$\begin{cases} M_{wD} = 95\,500 \\ M_{nD} = 81\,400 \\ M_{wH} = 1\,600\,000 \\ M_{nH} = 1\,300\,000 \end{cases}$	$5 \cdot 10^3$ s		same as II		idem	[9] [10]
IV Stretch-ed $\lambda = 3$	$\begin{cases} M_{wD} = 31\,900 \\ M_{nD} = 28\,500 \\ M_{wH} = 1\,600\,000 \\ M_{nH} = 1\,300\,000 \end{cases}$	100 s	113	30 s to 20 mn	6 s to 240 s	idem	[9]
V Shear $\gamma = 4$	$\begin{cases} M_{wD} = 780\,000 \\ M_{nD} = 555\,000 \\ M_{wH} = 760\,000 \\ M_{nH} = 660\,000 \end{cases}$	$5 \cdot 10^5$–$5 \cdot 10^6$ s	117 122 128	1 mn 1 mn to 20 mn 10 mn	60 s 400 s–8 $\cdot 10^3$ s $2 \cdot 10^4$ s	$7 \cdot 10^{-3}$, 10^{-1}.	[28]
VI Stretch-ed $\lambda = 4.6$	$\begin{cases} M_{,D} = 2.6 \cdot 10^6 \\ M_{nD} = 2.1 \cdot 10^6 \\ M_{wH} = 1\,600\,000 \\ M_{nH} = 1\,300\,000 \end{cases}$	$5 \cdot 10^6$–$5 \cdot 10^7$ s	116 150	1 mn 1 mn, 10 mn 30 mn, 40 mn	60 s $3 \cdot 10^4$ s, $3 \cdot 10^5$ s $1.2 \cdot 10^6$ s	$7 \cdot 10^{-3}$, $6 \cdot 10^{-2}$	[29]
VII Stretch-ed $\lambda = 4$	$M_{wD} = 400\,000$ $M_{wH} = 200\,000$ $\quad\;\;\; 400\,000$ $\quad\;\;\; 760\,000$ $\quad 1\,600\,000$	$4 \cdot 10^4$–$4 \cdot 10^5$ s for each blend:	122 128	1 mn 1 mn	400 s $2 \cdot 10^3$ s	$7 \cdot 10^{-3}$, $6 \cdot 10^{-2}$.	[48]
VIII γ crosslinked Matrix 1 600 000 Labeled path of mass $2.6 \cdot 10^6 = M_{wD}$ Stretched $\lambda = 4.6$	Rubber	$T_{ter} = \infty$	116 150 150	10 s 1 mn 30 mn	10 s $3 \cdot 10^4$ s $9 \cdot 10^5$ s	$7 \cdot 10^{-3}$, $6 \cdot 10^{-2}$.	[31]

range it follows that $t >$ or $\sim \tau(q)$ for $q > 1/R_g$ and $t \ll \tau(q < 1/R_g)$; for the second range we have $t \sim \tau(q)$, while $\tau(q)$ varies with q. In a third range, the data for large q are already very close to the isotropic value; the relaxation does not seem faster at large q than at small q. An inflexion point appears in both the parallel and perpendicular curves.

This behaviour will be explained in the following sections. Here a qualitative comparison with the variation of $G(t)$ is made. In Fig. 1 three behaviours also appear for $G(t)$, corresponding to the transition zone ($t < \tau_{tr}$), the plateau regime ($\tau_{tr} < t < T_{ter}$), and the terminal region ($t \sim T_{ter}$). Using a time-temperature superposition at 117 °C for $M_{wD} = 7 \cdot 10^5$ yields: $\tau_{tr} \sim 10$ s, $T_{ter}/10 \sim 5 \cdot 10^4$ s, and $T_{ter} \sim 5 \cdot 10^5$ s. These values may agree with the boundary values for the three time ranges of $S_t(q)$. Similar comparisons are also made below for other masses.

3.4 The Opposite Point of View of the Radius of Gyration and of the Intermediate Regime

Particularly in the polymer field, experiments have been mostly, and often very schematically, divided into two categories: studies of the global size of a chain, and measurement of the radius of gyration R_g, or studies of the inside of the chain, then considered to be infinite and with a behaviour $f(q)$. R_g is easily measured, tabulated and compared, while $f(q)$ can be difficult to interpret, unless a simple variation is predicted, as certain scaling concepts in polymer physics, which may lead to a power law as $f(q) \propto 1/q^\delta$.

The regime in which to study the global size corresponds to $qR_g \ll 1$ and is called the Guinier regime. In practice, allowing certain approximations, this can be extended to $qR_g < 2$ for a quasi-Gaussian chain. The regime of the inside of the chain, as it is infinite, corresponds to $qR_g \gg 1$ and is called the intermediate (or assymptotic) regime. In practice, for a quasi-Gaussian chain it starts at $qR_g = 5$ and is limited at high q by the size of the statistical unit b (see Sect. 13; $b \sim 12$ Å for PS).

Difficulties in quantifying the form factor. One difficulty of SANS is apparent: the q range is not a large as hoped for! For a radius of 200 Å, q will be at the extreme edge of the Guinier regime, while the molecular weight of the chain is not as large ($5 \cdot 10^5$ for PS melt). For a radius of 100 Å, the intermediate regime starts at $q \sim 5 \cdot 10^{-2}$ Å^{-1}, but is narrow as $1/b \sim 10^{-1}$ Å^{-1}. In the case of deformed melts, the complication is that the different boundaries vary with the direction φ of q with the stretching axis. Also, as the chain is not really Gaussian, their relation with R_g could be slightly different.

For those reasons the use of a calculated form factor derived from models and their comparison to the whole experimental form factor is an alternative to interpretating the data with different advantages and incoveniences. In the following sections we will thus use the three approaches: radius of gyration, intermediate regime, and calculated form factors.

4 Review of Neutron Investigations on Stepstrain Deformation

A first set of data was obtained by the CRM Saclay group in a range of rather large q from 10^{-2} Å^{-1} to $5 \cdot 10^{-1}$ Å^{-1} [23]. Only the effect of a weak deformation ratio λ (1 to 1.7) was observed, without taking into account the effect of other important parameters such as temperature of stretching, speed of deformation, duration of stretching, and quenching. No loss of anisotropy with increasing q was found, arising from the combination of the small values for λ, the narrow domain for q, and the rather large uncertainty. Thus it was possible to fit a totally affine deformation (see definition below). Values for larger λ (2, 3.5, and 5) are also given in Ref. [23] where the temperature assumed three different values for $\lambda = 3.5$. These data were analysed in the parallel direction in terms of a one-dimensional rubber elasticity model of deformation, where the mesh size was taken as an adjustable parameter depending on the temperature. The results were not systematic, mainly because of inefficient substraction of incoherent background values, consecutively to low fractions of deuterated chains, and low flux of the spectrometer.

A systematic variation of S(q) with the different parameters was obtained [24] with λ as 2, 3, and 4.5 and T as 115 °C, 120 °C, 130 °C, and 140 °C. The molecular weight of the chains was monodisperse before moulding and ranged around 150 000. A simplified model of rubber elasticity was again used; within this frame the data provided a systematic variation, assuming that molecular weight of the mesh — i.e. the part of chain between two consecutive crosslinks —, depends on the temperature. As no dynamics was included in the model, this mesh molecular weight had also to be time-dependent. A proposal to assimilate the melt to a temporary rubber in the so-called plateau regime will be discussed below. This regime, for which the mesh size is expected to be constant with time and very weakly dependent on temperature, was not explored by the first experiments, both because of too short times and too short chains.

Another published preliminary experiment was performed by the Imperial College group [25]. The obtained qualitative results were in agreement with the results of Ref. [23].

The most complete set of data, referred to below, is a thesis [9] published only in part [10, 20, 26]. Four mixtures with different molecular weights were used (set I, II, III, and IV of Table 1). The q range was wide, as various settings of spectrometers were used (at ILL, Grenoble); q ranged from $5 \cdot 10^{-3}$ Å^{-1} to $3 \cdot 10^{-1}$ Å^{-1}. Set I had the largest molecular weight ($M_{wD} = 7 \cdot 10^5 \sim M_{wH}$) over a wide range of time and temperature. Using the time-temperature superposition covers four and a half decades. The elongation ratio was mainly 3, while 2 and 1.5 have been used.

Other authors [27] published interesting results on the radius of gyration for coextruded samples; polystyrene of relatively high mass ($M_{wD} = M_{wH} = 6 \cdot 10^5$) was used. A strip of polystyrene was inserted into a cylindrical billet of polyethylene, extruded at T = 127 °C at a pressure of 280 kg/cm^2 and with a deformation rate of 0.1–0.2 cm/mn. The deformation ratio, measured from marks on the sample, attained very large values: $\lambda = 3, 4, 5,$ and 10. It is difficult to give a corresponding value of t for this deformation, but an indication is that the recovering was tested as total. The results, obtained only from a comparison to the affine behaviour, are presented in Sect. 5.

In addition to sets I–IV, we performed experiments on a different type of deformation (shear [28] (set V)) and recently on higher molecular weights (set VI). For the latter we also studied the same mixtures but crosslinked (set VIII); this creates long labeled paths of high molecular weights linked to the network by numerous junctions and allows to compare rubbers and melts [29, 49] (Sect. 10). Finally, some studies on the effect of the molecular weight of the host chains, i.e. the matrix, were also performed; earlier, the four possible pair-combinations of molecular weights $1 \cdot 10^5$ and $6 \cdot 10^5$ (see Table 1) were compared [30]. More recently, we have investigated the relaxation of a labeled chain ($4 \cdot 10^5$) solved in four different matrices ($2 \cdot 10^5$, $4 \cdot 10^5$, $7 \cdot 10^5$, $1.3 \cdot 10^6$, set VII) [31]. Preliminary results exist also on the interesting case of high-molecular-weight specimens ($1.3 \cdot 10^6$) within a low-molecular-weight matrix (10^5) [32], for which the effect of the concentration of large chains was also studied [32].

5 Quantitative Behaviour of a Finite Chain: Radius of Gyration

Usually SANS does not provide a q range such that $qR_g \ll 1$, except for very small chains. Measurements at $q < 10^{-3} \text{Å}^{-1}$ are delicate and consume enormous beamtime. Measurements reported here are only for $0.5 < qR_g < 2$; the extraction of R_g in this range involves assumptions on the chain conformation (i.e. it was considered as Gaussian at large distances) and a careful use of the different methods (e.g. the Zimm plot and other plots which fit the Debye function, see Refs. [9] and [33]). It is actually not possible to measure the parallel radius for large extensions except in the case of rather small chains (as $M_{wD} = 32000$, see below). For $M_{wD} \sim 100000$, we give an "apparent" parallel radius. For chains of $M_{wD} \sim 600000$, this is no longer possible, except for a special extrapolation in the case of affine behaviour in all directions φ (see last part of Sect. 5.1).

5.1 Short Time Behaviour. Comparison to Affine Values

From early investigations it appeared natural to compare the measured value of R_g with the affine value:

$$R_{gx} = \lambda_x R_{g\,iso} \tag{5.1}$$

where x is a Cartesian coordinate of one principle axis of deformation. It occured that the earliest data checked grossly Eq. (5.1) for the following reasons: a streching at T between 115 °C and 130 °C followed by a fast air quenching, corresponding to $10 \text{ s} < t^{117} < 300 \text{ s}$, is possible for polymers of molecular weights of $M \sim 100000$ to 150000, which classically are obtained easily by anionic polymerisation. Equation (5.1) is, in fact, not perfectly obeyed for $R_{g//}$ as for $R_{g per}$ (in Ref. [23]) where $\lambda = 1.1$ to 1.7 ($M_{wD} = 1.2 \cdot 10^5$, $M_w/M_n = 1.1$, $(M/M_e) = 6$; the rapid stretching/quenching at $T = 102$ °C gives $t^{117} \sim 50$ to 100 s). The discrepancy is around 10 %, but the measurement of the macroscopic deformation λ is not accurate. A similar slight discrepancy appears in the data of Ref. [25] for equivalent t^{117} and slightly larger M_{wD} ($1.4 \cdot 10^5$, but $M_w/M_n = 1.7$): $R_{g//}/R_{g\,iso} = 1.74$, $\lambda_{//} = 1.9$, but the uncertainty is larger than

10 %. Better controlled data [24] show a better agreement for elongation at constant speed gradients, $s = 0.2 \, s^{-1}$, and $T = 115$ to $130 \, °C$ (corresponding to $t^{117} \sim 10$ to $1000 \, s$); we recall that the molecular weight $1.4 \cdot 10^5$ corresponds to the appearance of a plateau behaviour in the viscoelastic data.

More systematic data [9] allow to check Eq. (5.1) with three different molecular weights: $M_{wD} = 3 \cdot 10^4$, 10^5, and $7 \cdot 10^5$ (Table 1). For $3 \cdot 10^4$, even at the shortest time ($t \sim 5 \, s$), $R_{g \, per}$ as well as $R_{g//}$ (here measurable) are far from the affine value; for $M_{wD} = 10^5$, one obtains $R_{g \, per} = 60 \, Å$ instead of $R_{g \, aff} = 90/\sqrt{3} \, Å$; this corresponds to a slight relaxation: assuming a reptation model, it could be a relaxation up to about the tube size which could influence the global size in this range of molecular weight still small ($M/M_e = 5$) [9]. Finally, for a larger value $7 \cdot 10^5$, $R_{g \, per}$ was measured for three values of $\lambda = 1.5$, 2, and 3, and was found to be *precisely affine* [20]; Fig. 5 shows that it remains so for $7 \, s < t < 200 \, s$. In the parallel direction one can see from Fig. 3 (sample 71, $t = 7 \, s$) that the form factor, though not in the Guinier range, is close to the affine calculated one; from this we deduced that the parallel radius of gyration of that sample is most probably affine. For this molecular weight, M/M_e is now 30.

Another elegant test of the affineness was carried out by Hadzioannou et al. [27], as described in Sect. 4. As the deformation and coextrusion are different, the value of t is difficult to evaluate. M_{wD} and M_{wH} are similar, at around $6 \cdot 10^5$; thus, only the perpendicular or isotropic radius can be obtained (using, in principle, the same corrections as suggested above). Within the uncertainty, Eq. (5.1) is very well obeyed for $R_{g \, per}$. The parallel radius cannot be measured, but as the isointensity curves appear as ellipses with an affine ratio ($\lambda/\sqrt{\lambda}$), a kind of extrapolation is made along these ellipses, which leads to an affine value for $R_{g//}$. A similar approach, applied to results of set I, permits also to check Eq. (5.1) in the parallel direction for the shortest times.

5.2 Variation of the Radius of Gyration with Time

Figure 5 shows the variation of $(R_g^2/R_{g \, iso}^2)-1$ with log (t) for the data sets I–IV. At short t the value is affine (I) or close to affine (II, III) as seen before; it then changes as times increase in a monotonous way back to 0, the equilibrium value. The speed of decay depends strongly on the molecular weight of the labelled chain. In a sample of $M_{wD} = 95000$ in matrices of $M_{wH} = 105000$ and $1\,600000$, the effect of the matrix appeared weak, except maybe for the longest time (see Sect. 12). For a mixture with $M_{wD} = 32000$ in $M_{wH} = 1\,600000$ giving close to isotropic values for the longest time, it seems possible to use longer times than those applied here, to obtain a complete relaxation of R_g.

These variations are now used to extract two important behaviours:

The Lack of Contraction

The variation of R_g with time can be an important key test of the tube models if the chains are large enough, as in Ref. [9, 10]. The reason is that not only the complete variation of $R_g(t)$ is predicted by the Doi-Edwards model, but that it also predicts an effect of the tube, called contraction [4] (see Sect. 2.2). The contraction inside the tube leads to an important consequence: the radius of gyration will be reduced by the contraction

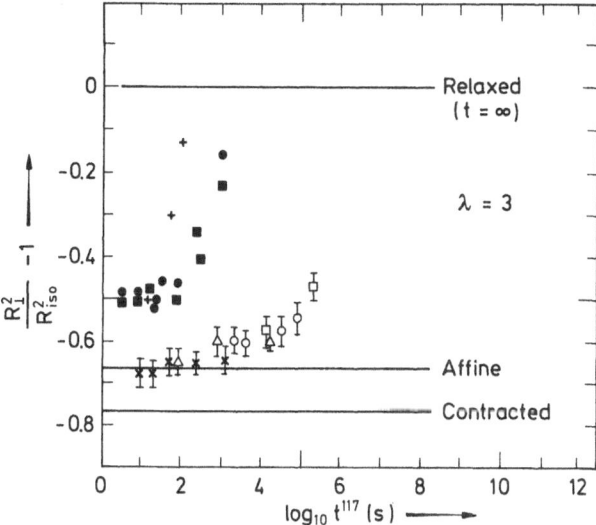

Fig. 5. Relaxation of the transverse radius of gyration after a stepstrain of elongation ratio 3. $+$, set IV $(35\,000_D/1\,600\,000_H)$; $\bullet$, set II $(105\,000_D/95\,000_H)$; $\blacksquare$, set III $(105\,000_d/1\,600\,000_H)$; $\times$, $\triangle$, $\bigcirc$, $\square$, set I $(780\,000_D/760\,000_H)$ at temperatures 117 °C, 122 °C, 126 °C, and 134 °C

factor [4] *in the parallel as in the perpendicular* direction. For the perpendicular case, as $R_{g\,per}$ is smaller than $R_{g\,iso}$, the relaxation process of contraction leads to a value even further from the equilibrium value, and the function $R_{g\,per}(t)$ exhibits a minimum.

A discrepancy between the data for the molecular weight $7 \cdot 10^5$ (set I), in Fig. 6a, is clearly shown in Ref. [20]. The data for $R_{g\,per}$, as previously mentioned, are monotonously increasing without any minimum. The calculated curves are for the ratio $T_{eq}/T_{dis} = 10^{-1}$ and 10^{-2}; for a mass M it is theoretically equal to $6M/M_e > 10^2$,

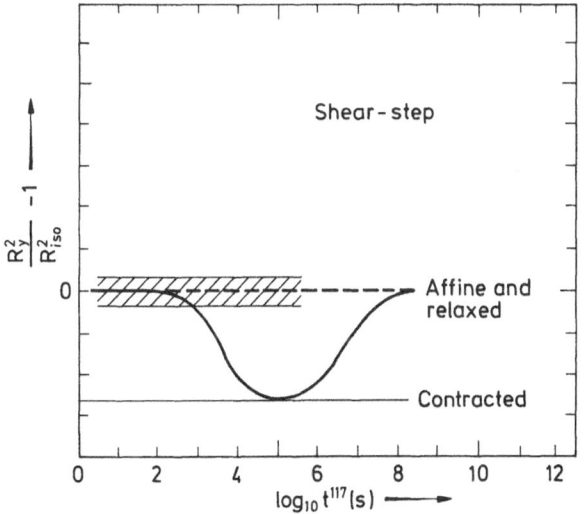

Fig. 6. Relaxation of the radius of gyration in a step-shear experiment: solid line predicted (DE model); dotted line, experimental (⬚)

and M is surely more than $5 \cdot 10^5$. The polydispersity certainly has to be taken into account ($M_w/M_n = 1.4$), but simple evaluations imply that this would not suffice to mix the two processes of equilibration and disengagement enough to cancel the effect.

The variation of $R_g(t)$ for small masses also does not show any contraction in Fig. 5. But in that case the molecular weights are too small for a conclusive test.

To increase the efficiency of the test, two complementary experiments were performed in collaboration with Osaki. The first [28] uses the same molecular weight ($7 \cdot 10^5$, set V), but a different deformation, namely shear instead of extension (see Sect. 2). In the case of shear there is one direction for which the macroscopic deformation ratio is $\lambda_y = 1$. Thus, the affine value of R_{gy} is equal to $R_{g\,iso}$; the contraction would again lead to a decrease and the disengagement would cause a return back to $R_{g\,iso}$ (Fig. 6). Such a variation could allow a more precise check. The results are the following: for short times ($t^{117} \sim 30$ s to 1000 s) R_{gy} is equal to $R_{g\,iso}$, as expected. For longer times a minimum is not observed and R_g always remains constant. The second complementary experiment uses uniaxial stretching but for much larger molecular weights (set VI, $M_{wD} = 2.6 \cdot 10^6$, $M_{wD}/M_{nD} = 1.15$ before moulding, and $M_{wD} \sim 2 \cdot 10^6$, $M_{wD}/M_{wH} = 1.5$ after moulding) with a larger stretching ratio, $\lambda = 4.6$, which enhances the theoretical contraction factor. Data are available only up to $t = 40$ mn at 140 °C, and thus approx. $t^{117} = 8 \cdot 10^5$. The Rouse time $T_{eq} = T_{Rouse}$ is estimated between 10^4 and $4 \cdot 10^5$ (Note 1). Thus, the contraction would be visible. On the contrary, $R_{gy\,per}$ increases monotonously with t.

Note 1: The Rouse time can be estimated: (i) from the value of T_{ter} (estimated in Ref. (13)), divided by $6M/M_e$, i.e. $5 \cdot 10^6/6 \times 100 \simeq 10^4$, (ii) from the estimated Rouse time (JPhys82) for $M = 7 \cdot 10^5$, multiplied by $(2 \cdot 10^6/7 \cdot 10^5)^2$: a first set of estimations 15000 to 50000 for $7 \cdot 10^5$ gives here 10^5 to $4 \cdot 10^5$. A second, much lower estimate of ~ 3000 gives $2 \cdot 10^4$.

Terminal Time of the Radius of Gyration

Apart from the contraction effect, described above, it is difficult to determine differences between the Rouse model and the reptation model via the time variation of R_g. The predicted variations are:

$$R_{g\,Rouse}^2/R_{g\,iso}^2 - 1 = \sum_{all\,p} 1/p^2 \exp\left(-p^2t/T_{Rouse}\right)$$

$$R_{g\,rep}^2/R_{g\,iso}^2 - 1 = \sum_{p\,odd} 1/p^4 \exp\left(-p^2t/T_{dis}\right)$$

These are slightly different at short time, i.e. in a range $t/T_{dis} < 10^{-2}$; but unless the chain is extemely large, this is the range where other processes are involved. At long t ($t/T_{dis} > 10^{-1}$), the decay is close to exponential for both models. Figure 7, plotting $\log(R_g^2/R_g^2 - 1)$ versus t displays a fast relaxation at short t; for the case of set I ($M_{wD} = 7 \cdot 10^5$) this corresponds to the relaxation observable in Fig. 5 occurring in the time range of equilibration but in the opposite direction. Only the points at long t give a straight line, the slope of which gives the terminal decay. We must stress that the weakness of this extrapolation is the too short times used compared to the obtained terminal time. These will be compared to the terminal time in the relaxation of the stress.

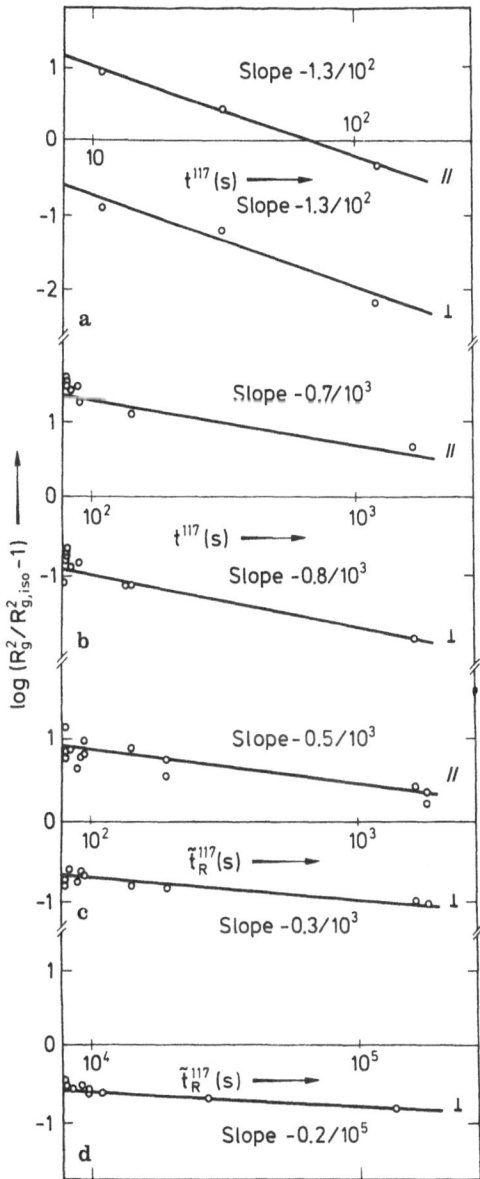

Fig. 7 a–d. Relaxation of radius of gyration for set IV (**a**), II (**b**), III (**c**), and I (**d**) with time in log-lin representation

In the case of the molecular weight $7 \cdot 10^5$ (set I) we measured the decay of the stress for different samples during their preparation, and used the time-temperature superposition to obtain a mastercurve [20]. By comparing the results with data from Tobolsky [34], a terminal time of $T_{dis} = 5 \cdot 10^5$ s was obtained. This is within estimations based on the extrapolation of other data of the latter author ($5 \cdot 10^5$ s $< T_{ter} < 2 \cdot 10^6$ s). The terminal time value for R_g is $T_{ter} = 5 \cdot 10^5$ s, which is equal to the value of the stress. For set II and III, an similar value was found for the perpendicular and the apparent parallel radius. Thus, $1.5 \cdot 10^3$ s for the matrix $M_{wH} = 105000$ and $3 \cdot 10^3$ s

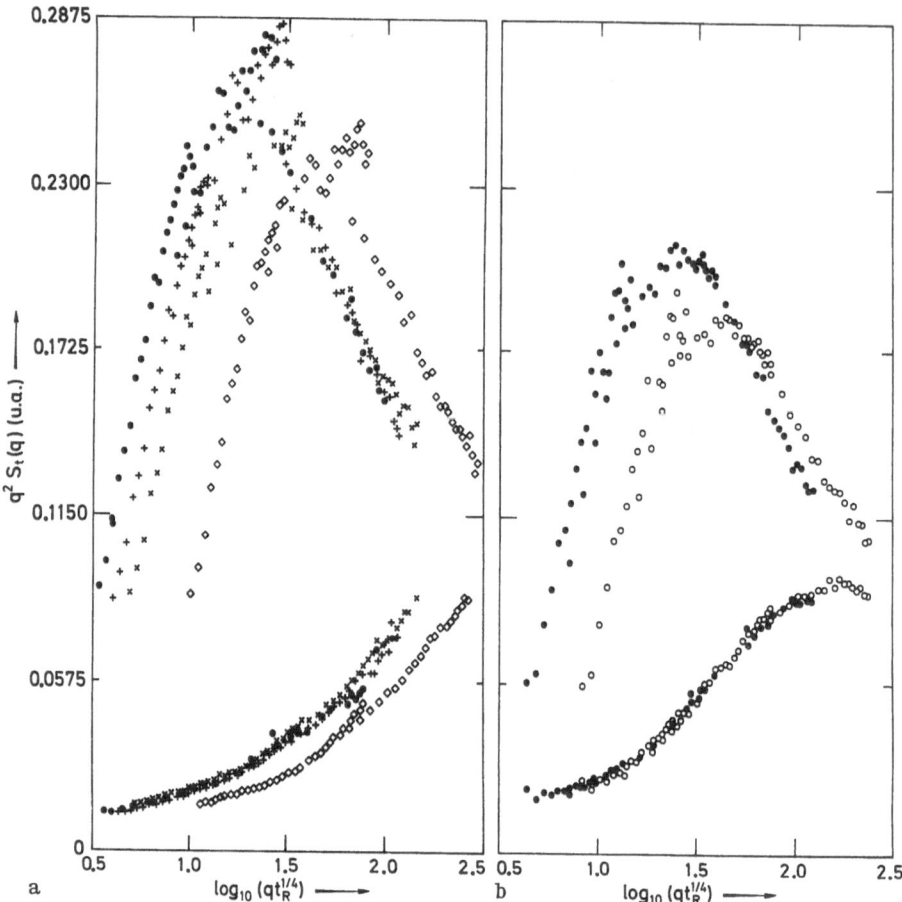

Fig. 8a. Test of a $qt^{1/4}$ using horizontal shift in $\{q^2 S_t(q), \log qR_{g\,iso}\}$ representation for short times, $t^{117} = 10$ s, 20 s, 60 s, and 240 s; **b** Same for long times 9000 s and 130000 s

for $M_{wH} = 1\,600\,000$. Table 1 gives $T_{ter} = 3 \cdot 10^3$ s. Finally the value for $M_{wD} = 32\,000$, from the perpendicular and parallel radius, is 70 s. Table 1 gives (from multiplying T_{Rouse} for $7 \cdot 10^5$ by $(32000/710^5)^2$) 30 s to 100 s, i.e. again a very good agreement.

5.3 Summary

SANS appears to be the only method to provide the time variation of a chain's global size in a melt, and this leads to three important observations:
— the radius of gyration is affine at short times, for large enough chains; the shorter the chain, the faster the departure from the affine value.
— at intermediate times, the relaxation of the radius is visible, and is in the opposite direction to the contraction predicted by the Doi-Edwards model.
— at longer times, the value of the radius terminal time is equal to the stress terminal time, obtained either from measurements during samples preparation or from estimations based on literature data, the comparison remaining unprecise.

6 Simple Asymptotic Laws for an Infinite Chain.
The Intermediate Regime

For an isotropic melt, the single chain form factor can be fitted by the Debye function over the q range $5 \cdot 10^{-3}, 10^{-1} \text{Å}^{-1}$. At large q it is close to:

$$S(q) \propto \frac{1}{q^2a^2 + 1/M} \propto 1/q^2a^2 \quad (M\infty) \tag{6.1a}$$

with

$$q \gg 1/R_g \tag{6.1b}$$

which can be called the asymptotic law for a Gaussian chain. M tends towards infinity, and a is equivalent to the size of the lattice cell: then, no other physical characteristic size is involved in the problem, as in the scaling laws of critical phenomena at the critical temperature T_c ($T - T_c = 0$ corresponds here to $1/M = 0$). Experimentally, Eq. (6.1) must be checked for $qR_g \gg 1$; in practice, $qR_g > 5$ is satisfactory. Pioneer tests [35] were on $M_w = 10^4$ to 10^5; during our experiments we checked Eq. (6.1) for $M_w = 7 \cdot 10^5$ and $2.6 \cdot 10^6$ (with a careful annealing of the sample). We describe in Section 14 how Eq. (6.1a) should appear not valid at large q, where the persistence length is involved.

What could be an asymptotic law for a deformed chain? Only three asymptotic laws are known; these correspond to classical models not explicitly involving any size or time parameter.

6.1 Totally Affine Deformation

This model is static, since time is not involved; it assumes:

$$\underline{r} \to \tilde{\tilde{E}}\underline{r} \tag{6.2}$$

where $\tilde{\tilde{E}}$ is the deformation tensor. In that case Eq. (6.1) becomes via Eq. (6.2):

$$S_{aff}(q) = 1/(\lambda_q^2 q^2) \tag{6.3}$$

where $\lambda_q^2 = \langle q \cdot \tilde{\tilde{E}} \cdot q \rangle/q^2$. To confirm this law one simply must plot $q^2S(q)$ versus any function of q; which must give a plateau. This is represented in Fig. 3 by the dotted curves for the perpendicular and parallel directions. Most of the data do not fit these curves in the region of q where they display the plateau. In the perpendicular direction, they overlap at $S_{aff}(q)$ only at short t and for small $qR_{g\,per}$, ~ 1, not in the asymptotic regime. For very long chains, this q range would correspond to $qR_{g\,per} > 5$ ($M \sim 10^7$), such that it is expectable that at equally short t, Eq. 6.3 would hold in the asymptotic regime. In parallel direction the overlap exists only for the shorter t and the smallest q values; this is nearer on more recent data on large molecular weight ($M_w \sim 2 \cdot 10^6$); in these case, we are in the asymptotic regime, as $qR_{g\,par} \gtrsim 5$.

6.2 The Rouse Superposition Law

Using the complete calculation (given in Sect. 7), and including the conditions of Eq. (7.10a),

$$qR_g > 3, \qquad t \gg \tau, \qquad t \ll T_{Rouse}, \qquad q\varrho \ll 1,$$

where ϱ is the size of the Rouse subchain, leads to [36]:

$$S(q) = 1/q^2 f(qt^{1/4}) \tag{6.4}$$

where f depends on λ and the direction of q with the deformation axes, but does not explicitly involve any size or time.

We do not clearly know ϱ, i.e. the Rouse subchain size, neither the elementary hopping time τ_N. But the two of them are related by the same relation than the radius of gyration and the maximum Rouse time, both measured quantities:

$$\frac{\tau_N}{\varrho_4} = \frac{T_{Rouse}}{R_g^4} = cst.$$

Taking ϱ equal to the persistence length, $\varrho = b = 12\,\text{Å}$, $\tau_N \sim 10^{-2}$ s at 117 °C.

The relation between q and t of Eq. (6.4) appears by plotting $q^2S(q)$ as a function of $(qt^{1/4})$. An easy procedure is to use for the abscissa log $(qt^{1/4})$ a large (q, t) range. The superposition of the curves in the range corresponding to Eq. (7.10a) should work for any direction. For the most complete set of data (set I), Fig. 8 shows the accuracy of the test. First, samples relaxed at same temperatures were compared to avoid the uncertainty caused by the time-temperature superposition. The largest uncertainty then results from the form factor, but would not artificially improve the fit in both directions: e.g., if an error reduces the form factor in the perpendicular direction, as would a supplementary relaxation, it would also do so in the parallel direction, which would then appear less relaxed. The result of this fit is:
— within a first time range, the superposition is satisfactory for both parallel and perpendicular directions in the q range between $2 \cdot 10^{-2}$ and $2 \cdot 10^{-1}\,\text{Å}^{-1}$. This corresponds to Eq. (7.10a).
— within a second time range there is no overlapping neither for the perpendicular nor for the parallel direction.
— within a third time range, of the largest times, the superposition is again surprisingly good in the parallel direction, while it appears unfavorable in the perpendicular.

A slightly different method, using the $q^2S(q)$ representation, is to search for the best superposition by translation of the log (q) abscissa, which gives Δ_t log q for each value of t (taking zero for a given t_0). Figure 9 allows to check whether this best superposition agrees with 6.4, i.e. whether Δ_t log q = 1/4 log (t/t_0). Equation (6.4) is checked

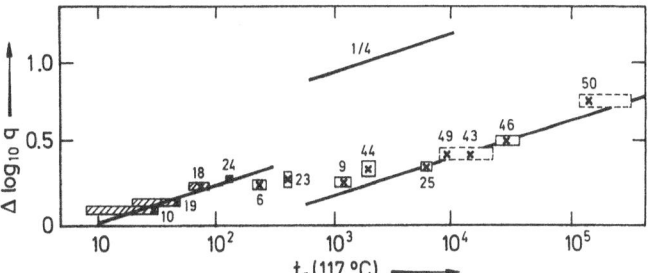

Fig. 9. Values of the horizontal shift $\Delta \log q(t)$ from superposition as in Fig. 8. Solid lines have the slope 1/4

at short times, where a departure is observed. A second slope 1/4 can be found at longer times.

It seems that the same behaviour on the same time scale also can explain the data for the lower molecular weights (100000 in a matrix of 100000 or 1600000).

In summary, a ($qt^{1/4}$) superposition is predicted by the Rouse model, and is checked within the uncertainty in the short time region for q large enough. It appears again at long times, but in the parallel direction only.

6.3 Asymptotic Law for the Disengagement Process

The genuine theory of reptation again does not involve any parameter, as it considers larger scales than the size of the tube. That condition is obeyed if

$$q \ll 1/D \tag{6.5}$$

where D is the tube diameter. Only the second and third process are testable. Here we consider only the third, i.e. at time t, chains during their disengagment after an initial deformation, with two ends already disengaged and a centre part still in its "old" tube. If the chain is long enough, it assumes a q range corresponding to the distances:

$$d(q) \ll R_{parts} \tag{6.6}$$

or chemical distances along the chain:

$$n(q) \ll N_{parts} \tag{6.7}$$

where R_{parts} and N_{parts} are the size and the chemical length of either the disengaged or the still engaged parts. Within this q range the interference terms between monomers of different parts is proportional to n(q), while the terms for monomers of the same parts are proportionnal to N_{parts}. Thus, the interference term is not significant and one can simply add the two terms of the engaged (S_{aniso}) and disengaged part (S_{iso}):

$$S_t(q) = \mu(t)S_{aniso}(q) + (1 - \mu(t)S_{iso}(t) \tag{6.8}$$

where $\mu(t)$ is the average fraction of the engaged part (memory function). Assuming a totally affine (contraction factor $c = 1$) or an affine plus contracted ($c > 1$) deformation for $S_{aniso}(q)$, we have:

$$S_{aniso}(q) \sim \left(\frac{c}{\lambda_q^2}\right)\left(\frac{1}{q^2}\right)$$

$$S_t(q) \sim \left(\frac{1}{\lambda_q^{*2}(t)}\right)\left(\frac{1}{q^2}\right)$$

$$1/\lambda_q^{*2}(t) = 1 + \mu(t)\,(c/\lambda_q^2 - 1)\,.$$

In this case there is a kind of uniform relaxation, valid only in Eq. (6.6) (in particular, the decay of the radius of gyration is different at short times; see Sect. 8).

Equation (6.9) is no more verified by the data than is Eq. (6.3) in the short time range which is due to the same reason, the impossibility to simultaneously obey Eqs. (6.5) and (6.1b). However, Eq. (6.8) is still claimed to be valid, with a different S_{aniso} than the one given by Eq. (6.9). In that case it follows that:

$$S_t(q) - S_{iso}(q) = \mu(t)\,(S_{short\,t}(q) - S_{iso}(q)) \tag{6.10}$$

where $S_{short\,t}$ represents the form factor at the beginning of the disengagement, admitting that the different processes are well separated in time. Then Eq. (6.10) reflects the important feature of reptation, i.e. that small distances are affected by the longer times of the disengagement process which actually involves larger scales much more strongly than for the Rouse model. This was also proposed for the dynamic form factor [35]. For that case we assumed that [26]:

$$\mu(t) \sim \exp\left(-t/T_{ter}\right) \tag{6.11}$$

and

$$(\log\,(S_{t1}(q) - S_{iso}(q)) - \log\,(S_{t2}(q) - S_{iso}(q)))/(t_1 - t_2) \tag{6.12}$$

was plotted versus q. A plateau at high q was approximately obtained for different combinations of four values of t. Here, the largest values of t ought to be used. A quan-

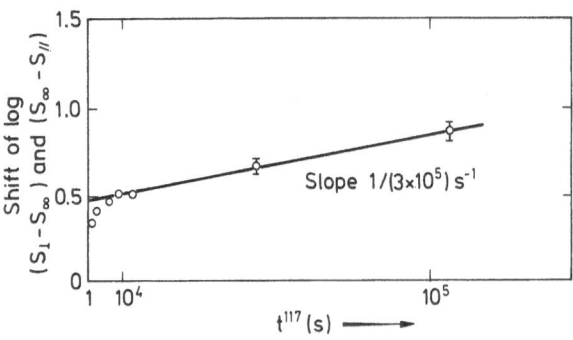

Fig. 10. Variation with time (linear scale) of the shift in $\log\,(S_{t\,per}(q) - S_{t\,iso}(q))$ (resp. $\log\,(S_{t\,para}(q) - S_{t\,iso}(q))$ allowing best overlapping of theses curves for the different values of t

titative calculation of Eq. (6.10) was given by Noolandi [37] for different pairs of times (t_1, t_2) both shorter or of the same order as T_{dis}, showing that an apparent plateau could be obtained even at small t/T_{ter}.

It is also simple to avoid the assumption of Eq. (6.11). By an alternative procedure we tested an overlapping of the curves log $(S_t(q) - S_{iso}(q))$ vs. q, at large q, by a simple transition $\Delta_t S$ along the ordinate axis [9, 10, 26]. Figure 10 shows the variation of $\Delta_t S$ with t. The value of the slope obtained from a straight line through the last four points gives $T_{ter\,h.q.} = 3 \cdot 10^5$ at 117 °C. The terminal times for the stress and for the radius of gyration are given in Sect. 5. The agreement between the three times is excellent, although we recall that the uncertainty in this case is very large. There is still a need for experiments at longer t; these are currently pursued.

7 Calculations of Form Factors

Several calculations have been given for the form factor of the chains, after a step-strain of given deformation tensor. These are restricted to a uniaxial deformation (extension to other cases is simple) and to a melt (no solvent): this latter condition allows to admit that, similarly to the isotropic case, the chains behave as Gaussian chains. Then the equality

$$S(q) = \left\langle \sum_i \sum_j \exp\left(i \cdot q \cdot (r_i - r_j)\right)\right\rangle = \sum_i \sum_j \exp\left(-q^2 \langle(r_i - r_j)^2\rangle/2\right) \quad (7.1)$$

generally holds for all other cases listed below in this section because all the probability distributions of $(r_i - r_j)$ are of the Gaussian type for all models used.

In the case of the Rouse model, the calculation first yields an equation of evolution for $\langle(r_i - r_j)^2\rangle$ (t), then leading to $S_t(q)$. This is similar in the case of the reptation model using the disengagement equation of motion, but this latter concerns only the third process of the Doi-Edwards model; this requires to take into account:
— the second process (contraction), which acts on the entire form factor, but which does not coincide with the data.
— the first process, which acts on a smaller level than the tube diameter D, which is fully attainable in our q range. The description of the state of the system by the Doi-Edwards model is close to that by the rubber elasticity models.
This is one of the reasons for the calculation of a form factor of a deformed rubber at an infinite time after a stepstrain.

7.1 The Rouse Model

The Rouse equation is given by:

$$\zeta \, \partial R_n/\partial t = \frac{3kT}{\varrho^2} \, \partial^2 R_n/\partial n^2 + \varphi_n \quad (7.2)$$

where ζ is a friction coefficient, n the index of one subchain, R_n its position, φ_n a random force acting on it, ϱ^2 the mean square length of one subchain, and the boundary conditions are $(\partial R_n/\partial n)_{n=0} = (\partial R_n/\partial_n)_{n=N} = 0$. A Fourier transformation gives:

$$\partial X_p/\partial t = -(3T/\zeta\varrho^2 p^2/N)\, X_p + \varphi_p(t) \tag{7.3}$$

solved as:

$$X_p = \exp\left(-t/\tau_p\right) \int_0^t \varphi_0(t')\exp\left(-t'/\tau_p\right) dt' + \int_0^N \cos\left(p\pi n/N\right) R_n(0)\, dn\,. \tag{7.4}$$

This allows to calculate:

$$\langle(R_m(t) - R_n(t))^2\rangle = 2/N \sum_p \sum_q \langle X_{p\alpha}(t)\, X_{q\beta}(t)\rangle\,(\cos\left(p\pi n/N\right) - \cos\left(p\pi m/N\right))$$

$$\times \{(\cos\left(q\pi n/N\right) - \cos\left(q\pi m/N\right))\} \tag{7.5}$$

which gives [9], along a given direction $\alpha\,(= x, y, z)$

$$\rangle(R_m(t) - R_n(t))^2\langle_\alpha = |n - m|\,\varrho^2/3 + \varrho^2/3(A_{\alpha\alpha} - 1) \sum_{p=1}^N 2N/p^2\pi^2$$

$$\times (\cos p\pi n/N - \cos p\pi m/N)^2\,\{\exp\left(-t/\tau_p\right)\} \tag{7.6}$$

with $A_{\alpha\alpha} = \lambda_\alpha^2$ for an initially affine deformation.

Equation (7.6) is directly inserted into Eq. (7.1) to calculate $S_t(q)$ by computer: Refs. [9, 10, 39] give plots in $q^2S(q)$ representation of the results, which are the plain and dotted lines of Fig. 11. An analytic derivation is given by Daoudi [36]. Though convinced to treat a reptation problem the latter author obtained, because of inversion in averaging, the same expression as Eq. (7.6), which can also be written as:

$$\rangle(R_m(t) - R_n(t))^2\langle_\alpha = \varrho^2\,|n - m| + \varrho^2(A_{\alpha\alpha} - 1) \sum_{k=n} \sum_{l=n} G_{kl}(2t) \tag{7.7}$$

where:

$$G_{kl}(t) = 2/N \sum_{p=1}^N \sin\left(p\pi k/N\right) \sin\left(p\pi l/N\right) e^{-t/c_p}\,. \tag{7.8}$$

This allows exact calculations of an infinite chain, by approximating the function $G_{kl}(t)$ as:

$$G_{kl}(t) = \sqrt{\pi^2/(4t/\tau_N)} \exp\left(-|k - 1|\,\pi^2/4(t/\tau_N)\right)\,. \tag{7.9}$$

Then $S_t(q)$ can be calculated by identifying the monomer and the subchain, i.e. $r_n = R_n$ (valid for $q\varrho < 1$). Using Eq. (7.1), leads, under the conditions: $qR_g > 3$, $t \ll T_{max}, t \gg \tau_N, N \gg 1$, to:

$$S_t(q) = 1/q^{*2} \int_0^\infty \exp\left(-(q^2/q^{*2})\, g(x)\right) dx \tag{7.10}$$

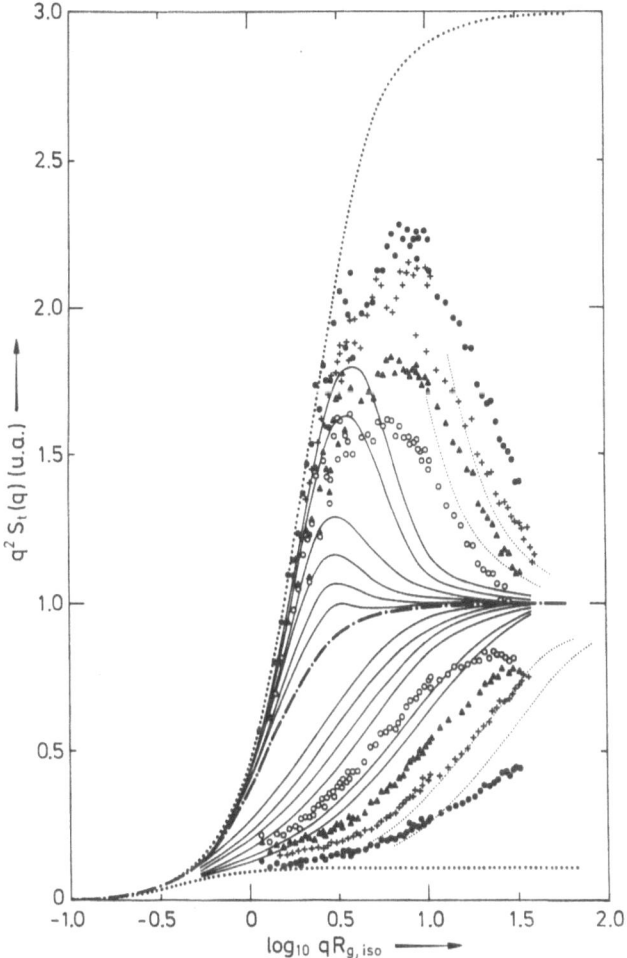

Fig. 11. Calculated form factors for isotropic Gaussian chains (dotted dashed line), completely affine deformation (dotted), Rouse.model ($t/T_{Rouse} = 5 \cdot 10^{-5}$, 10^{-4} (tiny dots), $5 \cdot 10^{-3}$, 10^{-2}, $5 \cdot 10^{-2}$, $5 \cdot 10^{-1}$, $2 \cdot 10^{-1}$ and $4 \cdot 10^{-1}$ (solid lines); data as in Fig. 3

or

$$q^2S_t(q) = (S_t(q)/S_{iso}(q)) = f(q/q^*) = f(qt^{1/4}) \qquad (7.11)$$

as $q^* = \pi/\sqrt{8}\,(1/R_g)\,(t/T_{max})^{-1/4}$. This equation shows what was called, in the preceding sections, a (q, t) superposition law. The function f depends on the direction of q with respect to the deformation.

7.2 The Disengagement Model

The calculation outlined here is based on the main process of reptation, i.e. the third process of the Doi-Edwards model, called the disengagement process [9, 39]. The two f̃irst processes are also included in Ref. [39] as the initial conditions for the third process. Therein it is seen that the three processes are well separated in time, which is possible only for long chains. Two averages are designed. The first is a thermal average of all chains with isotropic parts (i.e. already disengaged) from 0 to the linear abcissa along the tube line s_1 and from s_2 to L, but still anisotropic from s_1 to s_2. The distribution of vectors joining two points of the tube line is assumed to be Gaussian. This is obvious for the isotropic part. The anisotropic part of the chain remains under the initial conditions. An affine deformation of the tube line can be chosen. This allows the use of Eq. (7.1) with;

$$\langle (\underline{r}_m - \underline{r}_n)^2 \rangle = \varrho^2/3 \, |m - n| \left\{ 1 - \int_{m\varrho}^{n\varrho} (1 - \lambda_q^2/c) \, H(\sigma - s_1) \, H(s_2 - \sigma) \, d\sigma \right\}$$

where H(x) is the Heavyside function $H(x < 0) = 0$, $H(x > 0) = 1$, and λ_q^2 defined in 6.3 is used to calculate S(q) in a given direction of $\underline{q}$.

The second, is an average of the distribution at time t after stepstrain of the pair (s_1, s_2), $p(s_1, s_2, t)$:

$$p(s_1, s_2, t) = (\partial/\partial s_1) (\partial/\partial s_2) \, P(s_1, s_2, t)$$

where P is the probability that s_1 and s_2 are contained in the not disengaged part at time t. Noting that $g(s_1, s_2) = S_{s_1 s_2}(q) - S_{iso}(q)$, it follows that:

$$S_t(q) - S_{iso}(q) = \int_0^L dy \int_0^L dx \, g(x, y) \frac{\partial}{\partial x} \frac{\partial}{\partial y} p(x, y)$$

$$= \int_0^L dy \left\{ g(x, y) \frac{\partial p}{\partial y} (x, y) \right\}_0^y + \int_0^L dy \int_0^L dx \, dx$$

$$\times \frac{\partial g}{\partial x} (x, y) \frac{\partial p}{\partial y} (x, y)$$

$$= \int_0^L dy \left\{ g(0, y) \frac{\partial p}{\partial y} (0, y) - 0 \right\} + \int_0^{1.} dy \int_0^L dx$$

$$\times \frac{\partial g}{\partial x} (x, y) \frac{\partial p}{\partial y} (x, y)$$

(g(y, y) is equal to zero as it corresponds to one point in the old tube.)

$$= \{g(0, y) \, p(0, y)\}_0^L - \underbrace{\int_0^L dy \, p(0, y) \frac{\partial}{\partial y} g(0, y)}_{(p(0, y) = 0)}$$

$$+ \int_0^L dy \int_0^L dx\, dx \left\{ \frac{\partial g}{\partial x} (x, y)\, p(x, y) \right\}_x^L$$

$$\underbrace{\phantom{+ \int_0^L dy \int_0^L dx\, dx \left\{ \frac{\partial g}{\partial x} (x, y)\, p(x, y) \right\}_x^L}}_{T_2}$$

$$- \int_0^L dy \int_0^L dx\, dx\, p(x, y) \frac{\partial}{\partial x} \frac{\partial}{\partial y} g(x, y)$$

$$\underbrace{\phantom{- \int_0^L dy \int_0^L dx\, dx\, p(x, y) \frac{\partial}{\partial x} \frac{\partial}{\partial y} g(x, y)}}_{T_1}$$

The derivatives of $g(x, y)$ are calculated, following Eqs. (7.6), (7.14), (7.15) and Vaking $A = \lambda_q^2/c$,

$$T_1(q) = (1 - A)^2 \sum_{p\,odd} (8/\pi^2 p^2 L^2) \int_0^L ds(Q^2 s^3/Q^2 s^2 + p^2 \pi^2)$$

$$\times \exp(-AQ(1 - s)/L) - (p^2 L^2 t/s^2 T_{dis})$$

$$T_2(q) = (1 - A)\, Q(1 + e^{-Q}) \sum_{p\,odd} 8 \exp(-p^2 t/T_{dis})/(p^2 \pi^2 (Q^2 + p^2 \pi^2))$$

with:

$$Q = q^2 R_{g\,iso}^2 .$$

This not really simple algebra can be simplified by physical considerations [39]. First Eq. (7.14) combined with Eq. (7.1) gives the form factor of a isotropic-stretched-isotropic copolymer, which can be directly calculated. The second step then consists in calculating the average $S(q)$ of the copolymer (s_1, s_2) form factors, i.e. $\int p(s_1, s_2, t) S(s_1, s_2)\, ds_1\, ds_2$ instead of calculating crudely the form factor of the "average" copolymer $(S(\langle s_1 \rangle, \langle s_2 \rangle))$. The result is similar at small $qR_g(t)$ and large $qRg(t)$ but slightly different for intermediate values.

At short t, T_1 tends to zero, and $T_2(q)$ to $((1 - A)/6)\, Q \sum (96/\pi^4 p^4) \exp(-t/T_{dis})$, so that:

$$S_t(q) = 1/2 \left[1 - \frac{q^2 R_{g\,iso}^2}{3} (1 - (1 - A) \sum (96/\pi^4 p^4) \exp(-p^2 t/T_{dis})) \right]$$

$$(7.19)$$

which is the Guinier law with the Doi-Edwards expression for the radius of gyration $R_g(t)$ [4].

At large q, $T_1(q)$ tends to $(1 - A)^2/q^2 R_{g\,iso}^2 \mu(t)$ and T_2 to $(1 - a)/Q\mu(t)$ so that finally:

$$S_t(q) = (1/Q)(1 - \mu(t)) + (1/AQ)(\mu(t)) \qquad (7.20)$$

with:

$$\mu(t) = \sum_{p\,odd} (1/p^2 \pi^2) \exp(-p^2 t/T_{dis}) \qquad (7.21)$$

(note that the time function $\mu(t)$ $(\Sigma\ 1/p^2)$ is slightly different at short times from $R_g(t)$ $(\Sigma\ 1/p^4)$).

Another elegant derivation of these results was given later by Hong and Noolandi [37]. These authors remarked that an equation of evolution can be derived for the quantity:

$$F_{nm}(t) = \langle e^{i\underset{\sim}{q}\,\cdot\,(\underset{\sim}{R}_m(t)\,-\,\underset{\sim}{R}_n(t))} \rangle . \tag{7.22}$$

directly from the Langevin equation of the reptation model:

$$\underset{\sim}{R}_n(t + \Delta t) = 1/2(1 + \eta(t))\,\underset{\sim}{R}_{n+1}(t) + 1/2(1 - \eta(t))\,\underset{\sim}{R}_{n-1}(t)$$

which gives:

$$F_{nm}(q, t) = 1/2\{F_{m+1,n+1}(q, t) + F_{m-1,n-1}(q, t)\} . \tag{7.23}$$

The boundary conditions for $R_o(t)$ and $R_N(t)$ give boundary conditions for F_{m0} and $F_{N+1,n}$. Taking the continuous limit of Eq. (7.22) and by Laplace transformation we obtain F_{mn} which can be directly used in $S_t(q) = \Sigma_m \Sigma_n F_{mn}(q, t)$. The behaviour at small and large q is the same as in Eqs. (7.19) and (7.20). The authors give additional calculations and discussions on the physical quantities measured in our work of Ref. [26] (see Sect. 6).

7.3 Initial Conditions for the Disengagement: Contraction and Tube Diameter

As mentioned above and by several authors [37, 40], the q range of our experiment largely occupies the domain $q > 1/D$, D being the tube size, estimated for polystyrene to be the radius of gyration of a chain of molecular weight $M_e = 2 \cdot 10^4$, $D = 0.275\sqrt{M_e} = 40\,\text{Å}$. The choice is then to:
(i)   discard the data and wait for available larger chains and smaller values of q.
(ii) use some laws independent of qD, as reviewed in Sect. 6.
(iii) include the tube size in the calculated form factor. We first used the crude assumption of a cut-off: above the tube size, the chain is affinely deformed; below, it is isotropic. The cut-off was actually based on the chemical length, so that:

$$i - j < N_e \qquad \langle(r_i - r_j)^2\rangle = (i - j)\,b^2 \tag{7.24}$$

$$i - j > N_e \qquad \langle(r_i - r_j)^2\rangle = (i - j)\,\lambda_q^2 b^2 . \tag{7.25}$$

This gives a sharp departure from the original disengagement calculation. A smoother form can be obtained with the same threshold value for the chemical length, by replacing Eq. (7.24) by a crossover expression of the "end to end pulled chain" type (see below):

$$i - j > N_e \qquad \langle(r_i - r_j)^2\rangle = ((\lambda^2 - 1)\,((i - j)/N)^2 + ((i - j)/N))\,Nb^2 \tag{7.26}$$

Figures are given in Ref. [39]. These calculations were compared with other data [9, 10].

However, we were later led to more precise calculations of the tube effect, especially as the description of the chain after the first process is exactly the one of a chain in a rubber using the classical models of the rubber elasticity theory.

7.4 Form Factor for a Deformed Rubber

Phenomenological Expression

The first attempt to calculate a deformed form factor was [41] to combine an end-to-end pulling with an affine deformation within the entire range in order to obtain the probability of finding the monomers i and j at the distance $r_{ij} = r_i - r_j$.

$$p(r_{ij}) = \exp\left[\left((x_{ij} - X)^2/\tilde{\lambda}_x^2 - (y_{ij} - Y)^2/\tilde{\lambda}_y^2 - (z_{ij} - Z)^2/\tilde{\lambda}_z^2\right)/|i - j| b^2\right].$$

(7.27)

Because only $(i - j)$ is involved in this exponential term, the sum over i and j reduces to a sum over $n = i - j$ only which leads to an expression of the same kind as the one for the isotropic form factor. The two limit behaviours are: the affine deformation corresponding to $X = Y = Z = 0$, and the end-to-end pulled chain $X \neq 0, \tilde{\lambda}_{x,y,z} = 1$ already treated by Kuhn and Katchalsky [42]. In plotting $1/S(q)$ versus q^2, the asymptotic expression $(qR_g \gg 1)$ yields for the first case a change in the slope of the straight line, and for the second a vertical shift, as:

$$S(q) \sim 1/(q^2 R_{g\,iso}^2 + X^2/b^2 + 1).$$

(7.28)

The general case is predicted to lie in between; some examples on the possibility of obtaining the different cases are given later, together with parallel orientation averages $\langle \cos \theta \rangle$ and $\langle \cos^2 \theta \rangle$ (see Sect. 15).

Classical Models

In fact, this derivation [41] contains two main weaknesses. The first is the assumption that all chains between two crosslinks have the same end-to-end vector for the direction (parallel to the stretching axis) and modulus. This could be the case for a one-dimensional material, but in an entangled or crosslinked melt the end-to-end vectors are expected to have all directions. Moreover, a distribution of the moduli is also also sensible. For the isotropic case, the end-to-end vector distribution is Gaussian, and integrating the total distribution leads back to the usual average $\langle r_{ij}^2 \rangle = |i - j| b^2/3$. The simplest assumption for a deformed material is an affine deformation of this end-to-end distribution:

$$P(R) = \exp -(3/2Nb^2(X^2/\lambda_x^2 + Y^2/\lambda_y^2 + Z^2/\lambda_z^2)).$$

(7.29)

Convolution of Eq. (7.27) (with $\tilde{\lambda}_{x,y,z} = 1$) by Eq. (7.29) leads to the conformation of a mesh inside a deformed network for the **junction affine model** often called Kuhn model [43], as shown by Ullmann [44] for the uniaxial extension, who gives:

$$p(x_{ij}) = \sqrt{(a_x/\pi)} \exp(-a_x x_{ij}^2),$$

$$a_x = 3/(2Nb^2 |i-j|/N(1 + |i-j|/N(\lambda_x^2 - 1))) \qquad (7.30)$$

and similar expressions for y and z. Here, the distributions are also Gaussian, which allows to use the Gaussian approximation of Eq. (7.1) to derive the form factor. Following the same scheme, Pearson [45] elegantly remarked the possibility to adapt it with a few modifications to the *phantom network model* often called James & Guth model [46], as the distribution P(R) is still Gaussian with a different second moment.

Labeled Path

The second restriction [41, 45] is that chains are attached to the network only by their two ends. In the case of a chain in a melt of mass M ≫ M_e, it would be more realistic to apply the constraints by M/M_e points on the chain. A simple possibility is to divide the sum Σ over n = |i − j| in two parts, one for n < n_c and one for n > n_c, and use Eq. (7.24) for the first and Eq. (7.26) for the second. This was stated in Ref. [24], where calculations and curves are given for the parallel direction while for the perpendicular Eq. (7.25) was used instead of Eq. (7.26), used later [9]. Numerical evaluations are plotted in Refs. [9, 10, 29]. This calculation introduces a cut-off in the linear abcissa *along the chain*, which we called sliding blob, as no point is particularly affected along the chain.

The classical rubber elasticity model considers, however, that the crosslink points are particular, such that the cut-off occurs by these points in real space. The corresponding calculations for a chain obliged to pass by several crosslinks are recalled in Ref. [29]. The calculation for the junction affine model was accomplished by Ullmann for R_g and by Bastide for the entire form factor; for the case of the phantom network model, this was achieved by Edwards and Warner [47] using the replica method. Defining the number of crosslinks as two actually leads back to the expressions for only one labeled mesh; numerical evaluations are given in Ref. [29].

8 Comparison of Data with Calculations for Rouse and Reptation Models [9, 10]

We will first discuss the data of set I. The comparison partially recovers the former comparisons carried out for the radius of gyration and intermediate regime. This comparison might be more quantitative while at the same time it requires, at one stage, knowledge of the characteristic times from other data.

The comparison with the *Rouse model* thus requires an estimation of the maximum time. That should be, in principle, the Rouse time $T_{Rouse} \sim M^2$. In fact, the calculated curve is compared and, if the fit is possible, one extracts from that a maximum time. The data of set I have been compared [9] as shown in Fig. 11. This yields:
— a first consistent possibility of comparison for short times (t^{117} = 10 s, 60 s). In

practice, the data are not sufficiently relaxed at high q. In the parallel direction, the data are too relaxed at small q, or not sufficiently at high q. The comparison, thus unperfect, leads to an estimate of T_{max}^{117} between 10^4 and $5 \cdot 10^5$ s. The estimated value $R_{Rouse} \propto M^2$ (see Sect. 4) gives T_{max}^{117} between $5 \cdot 10^3$ s and $5 \cdot 10^4$ s. The experimental value T_{max} appears slightly overestimated.

— an inconsistency for long t. Data for $t^{117} \sim 10^4$ s could overlap with theoretical curves for $t/T_{max} = 10^{-3}$ giving $T_{max} = 10^7$ s; and for $t^{117} = 10^5$ s, with curves for $t/T_{max} = 2 \cdot 10^{-2}$ giving $T_{max} = 5 \cdot 10^6$. These values of T_{max} are much larger than the estimations for T_{Rouse}, and even for T_{ter}. Moreover, using the same maximum time the disagreement is total in the perpendicular direction.

In summary, it is possible to fit the short time data with a constant reasonable value of T_{max}, but it is not possible to take into account the complete set I with the same value of T_{max}. Beyond the short time region, T_{max} always increases as t increases: i.e. the Rouse model is too fast for the data. Moreover, it is not possible to account both for the parallel and perpendicular directions (this corresponds to the observation in Sect. 6 that a $qt^{1/4}$ superposition is obtained for long t only in the parallel direction, while it is obtained for both directions at small t).

A comparison with the reptation model was performed [9, 10] in two steps. A first dogmatic comparison assumed that $qD \ll 1$, which yields complete disagreement with the data. This led to the idea that the form factor might be much closer to the Rouse type. But accounting for the real value of qD ($qD > 1$ on a part of the q range), there remains only one fundamental disagreement, distributed over the entire q domain: the lack of observable contraction. We will finally see that suppressing the contraction effect — although it is completetly predicted by the model — allows the calculations to agree with the data better than using the Rouse model.

The dogmatic comparison is shown in Fig. 12. The form factor still reflects Eq. (6.9) at large q. The anisotropy of the function does no longer depend on q: this gives a plateau which decreases only with t, in sharp contrast with the data. A second discrepancy is visible at $q < 2 \cdot 10^{-2} \text{ Å}^{-1}$: the data lie clearly below the calculated curves.

The first discrepancy at large q can be suppressed within the framework of the reptation model, but with a realistic value of the tube diameter corresponding to Eq. (7.24). The result will be the disappearance of the plateau in the calculated curve $q^2S(q)$. A finite value of D is in fact the result of a Rouse motion during the period $(0, T_e)$, where $T_e \propto M_e^2$ is the Rouse time of a chain of mass M_e. Thus it resembles the Rouse model at this short T_e. At longer t the resemblance remains but the time decay is now slower and independent of q at large q, which is different from the Rouse model.

However, the second discrepancy still remains: the calculated curves still considerably exceed the data in *both* directions. That is, the contraction *should* be observable within the entire q range, while it is not. This confirms the result for the radius of gyration discussed in Sect. 5 (also see Sect. 13). It has been proposed [10, 39] to take a contraction factor c equal to one. Thus, the third process starts from the situation already obtained at $t \sim T_e$. The initial form factor should be the one calculated through the Rouse model, at $t = T_e$; this is suitable when taking into account the tube size in Eq. (7.24) for $\mu(t) = 1$ with $N_e = M_e/m$ (i.e. $D \propto \sqrt{M_e}$). The experimental form factor at $t \sim T_e$ agrees in a first approximation with both calculations (q). For t longer ($\mu(t) < 1$), Eq. (7.24) shows reasonable agreement with data for large experimental values of t using constant values of N_e (200, i.e. $M_e \sim 20000$) and of T_{ter}

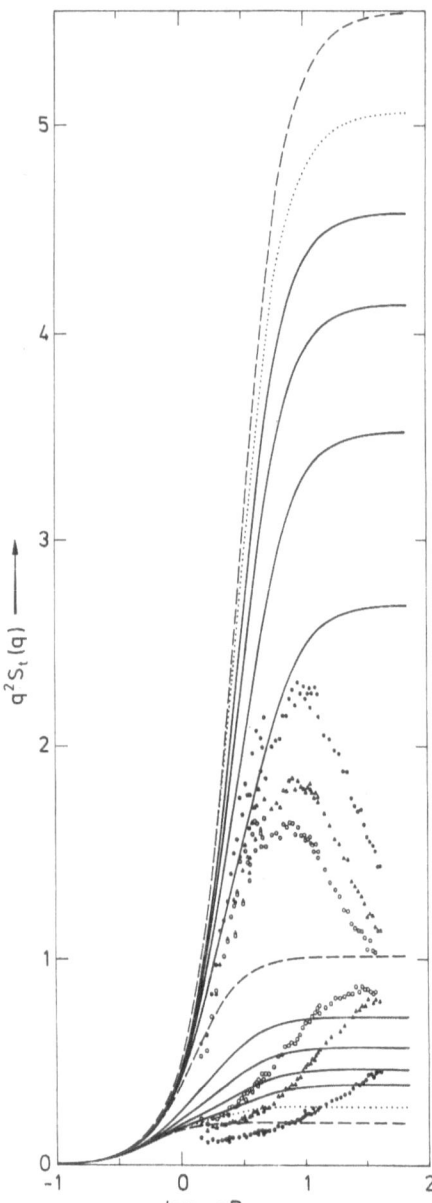

Fig. 12. Calculated form factor by assuming a disengagement of a contracted chain in a tube of zero diameter for $t/T_{ter} = 0$ (dashed line below and above), 0.1, 0.2, 0.4, and 0.8 (solid lines), and infinite (dashed line in the middle); data as in Fig. 3

$(5 \cdot 10^5$ s) for the different t's (Fig. 13). This corresponds to the behaviour expressed in Eq. (6.8), also shown in Fig. 10. However, in the parallel as in the perpendicular direction the data seem more relaxed at intermediate q than the calculations: it is seen that not only the contraction did not occur, but that another process of relaxation took place instead. A similar effect is discussed in the following section, which may serve as a generalization.

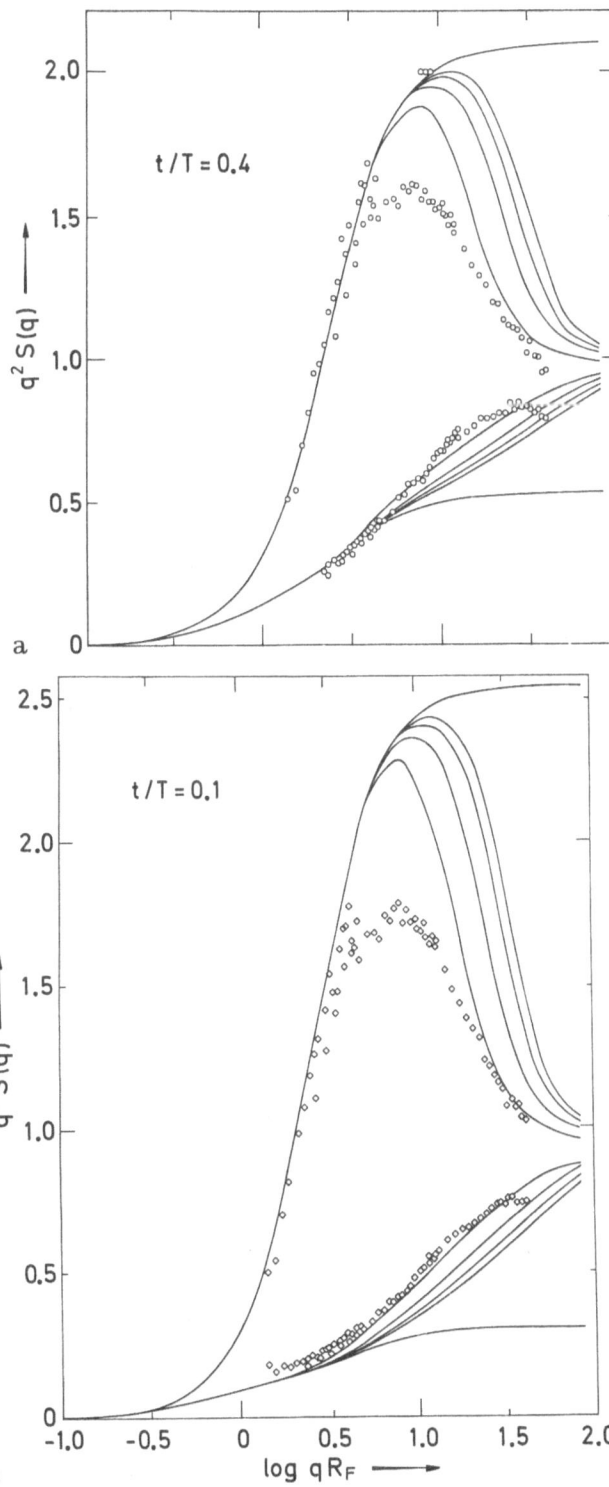

Fig. 13a and b. Calculated form factor by assuming the disengagement of a noncontracted chain in a diameter of finite value corresponding to $M_e/M = 30,\ 60,\ 90,\ 120$. **a** $t/T_{ter} = 0.4$, compared with data from sample 50; **b** $t/T_{ter} = 0.2$, compared with sample 46

9 Comparison of Data with Calculations Based on Classical Rubber Models

It is possible to investigate the initial state for the disengagement, or terminal process, in the form factor by exploring only a range of time t obeying both conditions:

$$t \gg T_e , \qquad t \ll T_{dis} . \tag{9.1}$$

For linear deformation, rheologists call this the plateau regime, because no important relaxation can be recorded. However, at large deformations, the rheological behaviour in a certain range is modified by an additional relaxation of an overstress, as observed by Osaki and others. This additional relaxation has been attributed to the contraction.

These conditions can be fulfilled by using high-molecular-weight specimens [29]: the labeled chain has a molecular weight of $2.6 \cdot 10^6$ dissolved in a $1.3 \cdot 10^6$ matrix (i.e. data set V). *If again we ignore contraction* (as we do not see it), we can compare the form factor in the regime of Eq. (9.1) to the one of a deformed rubber: this is indeed predicted by the Doi-Edwards theory at the end of the first and before the second process. The calculated curves were derived from the two classical models of Sect. 7,

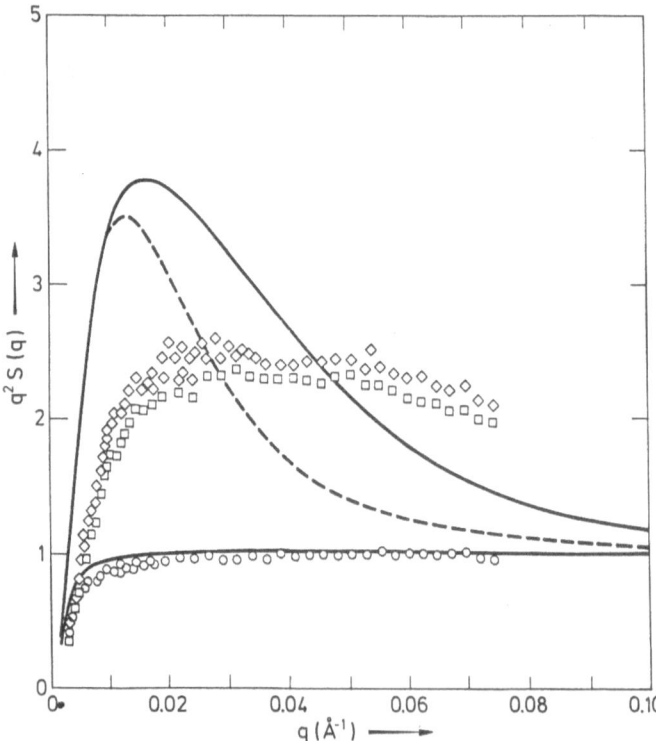

Fig. 14. Comparison of data for long chains in the plateau regime with the calculated form factor for a labeled chain ($M_{wD} = 2.6 \ 10^6$) crosslinked in a rubber of mesh $M_e = 20\,000$. Solid line below, isotropic; above, junction affine model; dashed line, phantom network. Data: set VI (Table 1): $\diamond$, $t^{140} = 10$ mn; $\square$, $t^{140} = 40$ mn; $\bigcirc$, isotropic

the junction affine [43] and the phantom network model [46] instead of the crude cut-off of Eqs. (6.23), (6.24). The molecular weight of the mesh was taken as $M_e = 2 \cdot 10^4$; the stretching ratio was $\lambda_{//} = 4.6$, leading to $\lambda_{per} = 1/\sqrt{4.6} = .46$. Two disagreements appear (Fig. 14): one at large q (data exceed the values on the curve), and one at small q, in which we are mainly interested:

— the data at small q are clearly below the calculated curves: even at $q \sim 5 \cdot 10^{-3}$ $Å^{-1}$ the chain is less deformed than predicted by the phantom network (dotted line), which already predicts a lower deformation than the junction affine model (solid line); the Flory-Erman [48] model which is intermediate, would then also disagree. It is as if the loss of affineness does not start at q of the order of the inverse of the mesh size, but that it spreads on a wider q range down to values smaller than the ones observed. The lack of deformation can be seen as an additional relaxation, not predicted by the models; no hope of improved agreement can be expected from introducing the contraction process in the calculation; this would enhance the discrepancy at least as far as the perpendicular direction is concerned.

— the data at large q are more deformed than the calculated values as if less freedom than predicted is actually available.

It is possible for this experimentally observed hyperrelaxation to correspond to a process occurring in a time $t \sim T_{Rouse}$. It could then produce the decay of the over-stress, observed by Osaki, by other means than by retraction. In that case, the Doi-Edwards model would have to be modified (see last part of Sect. 13). We will now see that excess deformation also occurs for deformed crosslinked materials, and melts and rubbers can appear to behave similarly under certain conditions.

10 Direct Comparison of Melt and Rubber

10.1 Deswollen Gels and Stretched Melts

The classical rubber models have also been used to compare data from a deformed *crosslinked* polystyrene [29]. This was done starting from a gel, crosslinked by gamma rays from a solution containing 10 % large PS chains of $M_{wD} = 2.6 \cdot 10^6$ of which 1 % were deuterated, in a radioreactive solvent. In the swollen state the chain appeared Gaussian (the solvent being a theta-solvent); after deswelling by drying the sample experienced a compression of $1/\sqrt[3]{10} = .46$ (close to the extension ratio of the melt of set VI). The deuterated chains provide the form factor of a long path in the network. The molecular weight of the mesh is carefully extrapolated from a complete set of diffusion coefficients and swelling ratios for calibrated gels; M is found to be close to 35000. Again, as for melts in the rubber plateau, the calculated curve, now for M = 35000, shows an important discrepancy with the data. The data appear also less deformed at low q, and more deformed at large q. The conclusion of Ref. [29] is that the two behaviours, for melts and for rubbers, appear to be physically similar. This last point led to another closer comparison between crosslinked and uncrosslinked melts, briefly outlined in the following.

10.2 Stretched Melt and Stretched Rubber

For a direct comparison one must have samples of the melt and dry rubber containing the same labeled chains. This was achieved by preparing samples of melt containing 10 % of M = 2.6 · 10^6 deutered chains in 90 % of M = 1.3 · 10^6 non-deuterated chains. Half of the samples were then slightly swollen (Q = 1.2), gamma-crosslinked, and then washed and dried. Half of the samples were not irradiated and let in the uncrosslinked state, and half was prepared in the crosslinked state, all of them in the same final shape. They were subsequently deformed under the same conditions of stepstrain at T > T_g, and relaxed for equal times before being quenched. In the preliminary experiments [49, 49 bis, 60], three values of time and three values of λ (1.46, 2.14, 4.6) were used; representative results are given for λ = 2.14 in Fig. 15. It is seen that the evolution of the form factor is very similar in the two cases. For very short t (1 mn at 116 °C) there is a difference at large q which can be related to a difference in T_g, suggested also by the stress behaviour. At higher temperatures (1 mn at 150 °C) the form factors overlap. For the rubber, the longest t (30 mn at 150 °C) — which has only recently been experimentally attained for the melt — leads to a form factor again less oriented than the prediction by the phantom network model.

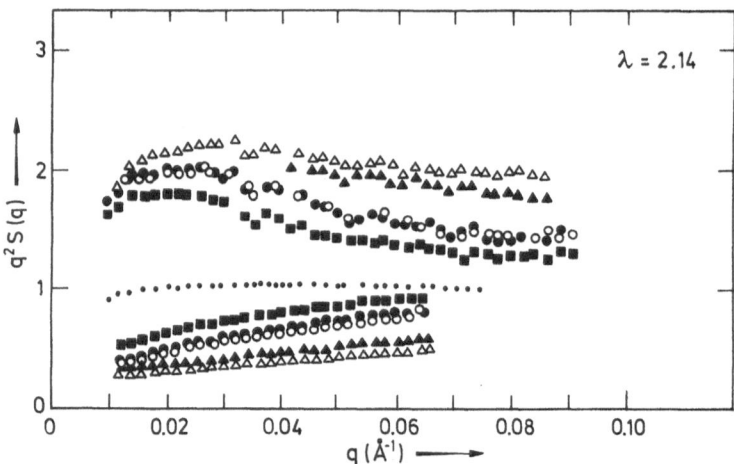

Fig. 15. Form factor from the same labeled path (M_w = 2600000) in melt (hollow symbols) and crosslinked melt (full symbols); t^{16} = 30 s (△), t^{115} = 1 mn (○) and 30 mn (□)

The same features are observed for λ = 1.46 and 4.6; however, while rubber and melt still precisely overlapp, the discrepancy with models is enhanced by increasing λ. A detailed comparison with models [49 bis, 60] shows that
— it is possible to fit grossly data in the perpendicular direction with a junction affine model and a crosslinking rate somewhat larger than the one expected from other characterizations (swelling ratio).
— the same sample in the parallel direction will scatter much more following the phantom network model, and with a crosslinking rate now 5 times weaker, thus noticeably smaller than extracted from the swelling.

11 Effect of the Molecular Weight. Alternative Tests of Reptation

In this section we directly compare the data for different molecular weights without using any predicted law or any calculation; this is an alternative for testing some intrinsic concepts of the current theories.

11.1 Comparison Between Mixtures of Different Molecular Weight, Keeping the One of the Host Chains Equal to the One of the Labeled Chain

Available data allows us to compare between shorter chains ($M_{wD} = 95000$) in a similar matrix ($M_{wH} = 105000$) and between larger chains ($M_{wD} = 780000$) in a similar matrix ($M_{wH} = 760000$) for all times. For short times, we have a satisfactory overlapping within the uncertainty of the two form factors for several equal values of t, at large q. This is in agreement with all models: at short times the conformation changes only for small sizes along the chain, independent of the size of the chain. For longer times, the form factors for the different molecular weights separate within the entire range of q: even the evolution at large q is slowed down at the longest time of the macromolecule. This is consistent with the tube model, but does not coincide with the Rouse model, for which the rate of decay depends only on q, whatever the size of the chain. The time at which the separation begins is of the order of 100 s at 117 °C. This needs to be compared to T_e^{117}, estimated as 10–50 s. Again we find, without any delicate model comparison, an agreement with the idea that the free chain behaviour ceases at T_e.

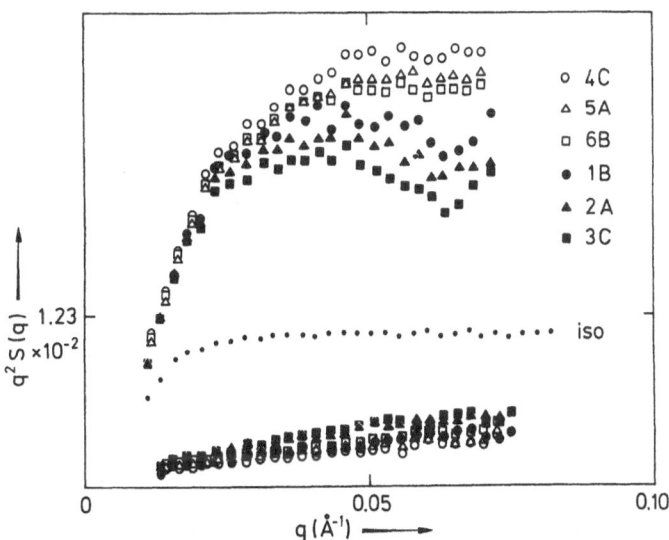

Fig. 16. Form factor from the same labeled path ($M_w = 400000$) in matrices of different molecular weights: ○, 1 600 000; △, 400 000; □, 200 000. Hollow symbols are for time 1 mn at 122 °C, full symbols for 1 mn at 126 °C

11.2 Effect of the Matrix: Variation of the Molecular Weight of the Host Chains without Changing the One of the Labeled Chain

A first insight into this effect is allowed by the data of set II and III, obtained at the same time, under same conditions of stretching and scattering. The labeled chain was $M_{wD} = 95\,000$ and the matrices $M_{wH} = 105\,000$ and $1\,300\,000$, ten times more (one thousand more for the terminal times). Comparison indicates a very good coincidence over the entire time range explored up to about $T_{ter}/50$ (T_{ter} being the terminal time $M_{wH} = 105\,000$). Only the longer times produce a considerable difference; for this case the matrix of the sample being essentially constituted of small mass material may have slightly changed of shape during the long relaxation period.

Experiments were carried out with four blends: $100\,000_D/100\,000_H$, $100\,000_D/600\,000_H$, $60\,000_D/600\,000_H$, and $600\,000_D/100\,000_H$. A matrix effect was noticed for both deuterated chains [30]. A more complete experiment investigated the matrix effect at longer times [31]. The same labeled chains, of medium molecular weight $4 \cdot 10^5$ ($M/M_e = 20$), were dissolved in four different matrices: $M_{wH} = 2 \cdot 10^5$, $4 \cdot 10^5$, $6 \cdot 10^5$, and $1.6 \cdot 10^6$ (see Table 1), with narrow polydispersities (1.15). The samples were then stretched under the same conditions, and relaxed for equal times. For $T = 122\,°C$ and $t_R = 1$ mn, the form factors are located close to each other in Fig. 16 (the largest molecular weight being a little less relaxed). For $T = 128\,°C$ and $t_R = 1$ mn, a systematic effect of the matrix is seen: the chain appears more relaxed for smaller molecular weights of the matrix. This also appears in the values of the radius of gyration (detailed data to be published). Here, only two general remarks:

— in such an experiment great care must be taken of the polydispersity of the samples: wide distribution could lead to a non-neglectible amount of very small chains; these could be more for smaller average molecular weights and would lead to faster relaxation.

— the fluorescence measurements of Tassin et al. [16], who used PS chains with an anthracene probe grafted in the middle, dissolved in different PS matrices, show a matrix effect. We chose the same samples as matrices, a deutered chain of about the same length as their grafted one, and corresponding temperatures and times to allow a comparison between the two experiments. The tests appear consistent, at first sight; neutron data appear however more sensitive to a matrix effect even when M_{wH} is very large. We do not see such a simple saturation effect when increasing M_{wH} as claimed by Tassin. Direct quantitative comparison is difficult without involving a model, as will be briefly pointed out in Sect. 15.

12 The Observation of Anisotropy

Up to here we have mainly considered the data grouped in angular sectors, slightly narrow, centred on the direction of q parallel or perpendicular to the stretching axis, š. We now want to focus on:

(i) the possible errors arising from this regroupment method

(ii) the potential information content of considering *all the data*, especially via isointensity maps.

12.1 Error from Angular Regrouping

When the cells are regrouped in an angular sector centred on 2 given φ — assuming an accurate control of sample alignment [9] — an uncertainty results from the width $\delta\varphi$ of the sector. The usual situation, at large deformation ratios, is the following: around the parallel direction, the isointensity curves slowly vary with φ; thus, the error when using large $\delta\varphi$ (20 deg.) is small. On the opposite, the φ dependence around the perpendicular direction is much more acute, and care must be taken. We usually find that $\delta\varphi = 5$ deg and $\delta\varphi = 10$ deg do not give any relevant difference. A computerised extended procedure of control was designed at ILL (P. Lindner): the value at each q can be plotted as a function of $\delta\varphi$ and extrapolate at zero value. Linear extrapolation appears to be possible, while this is not justified completely, and thus maybe confusing. On the other hand, it can be especially useful for poor counting as it is a way of using more cells of the detector. We will now describe other possibilities of using the detector.

12.2 Analysis of the Dependence on Angle φ: Isointensity Curves. The Strange Losanges. Extinction Angle

Instead of restricting ourselves to $\varphi = 0$ deg and 90 deg, corresponding to the perpendicular and parallel directions of q with respect to the stretching, one should explore all values of φ. A first procedure would be to make angular regroupments for a reasonable number of φ values (about ten for a width of 10 deg). This treatment, which is long, can be used for two purposes:
— If we know in advance the shape of the isointensity curves, testing and then fitting it on the whole detector will provide more accurate measurements. Two levels of exactness can be used. The most exact will impose a shape: one example is the ellipse with a ratio between the two axes of $\lambda/\sqrt{\lambda} = \lambda^{3/2}$; then we take an elliptic ring and group all cells inside this ring: the accuracy is then the same as for isotropic spectra when one regroup the cells inside a circular ring! Earlier [23], we extracted a given value S for each value of q and for each φ, in the X, Y plane, by smoothing the data, drew each isointensity curve and fitted it to an ellipse in order to extrapolate more accurate values of S(q) in parallel and perpendicular directions. We later ran a program for an analytical fit to the ellipse (at ILL, Grenoble). Improving the accuracy is delicate and there might be a distorsion of data by these methods. One reason might be simply that assuming a priori an analytical shape is more than unusual.

The small t_R/T_{ter} data — including data of earlier work — give elliptic shapes. They are characterized by their axis ratio, which can be compared to the affine value $\lambda^{3/2}$. In the q, t range where both the parallel and perpendicular S(q) are close to the affine value, the axis ratio has the affine value (even in the intermediate regime for the parallel or in the Guinier regime for the perpendicular). As soon as the affine variation is abandoned in one direction, the axis ratio decreases with time and with q, but curves remain elliptic [20, 23, 24] (also see Fig. 17a). A second set of data, obtained recently, involves large durations, for various situations, e.g.:
— For molecular weights equal for labeled and matrix chains, when durations are about one tenth of the terminal time (sample #50 of set IV).
— For labeled chains in a matrix of larger chains, when times are attained which are

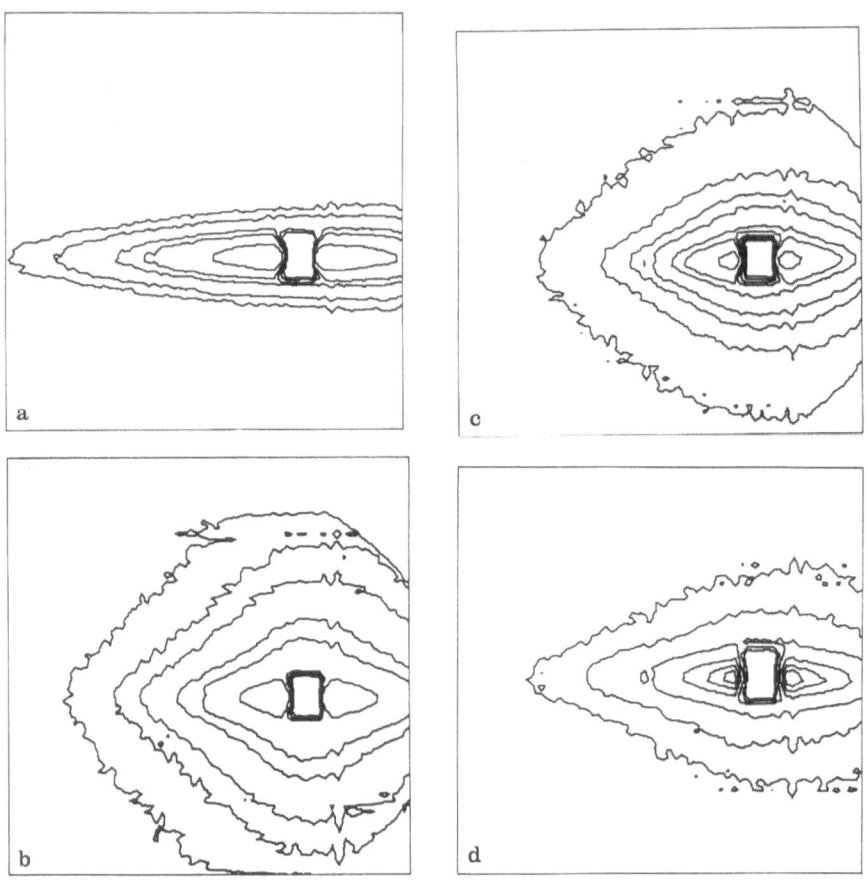

Fig. 17 a–d. Isointensity lines on the two-dimensional detector. **a** $\lambda = 4.6$, $T = 110\ °C$, $t = 1$ mn, crosslinked sample (uncrosslinked gives the same pattern); **b** same crosslinked sample, $\lambda = 4.6$, $T = 150\ °C$, $t = 30$ mn; **c** same sample, same λ, T, t but in addition return to $\lambda = 3$, at $T = 110\ °C$, which took $t = 7$ mn; **d** melt, $\lambda = 4$, $M_{wD} = 400\,000$ in $M_{wH} = 1\,600\,000$, $T = 134\ °C$, $t = 5$ mn

comparable to the terminal time of the labeled species, while the sample still behaves like a solid.

— For a crosslinked polystyrene (see Sect. 10) when the sample is relaxed indefinitely, in principle.

Representative curves are displayed in Fig. 17b, c, d. One clearly sees a losangic shape, instead of an elliptic one. Our explanations for that experimental fact are yet insufficient. A losange appears as a combination of ellipses of different axis ratios. Some chains could be more relaxed than others. The sample could be inhomogeneous on a very small scale, as predicted by some theories [50]. Also for rather polydisperse samples, the smallest chains could have become quasi-isotropic, while the largest being still very anisotropic (this is possible even for not too high polydispersity because of the $M^{3.4}$ law for the terminal time); a naive explanation could involve reptation which during disengagement provides a mixture of deformed chain parts in the middle of the chain and of isotropic parts at its ends. For cross-linked systems these proposals have to be replaced by an effect of pendant chains. We have done recently calculation for rubber deformation models including this effect: a lemon shape may be observed, never a losange shape, which we obtained only in a particular combination of orientation of end to end vectors R, only parallel or perpendicular to the elongation [49 bis].

"Extinction" angle. It is interesting to note that the assumption of affine displacement of the mean positions of the junctions lead always to a dependence on λ and θ of the type

$$\lambda^2 \cos^2 \varphi + 1/\lambda \sin \varphi$$

for both junction affine and phantom-network models. Then an extinction value $\varphi^*(\lambda)$ always exist for which the form factor should be $S(q, \varphi^*) = S_{iso}(q)$. Disengaged parts of chains, pendant chains, will give contributions, of isotropic smaller Gaussian chains. The resulting signal could always be analysed as the one of isotropic polydisperse distribution of chains, and in particular $q^2 S(q)$ should be a plateau at large q. In our data for rubber, an agreement with that behaviour was found for $\lambda = 1.46$, but *not for 2.14 and 4.6.*

We propose finally another way of exploring the φ dependence, which is by analysing, during the neutron experiment, the ω dependence. Namely, we vary the angles of the axis of the incident neutron beam and the stretching axis: this is, e.g. a way to obtain access to the large values of the distances $(R_{g//})$ in the parallel direction.

13 Summary of the Test of Reptation

The reptation model in the melt predicts three regimes. Let us recall the predictions for $S_t(q)$ for each regime by estimating each time range for our experiment, and then — keeping in mind the large uncertainty of the estimates — search for any predicted behaviour in the corresponding range.

The first predicted regime is a Rouse relaxation, with the normal Rouse time distribution T_{Rouse}/p^2, where T_{Rouse} is the maximum Rouse time of the whole chain. At any time t, it has acted only upon sizes smaller than $r^*(t) \propto R_g (t/T_{Rouse})^{1/4}$. The relaxation must stop when $r^*(t) \sim D$, i.e. when t is equal to the maximum Rouse time of a chain of mass M_e, $T_e \propto M_e^2 \propto D^4$. At small distances it must give exactly the same conformation whatever the value of the mass M of the chain, given that $M > M_e \sim 2 \cdot 10^4$. Our estimate for T_e^{117} (i.e. T_e at ^{117} °C) is 10 s to 100 s. For the range from 5 s to 200 s, the data show a rapid relaxation at large q, and a quasi nil relaxation at small q: R_g varies a few percent for $M_D = 10^5$, and not at all for $M_D = 7 \cdot 10^5$, and the same q range shows no relaxation for $M_D = 2 \cdot 10^6$. This is only contradicted for $M_{wD} = 3 \cdot 10^4$, close to M_e. In the q range of fast relaxation, the data for the masses 10^5 and $7 \cdot 10^5$ overlap (not tested for $2 \cdot 10^6$ because of a different stretching ratio). The intermediate regime appears to comply with a superposition law (Eqs. (6.4), (7.11)) of a unique parameter $(qt^{1/4})$. The fit to a form factor calculated form the Rouse model is acceptable at large q using a Rouse time for the whole chain within the uncertainty range of the estimates. In the case $M_{wD} = 7 \cdot 10^5$, for $t^{117} > 500$ s, this fit is no longer possible without changing the value of the Rouse time and progressively removing it out of the uncertainty range. The relaxation at large q appears to become slower while the radius of gyration starts to decay; the $qt^{1/4}$ superposition is not possible. For $M_{wD} = 1 \cdot 10^5$, the Rouse and reptation models give more similar behaviours, so do the data. In summary there is a good agreement with this first regime, inside and only inside a range of q, t, M corresponding to the theoretical expectation.

The second predicted regime is an equilibration inside the fixed tube. The chain feels only the finiteness of its own length, not the one of the tube. Thus, inside the contorted tube, this is a one-dimensional Rouse motion of maximum time $T_{Rouse}(M)$ $\propto M^2$. The motion should follow $r(t) \propto t^{1/8}$. For a deformed sample the equilibration corresponds to a contraction inside the tube. As this latter is three-dimensionally contorted, it corresponds to an increasing by a factor $c(\lambda)$ of the density correlation function $\langle c(0)c(r) \rangle$ for any direction of r. In terms of scattering this will produce, in particular, a decreasing of the radius of gyration in any direction, and, more generally, the form factor of a labeled path through a deformed phantom network, but with a density increased by a factor $c(\lambda)$. This will be the case when the equilibration is completed, at the end of the process, i.e. when $\exp(-t/T_{Rouse})$ is neglectible ($t/T_{Rouse} > 5$).

For $M_{wD} = 7 \cdot 10^5$, $T_{Rouse}^{117}(M)$ is estimated to be $(5 \cdot 10^{-3}–5 \cdot 10^4 \text{ s})$; for the range $t^{117} = 200 \text{ s}$ to 10^4 s, the data for both uniaxial extension and shear deformation do not agree with the theory. A slowing down is observed around the reduced time at 117 °C of $t^{117} = 200 \text{ s}$, and then the motion $r(t) \propto t^{1/8}$ could agree with the experimental variation (see Fig. 10). However, S(q) relaxes by a continuous increase of $S(q_{//})$ and continuous decrease of $S(q_{per})$; it is very clear on measurements of $R_{g\,per}(t)$ and coincides with the behaviour within the entire q range; contraction is not observed.

Moreover, a relaxation in the opposite direction *is* observed. This one is sufficiently significant to establish the following situation: not only does the form factor not follow a contracted labeled path (c > 1) in a phantom network, but it does not either comply with the Gaussian density (c = 1); i.e. it is even more relaxed than predicted by the model of rubbery deformation with no contraction, assuming the less oriented, so-called phantom network. The argument that the third process could have already started is extremely weakened by more recent experiments: the same effect is also observed for larger M, $2 \cdot 10^6$, for which the ratio M/M_e is much larger, ~ 100, theoretically corresponding to a strong decoupling of the second and third process. Finally, the form factor of a labeled path of same molecular weight in a *crosslinked melt* was found to be very similar, though no terminal process should occur. Also, still in that regime, we have observed an effect of the matrix on the relaxation of a considerably long chain $M_{wD} = 4 \cdot 10^5$ among chains of larger M. This effect is not predicted by the Doi-Edwards model.

The third predicted regime is the disengagement of the chain from the initial deformed tube: here, the finiteness of the tube becomes relevant, and the maximum time is $T_{dis} \propto M^3$. In practice, it is expected to be equal to the terminal stress relaxation time, though this latter is actually proportional to $M^{3.4}$. The main feature in the case of deformed materials is the inhomogeneous structure of the chain which appears as a triblock copolymer isotropic-anisotropic-isotropic. Quantities related to sizes smaller than $\beta(t)$ of the blocks behave as in the case of adding isotropic and anisotropic parts. This is the case for the theoretical behaviour at $q > 1/\beta(t)$, which gives $S_t(q) - S_\infty(q) \propto F(q) \mu(t)$. The rate of growth of the isotropic part is $\beta_{iso}(t) \propto \sqrt{1 - \mu(t)} \propto t^{1/4}$ at small t/T_{ter}. $\mu(t)$ is asymptotic to $\exp(t/T_{ter})$ at long times, and so is the variation of the reduced squared radius $(R_g(t)/R_{g\,iso})^2 - 1$.

Experimentally, the maximum achieved value for t is T_{ter} only for $M_{wD} = 3 \cdot 10^4$ chains, which is too short for the reptation test. For $M_{wD} = 10^5$ and $7 \cdot 10^5$, it is $T_{ter}/10$. The reduced squared radius variation appears close to an exponential in all cases, and the decay times are close to the estimated (and also measured for $M = 7 \cdot 10^5$) stress relaxation time. But this is not a key test of the reptation theory: for example, the Rouse model predicts also the same time variation of both quantities at long times (within a factor two for the terminal time).

A more exclusive test is the behaviour at large q. As both large enough t/T_{ter} and M/M_e are needed, only data for $7 \cdot 10^5$ can be used, and they agree with the factorisation law of Eq. (6.8) within the uncertainty of the measurement. This observation is strengthened by the fact that one obtains a terminal time equal to both the one of relaxation of the reduced squared radius and the one of relaxation of the stress. In summary, at that level, the few existing data agree with a disengagement process starting from the situation observed at $t \sim 10^4$ s, this latter disagreeing with the second process. This is also observed by comparing to calculated form factors for the disengagement.

A puzzling $qt^{1/4}$ superposition for this range of time has been reported. A possible explanation [9] may be that the threshold q value for the validity of Eq. (6.6) is $q^*(t) = 1/\beta(t) \propto t^{1/4}$ at short t. This was then observed only in the parallel direction because in the perpendicular, q^* is very close to $1/R_{g\,per}$. The curves calculated for $D = 0$ do not lead to a real superposition, which would only result from a finite tube size. In that matter, other experiments are needed at longer times; technically this would imply to remove the limitation of the chemical degradation which becomes as fast as the mechanical relaxation (see Sect. 2).

In summary, the experimental data contain two regimes which can coincide, both for their time range, and for the time dependence of $S_t(q)$ inside that range, with the first and third process of the tube model, respectively. In both cases, almost for the third regime of disengagement, the experimental range should, however, be extended for a better check.

A different regime, between the two former ones, should correspond to the second process of the tube model, with respect to the time range. This regime is widely explored in our experiment. But for this regime the data for $S_t(q)$ do not agree with the theory for the time dependence. This stands in contrast to the very good agreement with the tube model observed for rheological data [51, 52]. It is still necessary to find ways of combining the two experimental behaviours. We propose the following: At the end of Sects. 9 and 10 it was shown that deformed rubber (by deswelling a solution after crosslinking or stretching a dry rubber) exhibits a deformation inside the q range, lower than predicted by both the two classical models, i.e. the Kuhn model with affinely displaced junctions [43], and the model of James & Guth and Deam & Edwards [46] where only the mean positions of the junctions are displaced affinely. Moreover, the form factor of the same labeled path in a melt showed a behaviour close to the one of the rubber. This prompts us to propose that:

— The idea of similarity between a melt in the plateau regime and a rubber can be maintained.

— For both materials, the classical models for rubber elasticity predict a too high deformation on the observed scale. This is because additional rearrangement of the chains occurs on a larger scale than that of the size of the mesh.

— This additional rearrangement, when occurring in the melt, will produce a "tube rearrangement" in the time range of the second process. This would allow the linear density of the chain to equilibrate with no (or much less) retraction, and still the decrease in stress would be identical, the initial and the final value being the same.

The M^2 dependence found by Osaki et al. could have two explanations:

— Either the tube rearrangement leaves a low linear density to relax at the equilibration time $T_{Rouse} \propto (M/M_e)^2$;

— Or the rearrangement is also a Rouse process of longer times $T_{ter\,rearr}(M) \propto (M/M^*)^2$.

14 Local Effects

In the previous sections we always supposed $q \ll 1/b$, where b is the persistence length. For $q > 1/b$, all universal laws for polymers, such as scaling laws [53], are no longer valid. This is the same for our concern with the calculated form factors (Sect. 7); the asymptotic law (Sect. 6) for the Rouse (disengagement) model can be maintained if $\varrho > b\,(D > b)$. Recent progress is SANS underlined the limitation of universal laws, by improving not the spatial resolution ($1/q$) but the *chemical resolution*. Instead of using completely labeled polystyrene (monomer $-C_8D_8-$, called PSD8), Rawiso used polymers with only some of the hydrogen nuclei replaced by deuterium nuclei:

(PSD$_5$)   and   (PSD$_3$)

First experiments were performed on dilute solutions in good solvent: here, in the so-called intermediate regime (see above), the measurement of S(q) using PSD8 chains gave a plateau $q^{1/\nu}S(q)$ versus q, with $\nu_{exp} = 0.6$ [54]. This was predicted by the scaling law $S(q) = 1/q^\delta$, where δ, the fractal dimension, is here equal to $1/\nu$. Using PSD3 ("labeled backbone"), a surprising effect arises: the plateau $q^{1/0.6}S(q)$ completely disappears, hardly signed by an inflexion point after which the curve continues to increase. Using PSD5, on the contrary the plateau is replaced by a maximum after which the curve decreases. These two facts led to the idea that the plateau results from a fortunate compensation between two effects:

— The persistence length effect, which will increase $q^{1/\nu}S(q)$ as q increases, tending to a rod behaviour ($S(q) \propto 1/q$). A crude reasoning would predict the onset of this behaviour at $q \cong 1/b$, b being the persistence length. More exact calculations using a worm-like chain show that this effect would be visible by SANS at rather low q (0.3/b, i.e. $3 \cdot 10^{-2}$ Å^{-1} for PS, where b = 12 Å).

— The axial radius effect, due to the thickness of the chain. The form factor is then multiplied by the Patterson function of the density distribution perpendicular to the axis:

$$S(q) = S_0(q) F(q) \qquad (14.1)$$

where F(q) is a decreasing function of q. It can be expanded as:

$$F(q) = 1 - q^2 R_c^2/2 + 0(q^4 R_c'^4) \quad \text{for} \quad qR_c < 1, \qquad (14.2)$$

where R_c is the axial radius.

The D8, D5, and D3 curves result from different values of $R_{c\,app}$, the apparent radius due to the contrast between labeled and unlabeled nuclei (a complete description is given in Ref. [55]; one must note that the D3 chain does not strictly correspond to the naked backbone because of the difference between the benzyl ring and the solvent at the level of their "density of diffusion length" (the equivalent for neutron of the indice for light); it will be obtained by "contrast variation" using mixtures of H solvent and D solvent).

A similar effect occurs in the case of a melt ($\delta = 1/\nu = 2$), as observed for small chains. Onleading investigations employed larger D8, D5, and D3 chains ($M > 10^6$) dissolved in a deuterated matrix (this reduces the incoherent scattering by a factor of 40, the substraction of which is very delicate at high qs). Isotropic and deformed ($\lambda = 3$) melts have been observed. Representative data for D3 chains are given in Fig. 18. The $q^2 S(q)$ plateau is not obtained for isotropic melts containing D3 or D5 chains (in fact, even for any D8 data the plots show a decrease at $q > 10^{-1}$ Å^{-1}). For deformed polymers the pattern is also unusual, as the parallel and perpendicular form factors tend, as for D8, to return to the isotropic form factor at large q. This clearly shows that local effects at large q are not negligible.

According to Eqs. (13.1), (13.2), the ratio of the form factors for the two different species PSD_i and PSD_j will be:

$$\frac{S_{Di}(q)}{S_{Dj}(q)} = \frac{F_{Di}(q)}{F_{Dj}(q)} \sim 1 - q^2 \, \frac{(R_{ci}^2 - R_{cj}^2)}{2} + 0(q^4 R_c^4) \qquad (14.3)$$

The q dependence of the ratio is determined by using a Zimm plot or a Guinier plot as for the usual measurement of the radius of gyration (though the values of q are much larger here), and $R_{ci}^2 - R_{cj}^2$ is calculated.

This works for our preliminary results, as well on the $S_{Di}(q)$ than on the 3 quantities, which can be extracted from S_{D8}, S_{D5} and S_{D3}, quoted $S_{BB}(q)$, $S_{B\varphi}(q)$ and $S_{\varphi\varphi}(q)$ corresponding to the correlations backbone-backbone, backbone phenyl and phenyl-phenyl. An interesting result for example, is that $\dfrac{S_{BB}}{S_{\varphi\varphi}}$ — or, more directly $S_{D8}(q)/$ $S_{D3}(q)$ for example — is a different function of q for the isotropic melt, the deformed melt in the perpendicular direction (faster decreasing) and in the parallel direction (even faster). This means that the correlations would be modified at the local scale [61].

In the isotropic case it is possible to extract the R_{c_i}'s from a fit to (14.1) where $S_0(q)$ will be the form factor of a wormlike chain. Because correlations appear modified

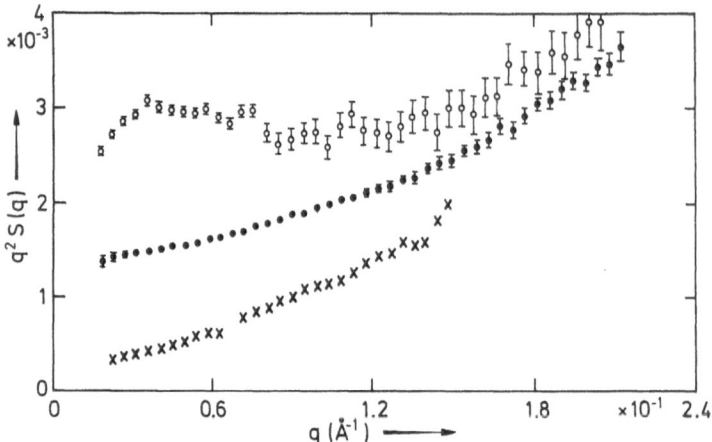

Fig. 18. Form factor of a PS chain ($M_w = 1\,000\,000$) labeled only on the 3 carbons close to the back-bone, in isotropic ($\bigcirc$) and stretched sample (per $\bigcirc$ para $\times$)

it is not possible to use these values of R_{c_i}'s for the deformed case: it follows that $S_0^{(deformed)}$ cannot be extracted that way. One would need a calculation of a deformed wormlike chain for further progress.

Here, the data of Fig. 18 can be used to evaluate the beginning of the local range in q: the effect is around 1 % at $q = 3 \cdot 10^{-2} \text{Å}^{-1}$ and 5 % at $q = 7 \cdot 10^{-2} \text{Å}^{-1}$. This latter limit will be assumed at the upper boundary of the subentangled regime. Its influence on the former conclusions is:

— In the case of the Rouse model another concept is involved, the "Rouse subchain" which is the elementary unit. It is not clear how large this is, compared to the persistence length; we assume it to be of the same order[1].

— In the case of the disengagement model, Eq. (6.8) holds *a priori* even at large q, thus also in the local regime, no matter whether $S_{short\,t}(q)$ and $S_{iso}(q)$ are affected by the local effects.

— of course, effects such as a local nematic field (see Sect. 15) would be sensitive at these q and modify $S_{short\,t}(q)$, but that implies a physical modification of the model.

15 Comparison with Other Techniques

15.1 Difficulties in Investigating the Melt

In the Introduction it was mentioned that most other techniques are unsuitable to study melts in the range T_e, T_{ter}. Welding studies ($t \sim T_{ter}$) and self-diffusion experiments ($t \gg T_{ter}$) were reviewed by de Gennes [26] together with results from other techniques for solutions and also for long times. Neutron Spin Echo suffers from an

---

* The same problem is encountered for inelastic neutron scattering. A Rouse behaviour ($\tau(q) \propto q^4$) is surprisingly observed for a q range larger than 10^{-1}Å^{-1}, sensitive to both the Rouse subchain and the persistence length.

available range of too short times, and also too large q. Only recently, a tube effect has been described (J. Higgins et al.) resulting from the very small tube diameter of PTHF, but this remains in the range of $t \sim T_e$. Apart from viscoelastic measurements, which still hold a great potential for research, another set of techniques may allow comparison to the neutron data in same time range: this technique employs no probe (birefringence) or does use probes in the absence of demixtion: very weak concentrations (10^{-6} for fluorescence polarisation [17]) or nuclear labeling (deuteration; infrared dichroism [17] and deuterium magnetic resonance [56]). These methods all measure the *orientation* of the chain on a *small scale*. It seems useful to add some comments on the relation which can be established between these methods and neutron scattering; especially considering the relation between $\langle \cos^2 \theta \rangle$ and S(q):

The optical methods measure the second moment $\langle \cos^2 \theta \rangle$ of the orientation distribution function f(cos θ) of the monomers or of a probe. A first difficulty is to relate this function to the orientation function of the statistical unit: one might need to use a model of the chemical arrangement of the chain, or to establish a proportionality coefficient in the case of a probe. Then there can not be, at our knowledge, any direct relation (i.e. not model-dependent) between $\langle \cos^2 \theta \rangle$ and S(q). This can be illustrated by three examples, corresponding to different models of chain conformation.

The first example is the totally affine deformation of the isotropic Gaussian conformation. In that case the deformation is the same on any scale, and thus; For a uniaxial deformation λ:

$$S(q) = S_{iso}(\sqrt{q \tilde{\tilde{E}} q}) = S_{iso}(\lambda_q q) = S_{iso}[\sqrt{(\lambda^2 \cos^2 \varphi + 1/\lambda \sin^2 \varphi) q^2}]$$

$$\langle \cos^2 \theta \rangle = 2/(1 - 1/\lambda^3) (1 - (\text{Arctg} \sqrt{(1 - \lambda^3)})/\sqrt{(1 - \lambda^3)}).$$

There exists a relation between the two quantities, though not simple, because $\langle \cos^2 \theta \rangle$ is a complicated function of λ.

The second example is the junction affine network model [43] for a chain fixed only by its two ends to the network. In Sect. 7 it was pointed out that the whole form factor can be calculated [44]. In particular, the radius of gyration is [32–], in a direction α = x, y, z,

$$R_{g\alpha}^2 = 1/(2N^2) \sum_i \sum_j \langle (R_i - R_j)^2 \rangle = 1/2(N^2/N^2) \sum_n (1 - n/N) r_n^2$$

$$= R_{g\,iso}^2 (1 + \lambda_\alpha^2)/2,$$

where $n = |i - j|$.

Ullmann [57] has also given the orientation function of the vector r_{ij} joining two monomers, which is for $i - j = 1$:

$$\langle \cos^2 \theta \rangle = 1/(\sin^2 \psi) (1 - \psi/\text{tg}\,\psi),$$

with $\sin \psi = (\lambda^2 - 1/\lambda)/(N - (\lambda^2 - 1)).$

It appears clearly that $\langle \cos^2 \theta \rangle$ is a very slow function of λ if N is large (note the presence of N in the denominator of sin ψ) while R_g is still a fast one. This reflects the well-known fact that the orientation of a monomer is much smaller than the orientation obtained for a totally affine deformation. Thus, $\langle \cos^2 \theta \rangle$ is much less anisotropic than R_g in this example.

The third example is a case where the monomers are highly oriented on the small scale, but give on the large scale a correlation function which is oriented in the perpendicular direction. Finally, all models are *a priori* possible and no general phenomenological relation can be written without strong model statements.

However, it remains possible to compare qualitatively S(q) and $\langle \cos^2 \theta \rangle$ on the same scale, i.e. for the unit for length b, i.e. $q \sim 1/b$. For the junction affine model (example 2) for instance, the order of magnitude of $S(q_{per})/S_{iso}(q)$ at $q \sim (1/10 \text{ Å})$ and of $P_2(\theta) = (3\langle \cos^2 \theta \rangle - 1)/2$ is comparable. However, we here enter the region where the orientation effect is weaker than the usual accuracy of the SANS experiment. We will now report a recent experiment which compares the two quantities and is related to these problems.

15.2 Experimental Comparison of a Free Chain in a Strained Rubber

Two experiments have been performed which allow a comparison between S(q) and $\langle \cos^2 \theta \rangle$ for same conditions and the same sample. The form factor is given by SANS [58] and the orientation $P_2(\theta)$ by deuterium magnetic resonance [59]. Both are sensitive to deuterated species and can be used in melts for large deuterated fractions. A small chain of deuterated PDMS ($M_{nD} \sim 10000$, $M_{wD} \sim 20000$) is dissolved in a matrix of a non-deuterated PDMS endlinked network (mesh size identical to the labeled chain, 10000), simply by spreading and waiting a convenient time. The network is then uniaxially stretched and one observes the form factor of the labeled chain in all directions between q and the stretching axis on the two-dimensional detector. The result is that the form factor does not change for $1 < \lambda < 1.45$ and $3 \cdot 10^{-2} \text{ Å}^{-1} < q < 4 \cdot 10^{-1} \text{ Å}^{-1}$. It remains isotropic within an uncertainty of 3% and equal to the one of the isotropic sample within 10%. It has the classical features of the form factor of the Gaussian chain. On the other hand, the DMR result is that the frequency peak splits into two peaks separated by $\Delta v = 80$ Hz, which is the signature of an orientation effect. The splitting value is within 10% equal to the one observed, at same λ, for a deuterated PDMS chain of same molecular weight *endlinked* inside the network. SANS and DMR results could appear contradictory, but may be explained following several remarks:

— the orientation given by DMR is weak ($\Delta v = 80$ Hz $\rightarrow P_2(\theta) \sim 1.5 \cdot 10^{-3}$), i.e. under the precision of the SANS experiment. If the anisotropy is the same at any scale, as in an affine deformation and as would give a nematic field, this will not be detectable by SANS.

— a local anisotropy effect may even give less orientation on the large scale than on the small scale [58].

— these two cases are opposite cases to the rubbery deformation for which the orientation is larger when the length scale is larger. This latter is very neatly observable by SANS, meanwhile $P_2(\theta)$ is of the same order of magnitude in the NMR results.

This explanation also implies, besides a comparison of techniques, that the local anisotropy effect in rubbers on the scale of the monomer is more than not negligible compared to the orientation by end-to-end pulling of the endlinked chains, as already suggested [17].

16 Conclusion

It is shown that it is possible to obtain reproducible data for the single chain form factor during the relaxation from a stepstrain, and that current models allow the calculation of this form factor. A comparison experiment-theory is then possible. The conclusion of this comparison is still incomplete: the models, though simple, involve numerous processes requiring a large amount of data, of which only a part is available. Also it appears that the real situation is even more complex, as several simple predictions are not fitted, or at least that a large crossover regime is involved, increasing again the research field. More precisely, the strong results are that:

— there is agreement with the Rouse model at short times and disagreement at long times.
— there is no really strong evidence for reptation. For example:
    1) the retraction predicted by Doi and Edwards is not observed.
    2) the disengagement, which would lead to a characteristic behaviour of the form factor, could explain the data but is not fully determined because this requires a very long duration of relaxation for rather long chains.
    3) a matrix effect seems to appear in the range of 100000–1000000 more strongly than expected.
    4) additional relaxation is observed, in particular for large masses.

Point 1 appears to be elucidated and confirmed by long chains and shear data. For point 4, the degradation of the long chains must be reduced even further, although this would not suffice to cancel the effect. The same long *nondegradated* chains must be used at long times for point 2. For point 3, the essential need is for more data.

The most important experiments seem to be those using very long chains $> 2.5 \cdot 10^6$; such measurements are very rare even in classical rheology. It is to be emphasized that experiments on crosslinked melts are of great interest: first, to compare with melts; and second, intrinsically, as rubbers are a particular state of matter, the dynamics of which are still widely unknown, and for which theory is still approximate.

Other extensions need to be purposed: it is easy to improve the calculation of the form factor, e.g. by combining the now-available correct calculation for rubber classical theories with the one for reptation, and this for any direction with respect to the deformation axes (aspect of isointensity curves). On the level of scattering techniques, further stimulations will result from the exploration at smaller q, in particular owing to the availability of "ultracold neutrons" (i.e. a wavelength of up to 40 Å, achievable at ILL, Grenoble) for SANS and Spin Echo Inelastic Scattering. A coupling with other techniques is certainly useful (fluorescence polarisation, DMR, diffusion coefficient), as well as systematic and precise measurements of the stress coupled to any experiment.

Finally, extensions of the techniques to other materials are obvious. Some are straightforward if they involve a melt close to a glass transition or if they are very viscous; this depends only on the possibilities of the chemistry of deuteration: star and ring polystyrene, blends, ionomers, and possibly semirigid polymers and compatible copolymers. The use of other techniques (Sect. 2) would be required for more liquid materials, as multicomponent solutions, where again neutron scattering has no equivalent.

17 References

1. Boué, F., Nierlich, M., Leibler, L.: Polymer 23, 9 (1982)
2. Edwards, S. F.: Proc. Roy. Soc. 9, 92 (1967)
3. de Gennes, P. G.: J. Chem. Phys. 55, 572 (1971)
4. Doi, M., Edwards, S. F.: J. Chem. Soc. Far. Trans. 2, 74, 1802 (1978)
5. Doi, M.: J. Pol. Sci. Letters 18, 775 (1980)
6. Oberthur, R. C.: Revue Phys. Appl. 19, 663 (1984)
7. de Gennes, P. G.: J. Chem. Phys. 72, 4756 (1980)
8. Hashimoto: IUPAC Proceedings, Washington, 1983
9. Boué, F.: These d'etat, Université Orsay Paris Sud, #2662 (1982)
10. Boué, F., Nierlich, M., Osaki, K.: J. Chem. Soc. Far. Symp. 18 (1983)
11. Boué, F., Cressely, R.: Lab. Leon Brillouin Report 1983
12. Ferry, J. D.: Viscoelastic Properties of Polymers, 3rd ed., Wiley, New York 1982
13. Daoud, M., de Gennes, P. G.: J. Pol. Sci., Pol. Phys. Ed. 17, 1971, 1979; Klein, J.: Macro-molecules 11, 852 (1978)
14. Viovy, J. L.: J. Physique, 46, 847 (1985)
15. de Gennes, P. G.: J. Phys. 36, 1191 (1975); Klein, J.: J. Chem. Soc. Far. Symp. 18, (1983); Helfand, E., Pearson, D. S.: Macromolecules, 17, 888 (1984)
16. Tassin, M., Monnerie, L.: J. Pol. Sci., Pol. Phys. Ed. 1983 or 1984
17. Monnerie, L.: J. Chem. Soc. Far. Trans. 18, (1983)
17 bis. Montfort, J. P.: These d'Etat Université de Pau et Pays de l'Adour 1984 (to be published)
18. Graessley, W. W.: Adv. Pol. Sci. 47, 68–117 (1982)
19. Marucci: Macromolecules 14, 434 (1981) (and paper submitted)
20. Boué, F., Nierlich, M., Jannink, G., Ball, R. C.: J. Physique 43, 137–148 (1982)
21. Allen, G., Higgins, J. S., Macconachie, A.: Far. Trans. 2, 72, 2117–2130 (1982)
22. Marin, G.: Thèse d'État, Université de Pau et Pays de l'Adour, 1978 and papers by Marin, G., Graessley, W. W., 1978–1979
23. Picot, C., Duplessix, R., Decker, D., Benoit, H., Boué, F., Cotton, J. P., Daoud, M., Farnoux, B., Jannink, G., Nierlich, M., de Vries, A. J., Pincus, P.: Macromolecules 10, 436 (1977)
24. Boué, F.: 3rd cycle Thesis, Université Louis Pasteur, 1977, and Boué, F., Jannink, G.: Physique, Colloques C2 (supplem. to #6) 39, C2–183 (1978)
25. Macconachie, A., Allen, G., Richards, R. W.: Polymer 22, 1157 (1981)
26. Boué, F., Nierlich, M., Jannink, G., Ball, R. C.: J. Physique Lettres L582–L589 and L590–L593 (1982)
27. Hadzioannou, G., Hui Wang, Li, Stein, R. S., Porter, R. S.: Macromolecules 15, 880–882 (1982)
28. Boué, F., Osaki, K.: (to be published)
29. Bastide, J., Boué, F., Herz, J.: J. Phys. 46, 1967–1979 (1985). More data on deswollen rubbers are in Ref. 29 bis.
29 bis. Bastide, J.: in Physics of Finely Divided Matter, Springer Proc. in Phys. 5, Daoud, M., Boc-cava, N., Eds., Springer (1985)
30. Boué, F., Nierlich, M., Jannink, G.: ILL Report 1978
31. Boué, F.: ILL report 1984 (and to be published)
32. Muller, Froehlich: (private communication)
33. Kirste, R. C., Oberthur, R. C.: "Small Angle X-ray Scattering", Chapt. 12, Glatter, O., Kratky, O. (Eds.), Academic Press, London 1982
34. Tobolsky, A. V.: J. Appl. Phys. 27, 673 (1956)
35. Wignall, G. D., Ballard, D. H., Schelten, G.: Eur. Pol. J. 10, 861 (1974); Cotton, J. P., Decker, D., Benoit, H., Farnoux, B., Higgins, J. S., Jannink, G., Ober, R., Picot, C.: J. des Cloizeaux, Macro-molecules 7, 863 (1974)
36. Daoudi, S.: J. Physique 38, 731 (1977)
37. Hong, Noolandi, I.: J. Physique 45, L149–L157 (1984)
38. de Gennes, P. G.: J. Physique 41, 735–740 (1981)
39. Boué, F., Osaki, K., Ball, R. C.: J. Pol. Sci., Polym. Phys. Ed., 23, 833–844 (1985)
40. de Gennes, P. G., Leger, L.: Rev. Ann. Phys. Chem., 1982
41. Benoit, H., Duplessix, R., Ober, R., Daoud, M., Cotton, J. P., Farnoux, B., Jannink, G.: Macro-molecules 8, 451 (1975)

42. Kuhn, W., Kunzle, O., Katchalsky, A.: Helv. Chim. Acta *31*, 1994 (1948)
43. Kuhn, V. W., Grunn, F.: Kolloid Zeitschrift, p. 248, 1942; Wall, F. T., Flory, P. J.: J. Chem. Phys. *19*, 1435 (1951)
44. Ullmann, R.: Macromolecules *15*, 1395 (1982)
45. Pearson, D. S.: Macromolecules *10*, 698 (1977)
46. James, H., Guth, E.: J. Chem. Phys. *19*, 1435 (1951); Deam, R. T., Edwards, S. F.: Phil. Trans. Roy. Soc., Ser. A *280*, 317 (1976)
47. Warner, M., Edwards, S. F.: J. Phys. A *11*, 1649 (1978)
48. Flory, P. J., Erman, B.: Macromolecules *15*, 800–806 (1982)
49. Boué, F., Bastide, J.: ILL report 1984 (and to be published). See also Reference 49bis.
49bis. Boué, F., Bastide, J.: Gomadingen meeting on Polymer Network, J. Coll. & Pol. Sci. (1987)
50. Marucci remarked that the application of the Doi-Edwards model for the relation between stress and deformation leads to an unstable situation and therefore inhomogeneities.
51. Osaki, K., Kurata, M.: Macromolecules *13*, 671 (1980)
52. Osaki, K., Doi, M.: Pol. Eng. Reviews *4*, 35–71 (1984)
53. de Gennes, P. G.: Scaling Concepts in Polymer Physics, Cornell University Press, Ithaca 1978
54. Farnoux, B.: Annales de Physique *1*, 73 (1976); Cotton, J. P., Decker, D., Farnoux, B., Jannink, G., Ober, R., Picot, C.: Phys. Rev. Lett. U32J, 1170 (1976)
55. Rawiso, M., Duplessix, R., Picot, C.: Macromolecules *20*, 630 (1987)
56. Dubault, A., Deloche, B.: (submitted to Polymer); Deloche, B., Lapp, A., Herz, J.: J. Physique Lettres *43*, L-763 (1982)
57. Ullman, R.: Macromolecules *11*, 987–989 (1987)
58. Boué, F., Bastide, J., Lapp, A., Picot, C., Farnoux, B.: Europhys. Lett. *1* (12), 637–645 (1986)
59. Dubault, A., Deloche, B.: (published in same issue as (58))
60. Bastide, J., Boué, F.: Physica *140A*, 251–260 (1986)
61. Rawiso, M., Boué, F.: to be published.
62. More precisely, χ is $2 \cdot 10^{-4}$ for the couple polystyrene monomer-deuterated polystyrene monomer. The upper limit for a stable one phase mixture (value at the spinodal) is

$$\chi_s = \frac{1}{2}\left[\frac{1}{N_H \phi_H} + \frac{1}{N_D \phi_D}\right].$$

where $N_H(N_D)$ is the number of monomers and $\phi_H(\phi_D = 1 - \phi_H)$ the volume fraction of the H(D) species. In the case $\phi_H = \phi_D = 50\%$, $N_H = N_D$, $\chi < \chi_s$ when $N_H < 10^4$, i.e. $M < 10^6$; but if $\phi_D = 10\%$, the stability condition for $N_H = N_D$ becomes $M \lesssim 2.5 \cdot 10^6$. The formula for the scattering is [1]

$$\frac{1}{S(q)} = \frac{1}{\phi_H S_H(q)} + \frac{1}{\phi_D S_D(q)} - 2\chi$$

so that χ can be neglected except for large mastes at $qR_g < 1$ ($S_H(q) \rightarrow N_H(1 - q^2 R_g^2/3)$). Limit cases are currently under study. Values for χ are given by: C. Strazielle, H. Benoit; A. Lapp, from data of Ref. 1 for PS); Bates and Wignall, Macromolecules, 1986.

Phase Relations and Miscibility
in Polymer Blends Containing Copolymers

Ryong-Joon Roe and David Rigby
Department of Materials Science and Engineering, University of Cincinnati,
Cincinnati, OH 45221-0012, USA

*This article reviews the reported work on thermodynamics of polymer blends in which at least one of the
components is a copolymer. Theoretical and experimental studies of miscibility and phase transition and
separation behaviors in such systems are summarized. Chapter 2 deals with blends involving random
copolymers, including the cases in which a random copolymer AB is mixed with either a homopolymer A,
or a homopolymer C, or two homopolymers A and B. Chapter 3 deals with blends involving a block
copolymer, and after giving a brief overview of systems containing a pure block copolymer alone, it
reviews the blend systems containing a block copolymer AB mixed with a homopolymer A or two homopolymers A and B.*

Advances in Polymer Science 82
© Springer-Verlag Berlin Heidelberg 1987

1 Introduction

The possibility of using block or random copolymers as compatibilizers in immiscible polymer blend systems has long been a subject of interest, with the first papers appearing in the patent literature during the 1940's in connection with the emerging synthetic rubber industry. Since then, and particularly during the past several years, a large number of publications has appeared. Of these, however, only relatively few studies deal with the basic thermodynamics of compatibility, though this is a subject of fundamental importance. In the present article, we attempt to review recent progress made in understanding the thermodynamic factors which govern miscibility. We are not concerned with studies involving mechanical behavior of copolymer-containing blends, except where the results have a bearing on the compatibility question, and instead refer the reader to the summary given in Tables 1 and 2 and to the references cited therein. These tables list all those blend systems, involving random or block copolymer, which have been examined in recent years either for their compatibility behavior or for their mechanical, rheological, and morphological properties. We note also that a number of earlier reviews contain material of relevance to the subject of copolymer-containing blends. For example, Krause [1] has given an extensive description of the blends known to be compatible up to 1977. Paul [2] has discussed the role of block copolymers as emulsifiers, while other more general reviews have been given by Schmitt [3] and Bywater [4], among others.

In the following two chapters, we address recent developments in understanding the thermodynamic behavior of mixtures containing random and block copolymers, respectively. In each case we will review the current state of theoretical understanding before describing the results of experimental studies, attempting to assess the extent to which present theory is able to explain (and predict) observed behavior. In our consideration, we include only mixtures containing non-crystallizable components, since the crystallization of any of the components introduces a whole new aspect of complexity which requires a separate, thorough study.

2 Systems Containing Random Copolymer

2.1 Binary Mixtures Containing One or Two Random Copolymers

2.1.1 Theory

2.1.1.1 Free Energy of Mixing

For discussion of polymer mixture thermodynamics, one can take the Flory-Huggins free energy of mixing as the starting point. The basic approximation embodied in the Flory-Huggins treatment is of the mean-field nature, ignoring the local concentration fluctuations around individual segments. It is well-known that for polymer solutions in low molecular weight solvent the mean-field approach becomes grossly inadequate near the critical point and at dilute concentrations. deGennes [5] has noted, however, that the mean-field approximation is fairly satisfactory when the mixture consists of all long chain molecules. In such mixtures, the chains assume ideal gaussian

Table 1. Blends Studied That Contain Random Copolymer

(a) *Binary blends containing one random copolymer.*

Copolymer	Homopolymer	Ref.
Styrene/butadiene	Polystyrene	8, 96)
Styrene/butadiene	Polybutadiene	169, 171)
Styrene/butadiene	Polypropylene	97)
Styrene/acrylonitrile	Polycaprolactone	98, 99, 101, 161)
Styrene/acrylonitrile	Poly(methyl methacrylate)	13, 22, 26, 27, 100, 102)
Styrene/acrylonitrile	Various acrylates & methacrylates	24)
Styrene/(methyl methacrylate)	Polystyrene	103)
Styrene/(n-butyl methacrylate)	Polystyrene or poly(n-butyl methacrylate)	14)
Styrene/(allyl alcohol)	Polycaprolactone	104)
Styrene/α-methylstyrene	Poly(2,6-dimethyl-1,4-phenylene oxide)	105)
Styrene/(halogenated styrene)	Poly(2,6-dimethyl-1,4-phenylene oxide)	10, 33−35, 37, 38, 40, 41, 106−108)
Styrene/(halogenated styrene)	Polystyrene	35, 37, 38)
α-methylstyrene/acrylonitrile	Poly(methyl methacrylate)	28)
α-methylstyrene/methacrylonitrile	Poly(vinyl chloride)	134)
Butadiene/acrylonitrile	Poly(vinyl chloride)	127, 169, 170)
Butadiene/acrylonitrile	Polychloroprene	169)
(methyl methacrylate)/(alkyl methacrylates)	Poly(methyl methacrylate)	109)
(methyl methacrylate)/esters	Polycarbonate	110)
(glycidyl methacrylate)/(ethyl acrylate)	Poly(2,3-dichloro-1-propylacrylate)	111)
(ethylene terephthalate)/oxybenzoate	Poly(butylene terephthalate)	112)
(ethyl acrylate)/(4-vinyl pyridine)	Poly(vinyl chloride)	113)
Ethylene/(vinyl acetate)	Polyethylene	164)
Ethylene/(vinyl acetate)	Chlorinated polyethylene	29, 114, 116)
Ethylene/(vinyl acetate)	Poly(vinyl chloride)	29, 30, 116, 117, 169)
Ethylene/(vinyl acetate)	Polychloroprene	169)
Ethylene/(vinyl acetate)	Poly(vinyl nitrate)	118)
Ethylene/(N,N-dimethylacrylamide)	Poly(vinyl chloride)	122)
Ethylene/propylene	Polypropylene	123−126)
Ethylene/propylene	Polyethylene	123)
Propylene/(vinyl chloride)	Polycarbonate	128)
(vinylidene chloride)/(vinyl chloride)	Polycaprolactone	129, 130, 132)

Various Sarans	Polyesters	130)
Various Sarans	Various polyacrylates and methacrylates	131)
(butylene terephthalate)/tetrahydrofuran	Poly(vinyl chloride)	163)
(ethyl acrylate)/ethylene	Poly(vinyl chloride)	119)

(b) *Binary blends containing two random copolymers.*

Copolymer	Copolymer	Ref.
Styrene/acrylonitrile	Styrene/acrylonitrile	13)
Styrene/acrylonitrile	Butadiene/acrylonitrile	171)
Styrene/acrylonitrile	(methyl methacrylate)/(alkyl methacrylate)	109, 133)
α-methylstyrene/acrylonitrile	(methyl methacrylate)/(alkyl methacrylate)	109)
Styrene/(methacrylic acid Na salt)	(ethyl acrylate)/(acrylic acid Na salt)	162)
Butadiene/acrylonitrile	Butadiene/acrylonitrile	169)
Styrene/butadiene	Styrene/butadiene	169)

(c) *Binary blends containing one random terpolymer.*

Terpolymer	Homopolymer	Ref.
Ethylene/(ethyl acrylate)/(carbon monoxide)	Poly(vinyl chloride)	119)
Ethylene/(vinyl acetate)/sulfone	Poly(vinyl chloride)	121)
α-methylstyrene/methacrylonitrile/(ethyl acrylate)	Poly(vinyl chloride)	167)
EPDM	Polypropylene	97, 123)
EPDM	Polyethylene	123)
Ethylene/(vinyl acetate)/carbon monoxide)	Poly(vinyl chloride)	119, 120)
Ethylene/(2-ethylhexyl acrylate)/(carbon monoxide)	Poly(vinyl chloride)	119)
Acrylonitrile/styrene/butadiene	Polypropylene	165)
Acrylonitrile/styrene/butadiene	Nylon 6 or Nylon 12	166)

(d) *Ternary blends containing one or two random copolymers.*

Copolymer	Homopolymer or Copolymer	Homopolymer	Ref.
Ethylene/propylene	Polyethylene	Polypropylene	135, 136, 138, 139)
Styrene/(methyl methacrylate)	Polystyrene	Poly(methyl methacrylate)	48)
Styrene/butadiene	Polystyrene	Polybutadiene	47)
Butadiene/acrylonitrile	(vinylidene chloride)/(vinyl chloride)	Poly(vinyl chloride)	137)

Table 2. Blends Studied That Contain Block Copolymer

(a) *Binary blends containing one block copolymer.*

Copolymer	Homopolymer	Ref.
Styrene/dimethylsilaoxane	Polystyrene	85, 140, 141)
Styrene/butadiene	Polybutadiene	17, 142)
Styrene/butadiene	Polystyrene	8, 17, 71, 72, 77–79, 143)
Styrene/isoprene	Polystyrene	73, 76)
Styrene/isoprene	Polyisoprene	73, 76)
α-methylstyrene/isoprene	Poly-α-methylstyrene or polyisoprene	144)
1,4-butadiene/1,4-isoprene	Poly(1,4-butadiene) or poly(1,4-isoprene)	49)
Dimethylsiloxane/styrene/dimethylsiloxane	Polystyrene	145)
Styrene/butadiene/styrene	Polystyrene	146, 149)
Styrene/butadiene/styrene	Polybutadiene	147–149)
Styrene/butadiene/styrene	Polyethylene	150)
Styrene/isoprene/styrene	Poly(2,6-dimethyl-1,4-phenylene oxide)	153)
Styrene/(ethylene-butene)/styrene	Polypropylene	151)
Styrene/α-methylstyrene/styrene	Polystyrene or poly(α-methylstyrene)	154)

(b) *Binary blends containing two block copolymers.*

Copolymer	Copolymer	Ref.
Styrene/butadiene	Styrene/butadiene	157)
Styrene/butadiene	Styrene/butadiene/styrene	146)

(c) *Ternary blends containing one block copolymer.*

Copolymer	Homopolymer	Homopolymer	Ref.
Styrene/isoprene	Polystyrene	Polyisoprene	73)
Ethylene/propylene/ethylene	Polyethylene	Polypropylene	168)
Styrene/(hydrogenated butadiene)	Polystyrene	Polyethylene	90–92)
Styrene/(ethylene-butene)/styrene	Polystyrene	Polyethylene	155)
Styrene/ethylene	Polystyrene	Polyethylene or ethylene/propylene	93, 94)
Butadiene/isoprene	Poly(1,4-butadiene)	Poly(1,4-isoprene)	156)
Styrene/(methyl methacrylate)	Polystyrene	Poly(methyl methacrylate)	48)
Styrene/Butadiene	Polystyrene	Polybutadiene	47)

conformation unperturbed by the excluded volume effect, thus resulting in further simplification. In fact, the method of random phase approximation, which is also of mean-field nature, was applied to the problem of polymer mixtures by deGennes [5, 6], and was found to lead to results that are identical to those obtainable from the Flory-Huggins approach.

The Flory-Huggins free energy of mixing, ΔG_M, per unit volume of the mixture can be written as [7, 8]

$$\Delta G_M = RT[(1/V_1)\, \varphi_1 \ln \varphi_1 + (1/V_2)\, \varphi_2 \ln \varphi_2] + \Lambda_{12}\varphi_1\varphi_2 \qquad (1)$$

where V_1 and V_2 are the molar volumes of the polymers 1 and 2, and φ_1 and φ_2 are the volume fractions of the two polymers in the mixture. The first term in Eq. (1) is the combinatorial entropy of mixing originally derived by Flory and Huggins, and the second term containing Λ_{12} includes all other contributions to the free energy of mixing that are not accounted for by the combinatorial term. Λ_{12} has the dimension of energy per unit volume, and can be called the polymer-polymer interaction energy density.

More traditionally the Flory-Huggins free energy of mixing is written in the form

$$\Delta G_M'/RT = [(1/N_1)\, \varphi_1 \ln \varphi_1 + (1/N_2)\, \varphi_2 \ln \varphi_2] + \chi_{12}\varphi_1\varphi_2 \qquad (2)$$

where $\Delta G_M'$ is the free energy of mixing per lattice volume (or segment), and N_1 and N_2 are the numbers of segment in polymer 1 and 2. If V_{ref} is the volume of a lattice (or a segment), then ΔG_M is equal to $\Delta G_M'/V_{ref}$, and it is seen that

$$\chi_{12} = \Lambda_{12}V_{ref}/RT . \qquad (3)$$

The interaction parameter χ_{12} is dimensionless, and its numerical value depends on the choice of the lattice volume V_{ref}. In the case of polymer solutions, V_{ref} is usually equated to the volume of a solvent molecule and no ambiguity arises, but in the case of polymer blends, the practice of implicitly equating V_{ref} to the volume of a monomer unit is not satisfactory since the monomer volumes of polymers 1 and 2 are usually significantly different from each other. For experimental evaluation the interaction energy density Λ_{12}, given in terms of a specific unit such as joules/cm^3, avoids such ambiguity.

In the model of "regular" solution (or its extension to polymer mixtures), the Λ_{12} term is purely enthalpic and Λ_{12} is thus a true constant. In a real mixture, Λ_{12} is a function of T, p, and the composition of the mixture, but the utility of Eq. (1) relies on the fact that the dependence of Λ_{12} on these variables is only moderate in most cases. Only in the case of the systems exhibiting LCST behavior is the temperature dependence of Λ_{12} appreciable. The strong concentration dependence of χ_{12}, often found with dilute polymer solutions, is not encountered with polymer mixtures. This is mainly due to the fact that the mean-field approximation, as stated earlier, is fairly satisfactory and the entropy of mixing two polymeric components is reasonably well represented by the combinatory entropy term.

2.1.1.2 Effective Interaction Parameter for Systems Containing Copolymers

The binary mixtures considered here can be classed into two categories. In the first, only two different monomers are involved in the mixture. This includes a mixture $(AB)_1/(AB)_2$ of two AB copolymers which differ in composition, and a mixture A/AB containing a homopolymer A and a copolymer AB. In the second category, three or more different monomers are involved. Its most general case would be a mixture AB/CD containing a copolymer AB and a different copolymer CD which has none of its segments in common with the other copolymer.

The free energy of mixing in such systems can still be written in the form of Eq. (1). What we desire is to be able to express the effective interaction parameter Λ_{12} in terms of the interaction between the constituent segments (i.e., in terms of Λ_{AB}, Λ_{AC}, etc.). If Λ_{AB} denotes the interaction parameter between homopolymers A and B, and f_{A1} denotes the volume fraction of A in copolymer 1, etc., then the following relationships hold:

$(AB)_1/(AB)_2$ mixture:

$$\Lambda_{12} = \Lambda_{AB}(f_{A1} - f_{A2})^2 \tag{4}$$

A/CD mixture:

$$\Lambda_{12} = \Lambda_{AC}f_{C2} + \Lambda_{AD}f_{D2} - \Lambda_{CD}f_{C2}f_{D2} \tag{5}$$

AB/CD mixture:

$$\Lambda_{12} = \Lambda_{AC}f_{A1}f_{C2} + \Lambda_{AD}f_{A1}f_{D2} + \Lambda_{BC}f_{B1}f_{C2} + \Lambda_{BD}f_{B1}f_{D2} - \Lambda_{AB}f_{A1}f_{B1} - \Lambda_{CD}f_{C2}f_{D2} \tag{6}$$

These relationships can be derived easily when Λ arises from a purely enthalpic effect contributed by nearest neighbor interactions. Then Eqs. (4)–(6) can be obtained by counting the number of contacts between various types of segment pairs A-B, A-C, etc., present in the mixture, and then by subtracting from it the number of similar pairs that were already present in copolymers 1 and 2 before mixing. Even when the Λ term contains the effect of non-combinatorial entropy of mixing, these relationships can be justified from a more general ground, and any interested reader is referred to the original literature [8] for the argument leading to this justification.

The necessary (but not sufficient) condition for mixing is that the free energy of mixing be negative. The first term in Eq. (1), arising from the combinatorial entropy of mixing is always negative, with its absolute magnitude diminishing with increasing molar volumes V_1 and V_2. Obviously, to attain a miscibility, Λ_{12} ought to be negative, or if it is positive, its magnitude should be as small as possible. In the case of $(AB)_1/(AB)_2$ and A/AB mixtures, Eq. (4) shows that Λ_{12} can be negative only if Λ_{AB} is itself negative. The magnitude of Λ_{12}, when Λ_{AB} is positive can, however, be made as small as desired, to achieve miscibility, by making the compositions of copolymers 1 and 2 as close as possible. In Eqs. (5) and (6) for A/CD and AB/CD mixtures, some of the terms have a negative sign in front, suggesting that a negative Λ_{12} value can be obtained even when individual Λ_{AB}, Λ_{AC}, etc., are all positive. Such a possibility has been

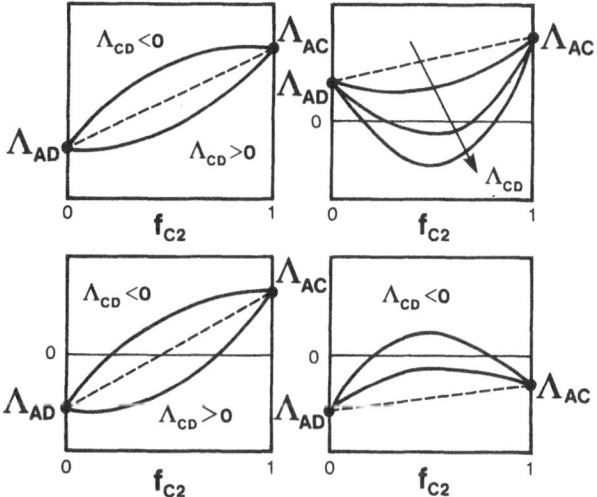

Fig. 1. Illustration of the various ways by which the effective interaction parameter Λ_{12} between homopolymer A and copolymer CD, given by Eq. (5), may vary with the volume fraction f_{C2} of co-monomer C in the copolymer, according to whether Λ_{CD} is negative, equal to zero (broken line), or positive. (From Paul and Barlow [9])

pointed out by a number of workers [9-11]. For example, Paul and Barlow [9] have discussed the various possibilities for Λ_{12} in the A/CD system. These are illustrated in Figure 1. The upper-left panel there shows the basic features, where Λ_{12} plotted against the composition of the copolymer is either concave or convex upward depending on whether Λ_{CD} is positive or negative, respectively. The upper-right shows the interesting situation in which both Λ_{AC} and Λ_{AD} are positive yet Λ_{CD} is sufficiently positive and large to make Λ_{12} negative over part of the composition range where miscibility might occur. Physically interpreted, this behavior would occur when there is a highly unfavorable interaction between segments C and D in the copolymer. Mixing with homopolymer A thus results in dilution of the unfavorable interactions between C and D and leads to a negative interaction parameter Λ_{12}. The lower-left panel shows that an overall positive interaction parameter Λ_{12} may result even when Λ_{AD} is negative, while the lower-right panel shows the opposite effect to the upper-right panel, namely that Λ_{12} can be positive over part of the composition range even when the homopolymer forms attractive interactions with both components of the copolymer. The situation described above is also clearly applicable to mixtures of two different copolymers AB and CD, although the possible combinations are more numerous. Thus there remains the possibility that, even if all intersegmental interactions are positive, mixing will occur when Λ_{AB} and/or Λ_{CD} are sufficiently large relative to Λ_{AC}, Λ_{AD}, Λ_{AB}, and Λ_{BD}.

While Eqs. (5) and (6) suggest the possibility of a negative effective interaction parameter even with all individual interactions positive, Paul and Barlow [9] have shown that such result is impossible if the interaction parameter depends exactly on the solubility parameter difference according to

$$\Lambda_{ij} = (\delta_i - \delta_j)^2 . \tag{7}$$

If this is the case, then, Eq. (6) for the AB/CD mixture can be rewritten as

$$\Lambda_{12} = (\langle \delta_1 \rangle - \langle \delta_2 \rangle)^2 \tag{8}$$

where

$$\langle \delta_1 \rangle = \delta_A f_{A1} + \delta_B f_{B1} \tag{9}$$
$$\langle \delta_2 \rangle = \delta_C f_{C2} + \delta_D f_{D2}$$

showing that Λ_{12} is always positive. Equation (7) results when the cohesive energy density C_{ij} between unlike segments is given by the geometric mean of the c.e.d.'s of the pure components,

$$C_{ij} = (C_{ii}C_{jj})^{1/2} \tag{10}$$

Paul and Barlow concluded, however, that if this restriction is relaxed by writing instead

$$C_{ij} = (1 - k_{ij})(C_{ii}C_{jj})^{1/2} \tag{11}$$

then only a small value of k_{ij} is sufficient to produce a negative interaction parameter Λ_{12}. (The authors considered the system A/CD with $f_{C2} = 0.5$, $\delta_A = 9$, $\delta_B = 8$, $\delta_C = 11$ (cal/cm^3)$^{0.5}$, and with $k_{AC} = k_{AD} = 0$. They then showed that Λ_{12} would become negative for $k_{CD} > 1/176$).

2.1.2 Experiment

2.1.2.1 A/AB and (AB)$_1$/(AB)$_2$ Systems

A number of blend systems in which only two monomeric species, A and B, are involved have been studied over the years. These include styrene/acrylonitrile [12, 13], styrene/ (n-butyl methacrylate) [14], styrene/butadiene [8], (methyl methacrylate)/(n-butyl acrylate) [15, 16], (methyl methacrylate)/(methyl acrylate) [15], (methyl methacrylate)/ (ethyl acrylate) [15], and (methyl methacrylate)/(butyl methacrylate) [15]. The techniques employed to determine the compatibility include visual observation of cloudiness/ turbidity, cloud point determination by light scattering [8, 17], determination of light transmission by visible spectroscopy [14], phase contrast optical microscopy [12], electron microscopy [15], differential scanning calorimetry [14] for determination of single or double T_g's, dynamic mechanical spectroscopy (torsion pendulum) [15, 16], dielectric relaxation spectroscopy [14], inverse gas chromatography [14], and neutron scattering [13]. When several techniques were employed at the same time, the threshold of compatibility determined by different techniques were often different [14, 16], making quantitative interpretation of the results difficult. Qualitatively, however, the expectation is borne out in all cases that the two polymers are compatible when their comonomer compositions are sufficiently close to each other.

Schmitt, Kirste, and Jelenic [13] utilized neutron scattering to determine the coil sizes and the second virial coefficients with mixtures containing two styrene/acrylo-nitrile copolymers, one of which was deuterated. The interaction parameter calculated from the second virial coefficient was indeed positive and increasing with the composition difference. The coils were found to be unpertubed in all mixtures except one in which it was slightly contracted.

Roe and Zin [8] investigated mixtures of polystyrene with styrene/butadiene random copolymer. Mixtures of polystyrene with styrene/butadiene block copolymer in the disordered state were also studied by them and by Rameau, Lingelser, and Gallot [17]. In these studies, cloud points were determined by light scattering as a function of temperature and concentration. Roe and Zin [8] fitted the resulting cloud point curve (see Fig. 2) with a calculated curve based on Eq. (1), thereby determining the best fitting value of the interaction energy density Λ_{12}. The values of Λ_{12} thus obtained were found to agree well with those calculated by Eq. (4) from the knowledge of the composition f_{A1} of the copolymer and the value of the interaction parameter Λ_{AB} between polystyrene and polybutadiene that were previously determined in a similar manner. The result thus demonstrates the essential validity of the thermodynamic considerations described in Sections 2.1.1.1 and 2.1.1.2.

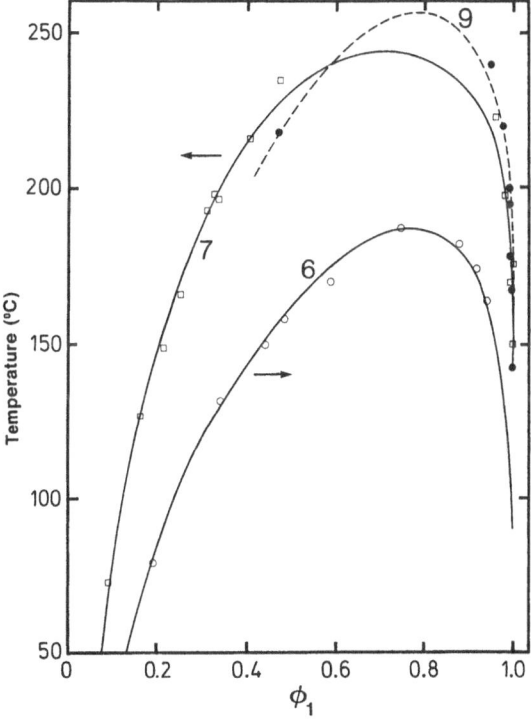

Fig. 2. The observed cloud points are plotted against the volume fraction φ_1 of the polystyrene for the mixture of polystyrene (M_w 5480) and styrene/butadiene random copolymer (50% styrene, M_w 24000) (curve 6), for the mixture of polystyrene (M_w 2400) and styrene/butadiene random copolymer (25% styrene, M_w 29000) (curve 7), and for the mixture of polystyrene M_w = 2400) and styrene/butadiene diblock copolymer (25% styrene M_w 28000, in the disordered state) (curve 9). The solid curves drawn are the results of the least-square fit using a temperature dependent Λ_{12} as an adjustable parameter. (From Roe and Zin [8])

On a more quantitative level, some of the detailed features in the experimental results were found to deviate from the predictions based on the Flory-Huggins theory. Fujioka et al. [14] studied mixtures of styrene/(n-butyl methacrylate) copolymer with either polystyrene or poly(n-butyl methacrylate). Although it is expected that A/AB mixture with the A fraction in AB equal to f should show exactly the same behavior as B/AB mixture with the A fraction in AB equal to 1 − f, such symmetry was not strictly observed. (The deviation from such symmetry might occur, aside from the possible inadequacy of Eq. (1), from the differences in the molecular weight or its distribution in the A and B homopolymers and in the compositional heterogeneity in the copolymer samples.) Probably a more serious deviation from the prediction of Eq. (1) is encountered in the observations of Kollinsky and Markert [15, 16]. As pointed out by Scott [18] long ago, it follows from Eq. (1) that the maximum difference $\Delta = f_{A1} - f_{A2}$ between the compositions allowed for a compatible mixture of two copolymers $(AB)_1$ and $(AB)_2$ will be independent of the mean $(f_{A1} + f_{A2})/2$ when the mixture concentration and the molecular weights of the copolymers are fixed. As is illustrated in Fig. 3 and 4, Kollinsky and Markert [15, 16] found that the maximum difference Δ

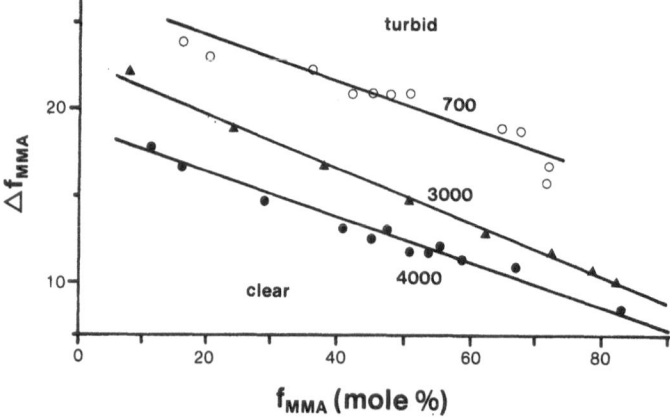

Fig. 3. The maximum composition difference Δf_{MMA} permissible for compatibility between two (methyl methacrylate)/(butyl acrylate) copolymers (in 50/50 mixtures) plotted against the average content f_{MMA} of methyl methacrylate. The degrees of polymerization are indicated. (From Kollinsky and Markert [15])

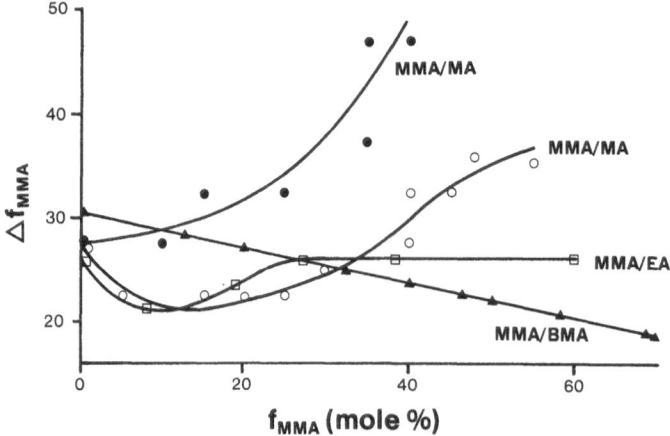

Fig. 4. The maximum composition difference Δf_{MMA} permissible for compatibility between two co-polymers, plotted against the average content f_{MMA} of methylmethacrylate. The types of copolymer are indicated. (From Kollinsky and Markert [15])

for compatibility of two copolymers consisting of methyl methacrylate and one of acrylate or methacrylate monomers drifted as the overall proportion of the methyl methacrylate in the mixture increased. Koningsveld and coworkers [19] noted two possible explanations for the observed behavior. It can arise [19a] if the polydispersity, either in molecular weights or in the comonomer composition, in the copolymer samples used changes with increasing methyl methacrylate content. It can also arise [19b] if the sizes of the A and B segments are different, so that the interaction parameter itself no longer depends on $f_{A1} - f_{A2}$ alone, as in Eq. (4), but depends also on the average composition.

2.1.2.2 A/CD Systems

A number of studies have uncovered instances of miscibility in mixtures of homopolymer A with copolymer CD when the corresponding homopolymer pairs A/C, A/D, and C/D are immiscible. The copolymer of styrene and acrylonitrile has been found to be miscible with a fairly large number of polymers, including poly(vinyl chloride) [20], poly(ε-caprolactone) [4], sulfone based polymers [21], poly(methyl methacrylate) [22-27], and poly(ethyl methacrylate) [24]. Copolymers of α-methylstyrene with acrylonitrile are also found [28] to be miscible with poly(ethyl methacrylate) and atactic and isotactic poly(methyl methacrylate). Other examples of A/CD compatible systems are copolymers of ethylene and vinyl acetate with poly(vinyl chloride) [29, 30] and copolymers of acrylonitrile and butadiene with poly(vinyl chloride) [31, 32]. In the list of compatible polymer pairs compiled by Krause [1], there are many cases which involve copolymers and some of them might belong to the A/CD systems discussed here.

In recent years, a systematic investigation was made by Karasz, MacKnight, and coworkers [10, 33-41] on the miscibility of halogen-substituted styrene copolymers

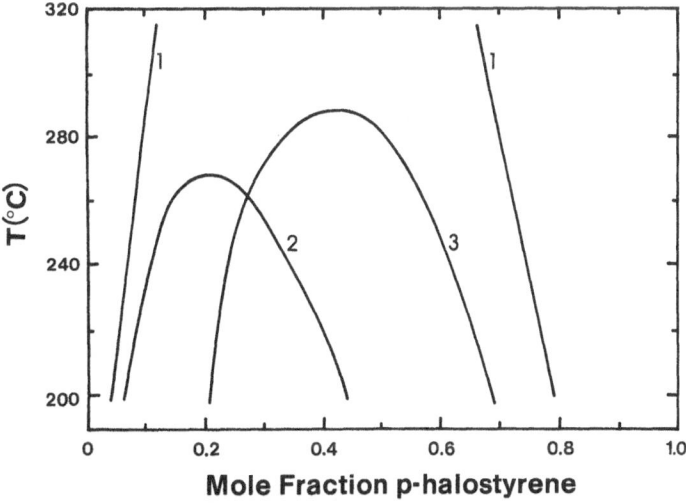

Fig. 5. Miscibility of PPO with random copolymer of o-fluorostyrene and p-chlorostyrene (curve 1), copolymer of o-fluorostyrene and p-fluorostyrene (curve 2), and copolymer of o-chlorostyrene and p-chlorostyrene (curve 3). The insides of the curves represent the miscibility regions. (From ten Brinke, Karasz, and MacKnight [10])

with poly(2,6-dimethyl-1,4-phenylene oxide) (PPO) (and also with polystyrene). The interesting aspect here is (i) that PPO and polystyrene are compatible with each other in all proportions, (ii) that neither PPO nor polystyrene is compatible with any of the halogen-substituted styrene homopolymers such as poly(ortho-chlorostyrene), poly(para-chlorostyrene), poly(ortho-fluorostyrene), poly(para-fluorostyrene), etc., and yet (iii) that many of the copolymers of these halogen-substituted styrenes (having a certain limited range of comonomer composition) exhibit compatibility with PPO. To cite a specific example, neither poly(ortho-chlorostyrene) nor poly(para-chloro-styrene) is compatible with PPO, but a random copolymer of these two types of chlorostyrenes is compatible with PPO, provided that the copolymer contains between 23 and 64 mole % of para-chlorostyrene [33]. The compatibility between polymer pairs was ascertained by the presence of a single T_g obtained by DSC, and was additionally confirmed by visual observation of the clarity of the films. The phenomenon of the miscibility "window" is well illustrated in Fig. 5, which [10] summarizes the miscibility limit of PPO when mixed with the three types of random copolymers of halogen-substituted styrenes. Only the copolymers having the composition of comonomers corresponding to the inside region of the curves are miscible with PPO. As the temperature is raised, the width of the miscibility "window" narrows, and this suggests a LCST behavior. In fact, all miscible pairs belonging to A/CD systems observed show a LCST behavior. Thus when a miscible pair at certain temperatures is annealed at a higher temperature, it was observed [33] that the mixture became cloudy and the DSC curve showed two T_g's. Figure 6 shows the cloud point curves obtained [28] with mixtures containing copolymer of α-methylstyrene/acrylonitrile and either atactic or isotactic poly(methyl methacrylate) or poly(ethyl methacrylate), and demonstrates the LCST behavior clearly.

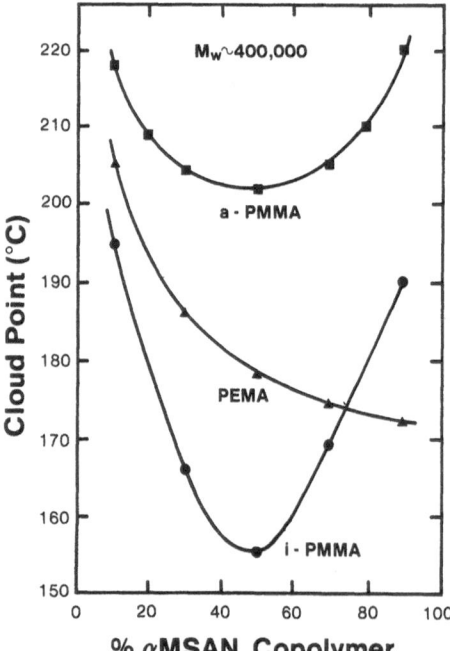

Fig. 6. Comparison of cloud point curves for blends of α-methylstyrene/acrylonitrile copolymer with isotactic poly(methyl methacrylate), with poly(ethyl methacrylate), and with atactic poly(methyl methacrylate). (From Goh, Paul, and Barlow [28])

The occurrence of the miscibility window can be explained readily [9, 10] by means of Eq. (5) and can be attributed to the strong "repulsion" between C and D being mitigated by A. The LCST behavior can then be ascribed to the same effect that gives rise to the LCST behavior of compatible blends of high molecular weight homopolymers. Paul and Barlow [9] extended the reasoning based on Eq. (5) to explain the miscibility behavior of a homologous series of aliphatic polyesters with polycarbonate [42], poly(vinyl chloride) [43], polyhydroxyether of Bisphenol-A [44], etc., by treating as if the homologous polyesters were copolymers.

2.2 Ternary Mixtures Containing a Random Copolymer and Two Homopolymers

2.2.1 Theory

Interest in A/B/AB systems arises from the potential utility of random copolymers as compatibilizers, i.e., their ability to produce a single phase mixture from two otherwise immiscible homopolymers.

Phase diagrams for the $(AB)_1/(AB)_2/(AB)_3$ system containing three AB copolymers of different compositions were considered by Leibler [45]. Manageable results could be obtained for the symmetric case where the molecular sizes of all three polymers are equal and the fraction of monomer A in the three copolymers are given by $f_{A1} = f + \Delta$, $f_{A2} = f - \Delta$, and $f_{A3} = f$. The A/B/AB blend considered here then becomes a special case corresponding to $f = 0.5$, $\Delta = 0.5$. He identified four distinct types of phase behavior, which for A/B/AB systems read as follows:

1) For $\Lambda_{AB}V/RT < 2$ (where V is the molar volume of the polymer molecules), the mixture is miscible at all compositions.
2) For $2 < \Lambda_{AB}V/RT \leq 6$, Fig. 7a, there is one critical point of demixing, located at $\varphi_1 = \varphi_2$, $\varphi_3 = 1 - RT/\Lambda_{AB}V$.
3) For $6 < \Lambda_{AB}V/RT \leq 8$, Fig. 7b, there are three critical points, two of which are physically meaningful. The third critical point is located in an unstable region, associated with the occurrence of a three phase triangle in the phase diagram.
4) For $8 < \Lambda_{AB}V/RT$, Fig. 7c, there are no critical points; all three of the binary pairs show a miscibility gap, and the three phase triangle remains.

Recently Koningsveld and Kleintjens [46] made some numerical analyses of possible phase diagrams of three random copolymers, in particular with the aim of identifying conditions necessary for a given copolymer to act as a compatibilizer. They deduced that addition of copolymer 3 with the composition intermediate between the other two copolymers 1 and 2 would not in general induce compatibilization unless components 1 and 2 were already of fairly low molecular weight.

The conditions necessary for the lowering of the critical, spinodal, or binodal temperature of binary mixtures of homopolymers A and B by addition of copolymer AB was also examined analytically by Rigby, Lin, and Roe [47]. They showed that under some special conditions, spinodal temperature T can be lowered linearly with the amount of added copolymer AB, so that

$$T/T^0 = (\Lambda_{AB}/\Lambda^0_{AB})(1 - \varphi_3) \qquad (12)$$

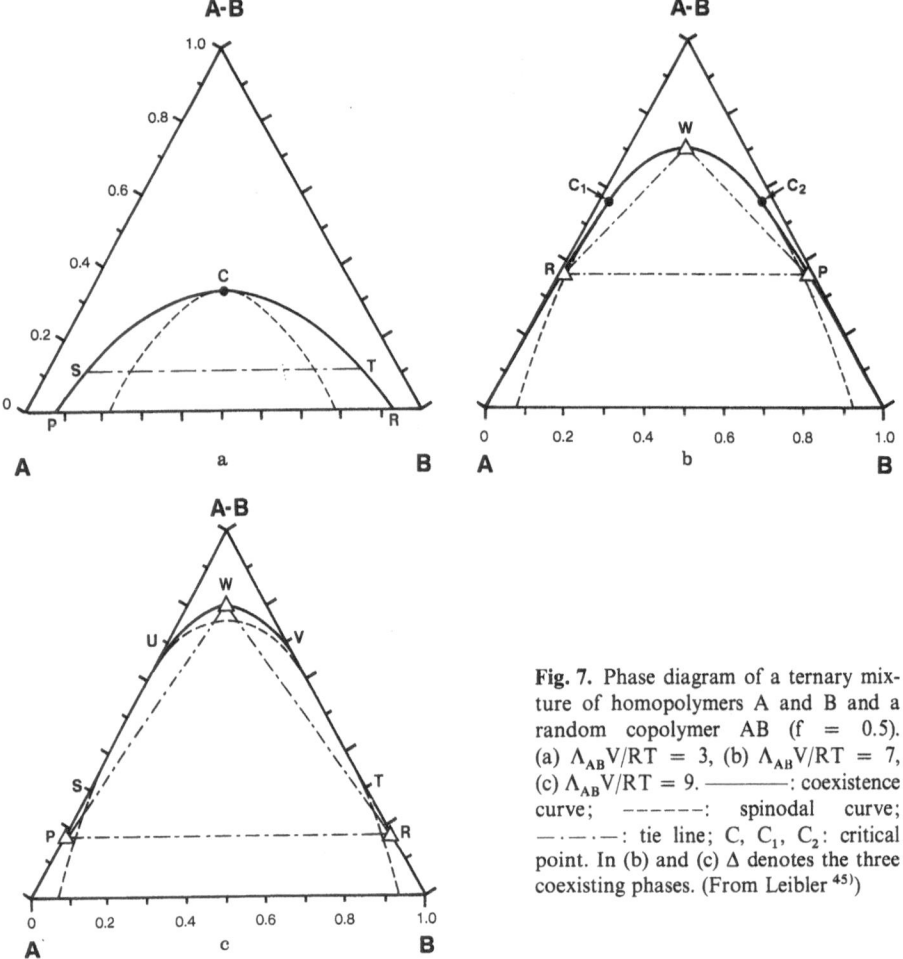

Fig. 7. Phase diagram of a ternary mixture of homopolymers A and B and a random copolymer AB (f = 0.5). (a) $\Lambda_{AB}V/RT = 3$, (b) $\Lambda_{AB}V/RT = 7$, (c) $\Lambda_{AB}V/RT = 9$. ————: coexistence curve; ------: spinodal curve; —·—·—: tie line; C, C_1, C_2: critical point. In (b) and (c) Δ denotes the three coexisting phases. (From Leibler[45])

where φ_3 is the copolymer volume fraction, T^0 is the spinodal temperature in the binary A/B blend and Λ_{AB}^0 is the value of Λ_{AB} at T^0. The condition under which Eq. (12) is obeyed is

$$\varphi_1 V_1/f_{A3} = \varphi_2 V_2/f_{B3} \tag{13}$$

where f_{A3} and f_{B3} are volume fractions of A and B in the copolymer and V_1 and V_2 are the molar volumes of the homopolymers A and B. Equation (12) represents the maximum possible lowering of spinodal temperature attainable with a given amount φ_3 of a copolymer AB. When the condition (13) is not satisfied, the lowering of spinodal temperature is always less than that given by Eq. (12). When the above condition is not satisfied, it is even possible for the spinodal temperature to increase initially on addition of copolymer AB, as is illustrated in Fig. 8. Rigby et al. [47] showed also that both binodal and critical temperatures can be lowered by addition of copolymer, but the condition under which Eq. (12) is obeyed for binodal or critical temperature

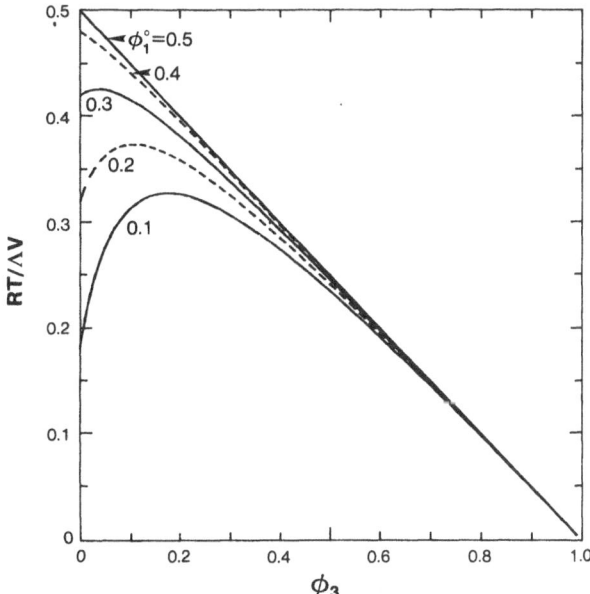

Fig. 8. Depression of spinodal temperature T with increasing volume fraction φ_3 of the added copolymer AB, calculated for the system homopolymer A, homopolymer B, and random copolymer AB with $V_1 = V_2$, $V_3 = 10V_1$, $f_1 = f_2 = 0.5$. The volume fraction φ_1^0 of homopolymer A in the initial binary mixture is indicated (From Rigby, Lin, and Roe [47])

is different from Eq. (13). Thus, for a linear reduction of critical temperature the two separate conditions

$$\varphi_1 V_1^{1/2} = \varphi_2 V_2^{1/2} \; ; \qquad V_1^{1/2}/f_{A3} = V_2^{1/2}/f_{B3} \tag{14}$$

have to be satisfied. For the linear reduction of binodal temperature the conditions to be satisfied are

$$V_1 = V_2 \; ; \qquad f_{A3} = f_{B3} \; . \tag{15}$$

It remains, however, to be established whether the lowering of cloud point according to Eq. (12) under conditions of Eq. (15) is the maximum possible.

Related to the compatibilization of homopolymers A and B by addition of copolymer AB is the question of so-called mutual miscibility enhancement [16] in heterogeneous copolymer samples; that is, whether the presence of molecules of a continuous spectrum of compositions in practical copolymer samples enhance the miscibility of components of extreme compositions in them. This problem was addressed by Koningsveld et al. [19b], who, on analysis based on Flory-Huggins equation, concluded that such mutual miscibility enhancement can indeed occur.

2.2.2 Experiment

Early experimental work of Riess et al. [48] substantially supports the above conclusion that the compatibilizing effect of added copolymer is only moderate in many cases.

These authors studied mixtures of polystyrene/poly(methyl methacrylate) with added copolymer, and evaluated in terms of optical clarity. Compatibility was found only when substantial amount of S/MMA copolymer was added.

Quantitative examination of Eq. (12) applied to cloud points of mixtures was reported by Rigby et al. [47] using low molecular weight polystyrene and polybutadiene (M_w = 1900 and 2650, respectively) with added S/B copolymer (M_w = 16300, 25000, and 270000). The linear relationship predicted by this equation was found to be obeyed on addition of either of the two lower molecular weight copolymers (see Fig. 9). The temperature dependence of the interaction energy density Λ_{AB} was determined from the slope in Fig. 9, and was found to give excellent agreement with the value previously obtained from a study of binary mixtures of the two homopolymers [8]. Lowering of the cloud point by added copolymer of molecular weight 250000 was much smaller than the values found for the other two copolymers. This was attributed to the occurrence of separation into three phases, calculated to occur at only 2% copolymer content according to the criteria by Leibler [45].

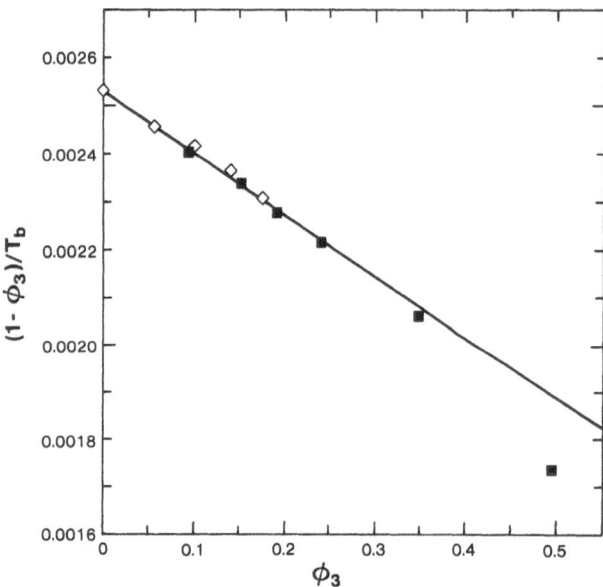

Fig. 9. The cloud temperature T_b of the mixture of polystyrene (M_w 1900) and polybutadiene (M_w 2650), to which styrene/butadiene random copolymer was added, is plotted against the volume fraction φ_3 of the copolymer, to show the conformation of the observed data to Eq. (12). The open squares were obtained with a random copolymer (52.5% styrene, M_w 25000) and the filled square with a random copolymer (46.7% styrene, M_w 16300). The slope gives the value of the temperature coefficient of the interaction energy density Λ_{12} that agree well with the value determined previously. (From Rigby, Lin, and Roe [47])

3 Systems Containing Block Copolymer

Among blends containing block copolymer, systems of primary interest to be reviewed in this section are binary mixtures of block copolymer AB with homopolymer A, and ternary mixtures containing immiscible homopolymers A and B combined with corresponding diblock copolymer AB. A study of the former is indicated by the need to understand to what extent and in what manner block copolymer structure and properties can be modified by the addition of homopolymer; while studies of the ternary system can lead to understanding of the emulsifying role of block copolymer which is exploited technologically.

3.1 Mixtures Containing a Diblock Copolymer and a Homopolymer

3.1.1 Overview

Before describing the behavior of these blend systems in detail, we here give a brief summary of the features exhibited by a block copolymer by itself, and of the types of phase behavior which are expected on addition of a homopolymer.

In pure block copolymer having two types of blocks, A and B, which are mutually incompatible, the blocks segregate out into their own microdomains. The number of like blocks that can aggregate to a microdomain is limited by the geometric constraint to form a space-filling structure, and consequently the dimension of the microdomains is usually of the same order of magnitude as the radii of gyration of the blocks themselves. When synthesized by an anionic polymerization mechanism, polydispersity of the block lengths is kept to a minimum and, as a result, the shape and size of the microdomains in a block copolymer sample can be very uniform. Depending on the relative lengths of the A and B blocks, and depending to some extent on the conditions of sample preparation, such as the solvent used, the collection of microdomains forms regular structure consisting of spheres, cylinders, or lamellae. Beautiful electron micrographs showing such ordered arrays of microdomains abound in the literature.

The segregated microdomain structure is not attained when the two types of blocks are mutually compatible. Such compatibility arises either when the blocks are very similar (e.g., 1,4-butadiene vs. cis-1,4-isoprene [49]), when the block lengths are relatively short, or when the temperature is altered (i.e., raised if the corresponding homopolymer pair exhibits UCST behavior). The structure of such disordered block copolymer is essentially liquid-like, forming a single homogeneous phase lacking any long range order. It differs, however, from the mixture of compatible homopolymers in that the two types of blocks, by virtue of the covalent linkage between them, maintains a fairly well defined distance between them. As a result, small-angle X-ray or neutron scattering from disordered block copolymer shows a peak [50] at a finite scattering angle, and this phenomenon has been termed the "correlation hole effect" by deGennes.

The transition between the ordered, microdomain structure and the disordered, homogeneous structure, induced by temperature change, can be oberserved by rheological measurement [51-53] or by small-angle scattering [54]. The transition, often called the order-disorder transition or microphase separation transition, resembles the solid-liquid transition, and should be of a first order according to the theory of Leibler [55], but experimentally its character has not been established clearly.

Next, we consider what will happen when we take the pure block copolymer in its ordered state and add to it increasing amounts of homopolymer A. Initially, the homopolymer will dissolve in the microdomains of block A of the copolymer until a solubility limit is reached, beyond which the excess homopolymer will separate out in a separate macroscopic phase. Even before the solubility limit is reached, as the relative volume of the A and B microdomains change, there might arise a change in the ordered structure, for example, transitions from spherical to cylindrical and to lamellar morphology. The solubility limit and these morphological transition points would depend on the relative lengths of the two types of blocks and the homopolymer molecules. The thermodynamic stability of the microdomains, and hence the microphase separation temperature (MST), will also be affected by the addition of the homopolymer. Unlike the melting point of a solid in the presence of a diluent, which is always depressed, the MST might go up or down, again depending on the relative lengths of the various species involved. If we look at the other end of the concentration scale, a small amount of block copolymer, added to a pure homopolymer, will initially dissolve into a homogeneous solution, but with increasing concentration of block copolymer, a critical micelle concentration (CMC) is eventually reached, beyond which any excess copolymer aggregates into micelles. Further addition of the block copolymer will increase the population of micelles, and may finally lead to agglomeration of micelles into ordered macroscopic structure. These are some of the features expected and the questions likely to be raised when we examine the behavior of the mixture containing block copolymer AB and homopolymer A.

3.1.2 Theory

3.1.2.1 Block Copolymer Theory

Before discussing the theories describing the mixture of block copolymer and homopolymer, it is necessary to review the theories dealing with pure block copolymer very briefly. The latter can be divided into three categories, the first dealing with ordered microdomain structure, the second dealing with the disordered phase and the third dealing with the transition between them.

The theories dealing with the ordered phase were first developed as soon as the fascinating properties of block copolymers became known. Main concerns of these theories were the delineation of the conditions necessary for the formation of spherical, cylindrical or lamellar morphology, the prediction of the size of the microdomains of these morphologies in relation to the block lengths, and the question of the thickness and density profile across the boundary between microphases. These theories were extensively reviewed over the years in monographs and symposium proceedings and will not be discussed further here.

Thermodynamic and scattering properties of block copolymer in the disordered state were evaluated by Leibler [55] by means of the random phase approximation method, which was introduced into the study of polymers by deGennes [6]. In the method, one starts with the segment density correlation function of ideal, independent polymer chains, and then calculates the modification to it arising from the requirement of uniform segment density in space; this is done by subjecting each segment to an extra potential determined in a self-consistent manner. The result gives the expression for the Fourier transform of the segment density correlation function, $\tilde{s}(q)$, (which is proportional to the scattered intensity of X-ray or neutron) as follows [55]

$$\frac{1}{\tilde{s}(q)} = \frac{s_{11}(q) + s_{22}(q) + 2s_{12}(q)}{s_{11}(q)\, s_{22}(q) - [s_{12}(q)]^2} - \frac{2\Lambda}{RT} \tag{16}$$

where $s_{ij}(q)$ is the Fourier transform of the correlation function between segments of type i and type j of ideal, independent chains in the sample, and q is the scattering vector equal to $4\pi \sin \theta/\lambda$. Equation (16) is applicable not only to disordered block copolymer but also to any amorphous polymer blend in which only two types of segments are present. For example, for a mixture of two homopolymers of molar volumes V_1 and V_2 and mean end-to-end distances R_1 and R_2,

$$s_{11}(q) = \varphi_1 V_1 g_D(x_1) \; ; \qquad s_{22}(q) = \varphi_2 V_2 g_D(x_2) \; ; \qquad s_{12}(q) = 0 \tag{17}$$

where x_i is equal to $q^2 R_i^2/6$, and $g_D(x)$ is the Debye function (the Fourier transform of the correlation function of a gaussian chain) given by

$$g_D(x) = 2(x + e^{-x} - 1)/x^2 \tag{18}$$

Substitution of Eq. (17) into Eq. (16) gives

$$\frac{1}{\tilde{s}(q)} = \frac{1}{\varphi_1 V_1 g_D(x_1)} + \frac{1}{\varphi_2 V_2 g_D(x_2)} - \frac{2\Lambda}{RT} \tag{19}$$

By letting $q \to 0$ in Eq. (19) we obtain

$$\frac{1}{\tilde{s}(0)} = \frac{1}{\varphi_1 V_1} + \frac{1}{\varphi_2 V_2} - \frac{2\Lambda}{RT} \tag{20}$$

which is equal to the second derivative (with respect to φ_1) of the Flory-Huggins free energy of mixing given in Eq. (1). The temperature at which $\tilde{s}(0)$ diverges gives the spinodal temperature signifying the limit of metastability of homogeneous single phase. Equation (20) implies the equivalence of the random phase approximation and the Flory-Huggins treatment, both being mean field theories. More recently, Benoit and Benmouna [56–58,] have shown that Eq. (16) can be derived by direct, methodical enumeration of correlations between segments belonging to the same molecule or to neighboring molecules, thus reaffirming the mean-field nature and equivalence of these three different approaches.

Although the possibility of the order-disorder transition was recognized in most of the block copolymer theories, it is Leibler [55] who has expressedly addressed this problem. He derived the free energy of a block copolymer system in a series expanded in powers of the order parameter ψ denoting the deviation of the local density from the mean. The coefficients of this expansion up to the fourth order term were evaluated by a method which is a generalization of the random phase approximation method described above (Equation (16) was, in fact, derived as the second order term in the free energy expansion). The MST, determined from the condition that the excess free energy due to the ordered structure should vanish, are shown in Fig. 10. Here, the abscissa gives the fraction f of one of the blocks in a diblock copolymer and the ordinate gives the value of χN at the MST, where N is the total number of segments in the copolymer molecule. Since $\chi\ (= \Lambda V/RT)$ is inversely proportional to T when Λ is independent of temperature (as is approximately true for polymer pairs exhibiting UCST behavior), the MST among molecules of the same total length is the highest when the lengths of the two blocks are equal to each other. The spinodal temperature, obtainable as the temperature at which $\tilde{s}(q)$ given by Eq. (16) diverges for some value of q, shows a closely similar dependence on f. When $f = 0.5$, the spinodal and the MST coincide, and at other values of f the spinodal temperature is somewhat lower than the MST, but the difference is so small and cannot be shown clearly in Fig. 10.

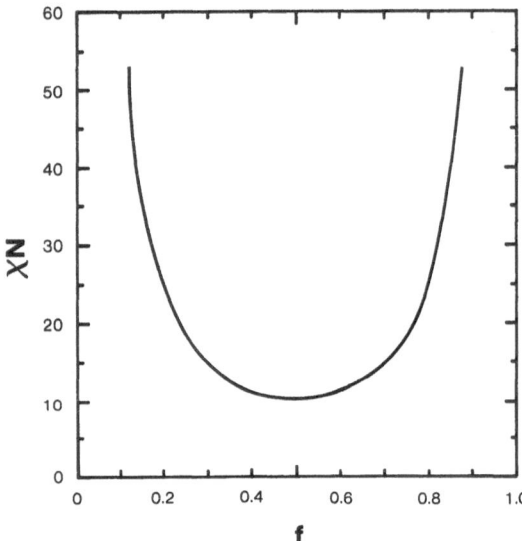

Fig. 10. The value of χN at the spinodal point (N is the number of segments per copolymer molecule) plotted against the composition f of the block copolymer molecule. The value of χN at the microphase separation temperature is slightly smaller than at the spinodal, but the difference between them is small and cannot be meaningfully displayed in this plot. (From Leibler [55])

3.1.2.2 Block-Copolymer Homopolymer Mixture Theory

Equation (16) derived from the random phase approximation is directly applicable to the mixture of block copolymers with homopolymers, provided that the correlation functions $s_{ij}(q)$ in it are duly evaluated for the collection of ideal, independent chains in the mixture concerned. The equation gives the X-ray or neutron intensity $\tilde{s}(q)$ scattered from such mixtures in the disordered state, and also leads to the prediction of the spinodal temperature. The MST itself is of more general interest, but because of the close relationship between MST and spinodal, the knowledge of the latter and its dependence on variables such as the size of the molecules and blocks involved is useful.

The prediction from Eq. (16) of the change in the spinodal temperature on addition of homopolymer A to diblock copolymer AB is qualitatively as follows [59]. Whether the spinodal temperature is raised or lowered by the addition of homopolymer depends on two factors: the symmetry of the block copolymer itself and the relative size of the homopolymer in comparison to the copolymer. The spinodal temperature tends to rise if the addition of the homopolymer shifts the overall monomer concentration in the mixture toward the 50/50 composition and also if it increases the overall average molecular weight of the chains in the mixture. Thus, with an unsymmetric copolymer having block A very much smaller than B, addition of homopolymer A will raise the spinodal except when homopolymer A is very small. With a symmetric copolymer ($f = 0.5$), the spinodal will rise if the added homopolymer is larger than about one-fourth of the length of the copolymer. With a unsymmetric copolymer having block A very much larger than block B, addition of homopolymer A will in general depress the spinodal, except when the homopolymer is very much larger than the copolymer.

Krause [60], some years ago, applied a macroscopic thermodynamic consideration to the problem of change in MST by addition of homopolymer, and came to the conclusion which, in very qualitative terms, agrees with the above predictions based on Eq. (16).

More recently Noolandi and Hong [61-63] undertook theoretical studies of this type of system by a method more rigorous mathematically. They expressed the free energy of the mixture as a series expansion in powers of the concentration fluctuation, in a manner analogous to Leibler's block copolymer theory [55]. The space filling requirement for the gaussian chains was likewise accommodated through the incompressibility condition and through the use of an effective, self-consistent potential. In the series, terms up to the fourth order were retained, and numerical results were obtained for some of the systems having lamellar morphology (for which the third order term vanishes). The phase diagrams calculated from the theory are illustrated in Fig. 11 and 12, where M denotes the region of stability of ordered microphase structure (mesophase), H the region containing a single disordered phase, and HH and HM the region in which two phases coexist. In these two examples, the MST is seen to increase with increasing amount of homopolymer. The complexity of the theoretical expressions does not allow ready evaluation of numerical results and it is, therefore difficult,

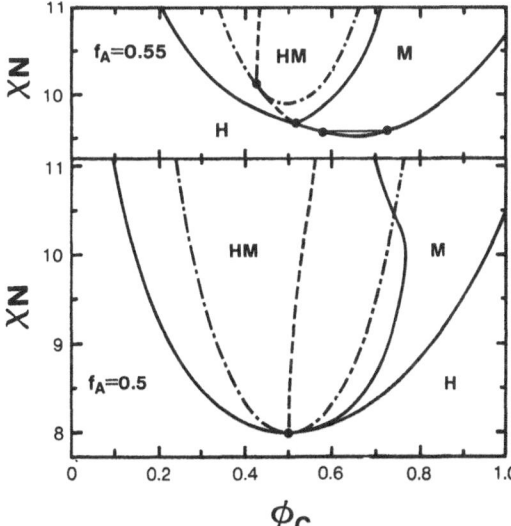

Fig. 11. Calculated phase diagrams of mixtures containing block copolymer AB and homopolymer A, where the number of segments per molecule of the copolymer and the homopolymer are equal to N, and the fraction f_A of monomer A in the copolymer is as indicated. (From Hong and Noolandi [63])

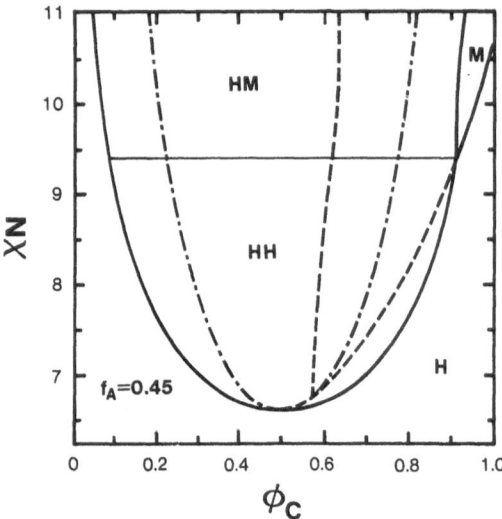

Fig. 12. Calculated phase diagram of a mixture similar to that in Figure 9, but here the fraction f_A of monomer A in the copolymer is equal to 0.45. (From Hong and Noolandi [63])

to deduce from the theory the conditions necessary for the MST to rise or fall with addition of homopolymer.

The solubility limit of homopolymer in the microdomains of ordered block copolymer phase was treated earlier by Meier [64], who, as in his earlier theories of block copolymer, enumerated the contributions to the free energy by separate factors, such as the interfacial energy, the entropy loss due to the confinement of joints in the interface and of blocks within their own microdomains, and the entropy loss arising from the distortion of conformations to achieve uniform density requirement. His theory, although approximate, has the virtue of making the contributions by various physical factors more intuitively visible. The result shows that the length of the homopolymer has to be of the same order of magnitude or smaller in comparison to the corresponding block length for it to be soluble in the microdomains of the copolymer. The solubility limit would decrease with increasing homopolymer molecular weight. It predicts, for example, that in lamellar morphology the volume ratio of homopolymer A to block A of copolymer in the swollen microdomains would be equal at most to 1.0 and 0.08, when the molecular weight ratio of homopolymer to copolymer block is equal to 0.1 and 1, respectively.

On the question of the micelles formed by a small amount of block copolymer present in a large amount of homopolymer, two theories are available, one by Leibler, Orland and Wheeler [65] and the other by Whitmore and Noolandi [66]. They are very similar to each other in their approach, and take a model of micelles consisting of a spherical core, mainly of block B, surrounded by a spherical shell (or corona) in which block A of the copolymer intermixes extensively with homopolymer A (Fig. 13). Contributions to the free energy from the core (in which conformations of block B chains are somewhat constrained), from the interface between the core and corona, from the corona (in which the effect of the entropy of mixing and the constraint to the conformation are taken into account) and from the bulk phase (in which some of the block copolymer remains dissolved) are all evaluated. Minimization of the free energy then gives the equilibrium values of a number of parameters characterizing the system, such as

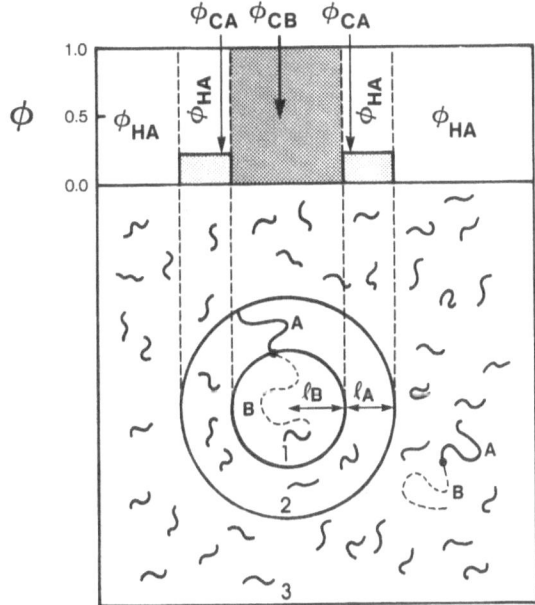

Fig. 13. Model of a spherical micelle consisting of diblock copolymer AB in the matrix of homopolymer A. (From Whitmore and Noolandi [66])

the critical micelle concentration, the radius of the core, the thickness and degree of swelling of the corona, and the number of copolymer chains per micelle. As the population of micelles in the system increases, copolymer chains in the coronae belonging to adjacent micelles may begin to interact with each other. Leibler and Pincus [67] considered such a situation and found that close approach of two micelles results in unfavorable entropic effect and thus develops a net repulsion tending to prevent

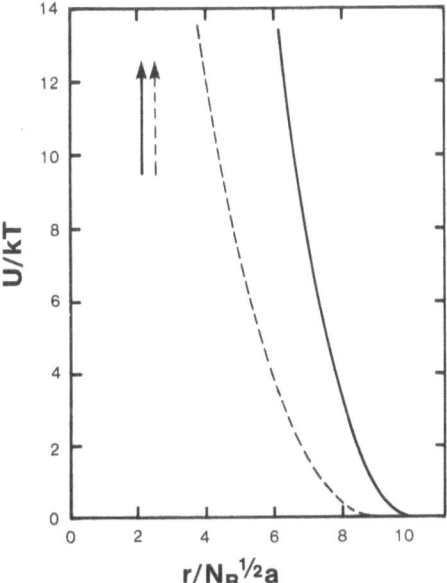

Fig. 14. Interaction energy U between two micelles as a function of the distance r between them, calculated for the system in which diblock copolymer (N_A = 1800, N_B = 200) is mixed with homopolymer N_h = 50 (solid line) or homopolymer N_h = 360 (dashed line). The interaction parameter χ is assumed to be 0.1. (From Leibler and Pincus [67])

extensive interpenetration. They calculate the effective potential U(r) between micelles as a function of their separation distance r (Fig. 14), and then predict the concentration at which the micelles will organize themselves into an ordered structure. This concentration will depend, of course, both on the length of block A and on the ratio of the chain lengths of homopolymer A and block A, but will depend only weakly on temperature.

3.1.3 Experiment

3.1.3.1 Microphase Separation Temperature

The influence of added homopolymer on the MST of block copolymer was studied first by Robeson et al. [68] with styrene/α-methylstyrene diblock copolymer (containing 50 mole % of each monomer) mixed with polystyrene. The presence of a single or double T_g's was determined by dynamic viscoelastic measurement to ascertain the occurrence of microphase separation. In the absence of added polystyrene, all three block copolymer samples of M_w equal to 80000, 150000, and 420000 exhibited single phase behavior. On addition of 30% or 50% (by weight) of polystyrene (M_w = 270000), the two block copolymers of lower molecular weights remained in a single phase state, but the third block copolymer of the highest molecular weight exhibited two separate T_g's. Krause et al. [69] also studied styrene/α-methylstyrene diblock copolymer (42% styrene, molecular weight 1.06×10^6). With DSC it was determined that the block copolymer possessed microphase-separated structure, and on addition of 25% by weight of polystyrene (M_w = 20000) or poly(α-methylstyrene) (M_w = 37000) the mixture could not be transformed into a single phase even though the homopolymer molecular weight was less than one-fourth of the corresponding copolymer block. In these studies relying on determination of T_g's, any change in MST could have been detected only when the effect was very large, and the results are therefore not necessarily in conflict with the prediction of Krause's own earlier theory [60].

Cohen and Torradas [70] studied a diblock copolymer of 1,2-butadiene and 1,4-butadiene (with block molecular weights of 30000 and 100000, respectively) mixed with either 1,2-polybutadiene (M_w = 30000) or 1,4-polybutadiene (M_w = 100000). On the basis of measurement of loss tangent peaks obtained with a Rheovibron, they found that the pure diblock copolymer was homogeneous, as were its blends with the 1,4-polybutadiene in all proportions, while the 1,2-polybutadiene induced microphase separation at concentrations above about 10%. Since the block copolymer has a shorter 1,2-butadiene block, the addition of 1,2-polybutadiene shifts the overall composition of the two types of monomers in the mixture toward 0.5 and thus induces the microphase separation.

Quantitative measurement of the variation in the MST with added homopolymer was performed by Zin and Roe [71]. These authors studied blends of a styrene/butadiene diblock copolymer (styrene weight fraction = 0.27, M_w = 28000) with polystyrene (M_w = 2400) using the small-angle X-ray scattering technique. The scattered X-ray intensity was determined at a number of temperatures and then $1/I_{max}$ was plotted against $1/T$, where I_{max} is the intensity of the characteristic low angle peak in the SAXS curves. Such a plot is suggested by Eq. (16). Extrapolation of the linear portion of the high temperature data (above the MST) to $1/I_{max} = 0$ (illustrated in Fig. 15)

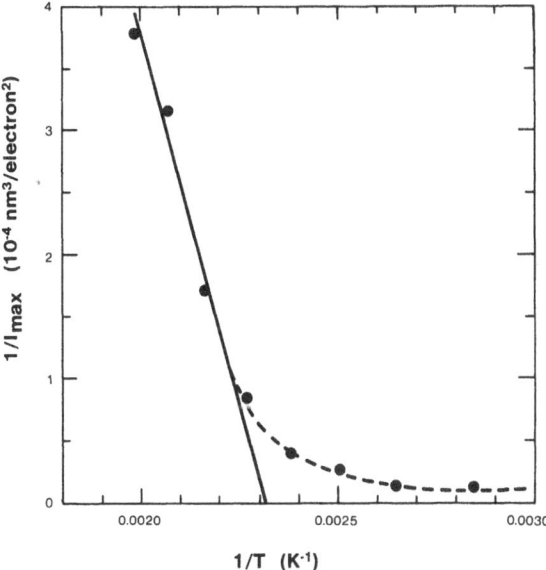

Fig. 15. Reciprocal of the peak intensity I_{max} obtained from SAXS measurement of styrene/butadiene diblock copolymer plotted against the reciprocal of temperature T. Linear extrapolation of high temperature data to zero gives the spinodal temperature, while the first deviation of the observed intensity from the straight line gives the microphase separation temperature. (From Zin and Roe [71])

gives the spinodal temperature for microphase separation, while the point of deviation of the observed intensity from the straight line gives the MST. The latter was thus found to increase from ca. 140 °C for the pure block copolymer to ca. 168 °C for a mixture with 20% polystyrene.

3.1.3.2 Phase Diagram

The phase diagram of a mixture of block copolymer with homopolymer, extending over an extended range of temperature and composition, was first reported by Roe and Zin [72]. The phase diagram shown in Fig. 16 refers to the same mixture of styrene/butadiene diblock copolymer and polystyrene mentioned in the above. For its construction the data for the MST obtained by SAXS and the data for the macroscopic phase separation (cloud points) obtained by light scattering were combined and were interpreted in the light of the thermodynamic principles governing phase equilibria and phase diagrams.

In Fig. 16, the left-hand ordinate gives the behavior of the pure block copolymer with its MST at around 140 °C. When polystyrene is added to the copolymer in its ordered state, the homopolymer dissolves into the microdomains consisting of styrene blocks. The area denoted as M_1 represents the region in which a mesophase (ordered microdomain structure) is stable. When the amount of the homopolymer is not large (below about 18%), on heating, the ordered structure is transformed to a disordered, homogeneous mixture denoted as L_1. The MST increases, as described in Sect. 3.1.3.1, with increasing amount of the added homopolymer. The area $L_1 + L_2$ depicts the region in which the mixture undergoes a macroscopic phase separation into two

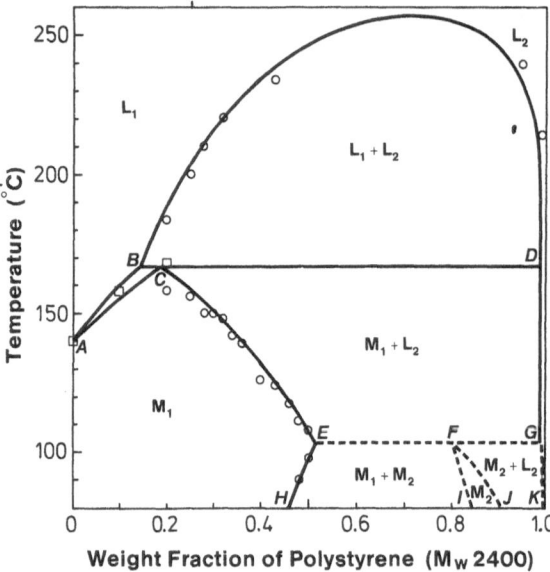

Fig. 16. Phase diagram of a mixture containing polystyrene (M_w = 2400) and styrene/butadiene di-block copolymer (27 % styrene, M_w = 28 000). Liquid phases L_1 and L_2 represent mixtures of disordered block copolymer and polystyrene. Mesophase M_1 consists of ordered microdomains of the block copolymer swollen with polystyrene. Mesophase M_2 probably contains aggregates of block copolymer micelles within the medium of polystyrene. The features on the lower right, drawn in broken lines, are more speculative. (From Roe and Zin [72])

coexisting, homogeneous mixtures L_1 and L_2. In this region, which is above the MST, the block copolymer behaves essentially the same as a random copolymer, and the phase behavior of the mixtures can be described by the usual Flory-Huggins treatment. Below this lies the area denoted as M_1 + L_2, and the mixture having composition in this region undergoes a macroscopic phase separation into a mesophase M_1 and a disordered phase L_2. The boundary between the area L_1 + L_2 and the area M_1 + L_2 constitutes a eutectic point (or a peritectic point, the latter being the terminology more in conformity with the traditional usage [95] of these words), at which three phases, L_1, M_1, and L_2 coexist. The lower right areas in Fig. 16, denoted as M_1 + M_2, M_2, and M_2 + L_2, are less well defined and more speculative. They probably involve the aggregation of block copolymer micelles into an ordered structure. The existence of the mesophase M_2 and the eutectic point associated with its upper extreme is contemplated mainly to satisfy the thermodynamic principles.

Roe and Zin [72] also gave the phase diagram of mixtures consisting of the same diblock copolymer and a polybutadiene (M_w = 28 000). The overall feature of this phase diagram is similar to that given in Figure 16, except that the MST is seen to decrease with increasing amount of added polybutadiene.

One can notice many similarities between the experimental phase diagram in Figure 16 and the phase diagrams in Fig. 11 and 12, calculated from the theoretical treatment of Noolandi and Hong. More detailed, quantitative comparison is, however, difficult because the molecular parameters used for the calculation of these predicted phase diagrams are very different from the experimental ones.

3.1.3.3 Solubility of Homopolymer

Solubilization of homopolymer by copolymer was reported in an early work by Inoue et al. [73], in which mixtures of styrene/isoprene diblock copolymer with polystyrene and/or polyisoprene were examined for optical clarity of toluene-cast films and for the microstructure by electron microscopy. The results, though not quantitative, suggest that the amount of solubilized homopolymer could be 2–3 times the volume of the like copolymer block when the corresponding molecular weight ratio was around unity, while films containing a much higher molecular weight homopolymer were invariably cloudy. Skoulios et al. [74] used SAXS and visual observation to determine the solubility of polystyrene of different molecular weights in the styrene domains of a styrene/(vinyl-2-pyridine) diblock copolymer in which the vinyl-2-pyridine block was swollen with octanol. On addition of a polystyrene with molecular weight equal to the copolymer styrene block, the solubility limit was reached when the volume ratio of polystyrene to the styrene block was roughly equal to unity, while cloudy macrophase-separated mixtures resulted when the polystyrene molecular weight was larger. Ptaszynski et al. [75] also used SAXS to study mixtures of polystyrene of varying molecular weight with a styrene/isoprene diblock copolymer with block molecular weights 40000 and 50000, respectively. Essentially corroborating the above results, they found that at fixed homopolymer concentration (15 % w/w) the mixtures were transparent until a homopolymer molecular weight of 60000 (i.e., 1,5 times the styrene block length) was reached and thereafter the mixtures were visibly cloudy. With polystyrene of molecular weight 10000, the solubility limit was reached when the polystyrene content was around 30 %. Thus it was concluded that the statement that the homopolymer molecular weight must be less than or equal to that of the corresponding copolymer block for solubilization to occur represents a good rule of thumb, but that a certain amount of solubilization occurs even at higher molecular weights.

Roe and Zin [72] investigated mixtures of styrene/butadiene diblock copolymer (block molecular weights 7600 and 20400, respectively) with either polystyrene or polybutadiene. They found the solubility limit to increase with increasing temperature in the case of mixtures with polystyrene of M_w 2400, but to be fairly independent of temperature in the case of mixtures containing polystyrene of M_w 3500 and polybutadiene of M_w 26000. The solubility limit at room temperature was about 48, 18, and 27 % for polystyrene of M_w 2400, polystyrene of M_w 3500, and polybutadiene of M_w 26000, respectively. These values are about an order of magnitude greater than those predicted by Meier's theory [64]. Roe and Zin [72] argue that the underestimation by Meier's theory arises because the theory assumes a model in which the homopolymer is uniformly distributed within the microdomain, whereas in practice it is more likely that the homopolymer will concentrate more toward the center of the microdomain in order to avoid overly stretching the block chains.

The morphology of blends of block copolymer with homopolymer was studied by means of electron microscopy and small-angle X-ray or neutron scattering by Hashimoto et al. [76] and by Bates et al. [77] The first group prepared blends of styrene/isoprene diblock copolymer and polystyrene cast from toluene solution. The second group studied blends of styrene/butadiene diblock copolymer and polystyrenes, cast from solution prepared with mixed solvents THF/MEK. Both groups noted that the long range order of microdomain packing, present with the pure block copolymer, was lacking in all the blends with added homopolymers. The SAXS results by the first group also showed that the radius of the spherical microdomains of isoprene block remained approximately constant but the distance between the spherical domains increased as the homopolymer content was increased (see Fig. 17).

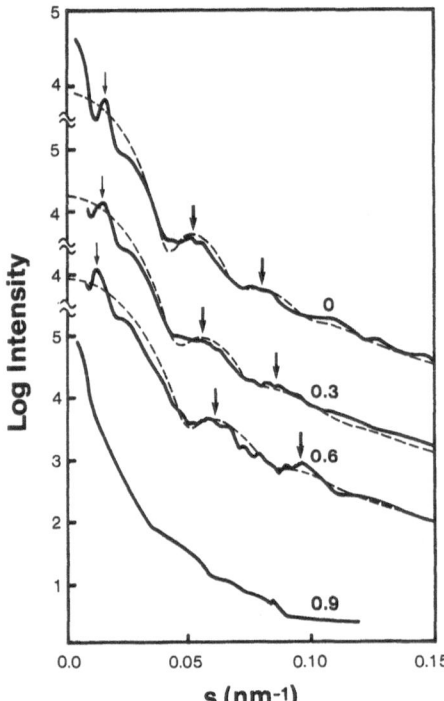

Fig. 17. Desmeared small-angle X-ray scattering intensities for a styrene/isoprene diblock copolymer (78% styrene, M_n 320000) and its mixtures with polystyrene (M_n 81000) (the weight fraction of the latter in the mixture is indicated on the plot). The dotted curves are best fitting curves calculated for scattering from isolated spheres. (From Hashimoto, Fujimura, Hashimoto, and Kawai [76])

A question naturally arises as to whether the solubility limits and the morphology observed with these blends correspond to equilibrium states. Meier [64] suggests that the apparent disagreement between the observation and his theory may arise from non-equilibrium effects. When the molecular weights of the components are fairly large and the blends are prepared from solution, the morphology of the samples obtained frequently depends on the types of solvents used. Equilibrium values of the domain size and the solubility limit can be obtained only when the condition during the sample preparation allows migration of the block copolymer and the homopolymer molecules through the continuous matrix in the blends. Such a condition is more likely to be met when the molecular weights of the components are fairly small and the blends are heated to temperatures approaching or exceeding the MST.

3.1.3.4 Block Copolymer Micelles

The micelles formed when styrene-butadiene diblock copolymer is mixed with a large excess of low molecular weight polybutadiene were studied by Rigby and Roe [78, 79] (by SAXS) and by Selb et al. [80] (by SANS). The first group employed three block copolymers in which the fractions of styrene are approximately equal to 25, 50, and 75%, respectively, and studied the effect of changing the temperature. The second group employed block copolymers in which the weight fraction of styrene ranged from 32% to 68% and blended them with polybutadiene of three different molecular weights and studied them at room temperature. Both groups report the radius of the spherical micelle core, determined by SAXS or SANS technique, which agrees fairly well [66, 81] with the value predicted on the basis of the theory by Leibler et al. [65] or by Whitmore and Noolandi [66]. The dependence of the critical micelle concentration on the relative lengths of the blocks and on temperature, reported by Rigby and Roe [79], is shown in Fig. 18. The CMC increases as the temperature is raised and as the butadiene block in the copolymer becomes longer, in accord with the accompanying increase in the compatibility between the block copolymer and polybutadiene. The theory by Leibler et al. [65] was shown to predict the overall trend and the order of magnitude of the CMC shown in Figure 18, but the agreement was not quantitative [81]. The results by Rigby and Roe [78, 79] also show that as the temperature is raised, the micelle core consisting of styrene blocks becomes progressively swollen with polybutadiene even before the micelles eventually dissolves into the homopolymer matrix at higher temperature. This aspect was not predicted by either of the theories.

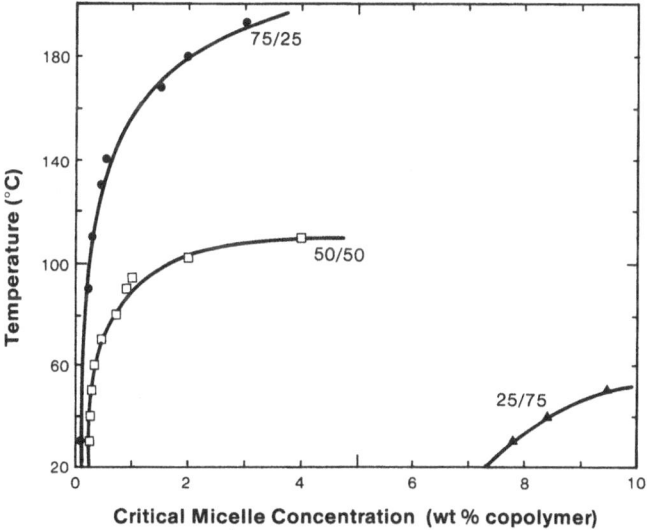

Fig. 18. The critical micelle concentrations of three styrene/butadiene diblock copolymer samples in the matrix of low molecular weight polybutadiene are plotted against temperature. The approximate compositions (stytene vs. butadiene) of the block copolymers are indicated. (From Rigby and Roe [79])

3.2 Mixtures Containing a Diblock Copolymer and Two Homopolymers

3.2.1 Theory

When a diblock copolymer AB is added to a phase-separated mixture of homopolymers A and B, one of the following three events is likely to occur.

1) When the homopolymer pair is only moderately incompatible, addition of the copolymer may actually cause the two-phase system to become a single homogeneous phase. The copolymer may then be termed a compatibilizer in the thermodynamic sense of the word. In this case, the block copolymer is acting in essentially the same manner as does a random copolymer of the same composition as described in Sect. 2.2.

2) Homopolymers A and B become solubilized in the microdomains of the like components of the copolymer. This situation is more likely to be found at high copolymer content.

3) The mixture may remain macroscopically demixed, but the copolymer is located preferentially at the interface, with its A block in the A-rich phase and B block in the B-rich phase. The main effect in this case is to lower the interfacial tension between the two phases.

Two theoretical treatments, published so far, address themselves on the aspect (3) in the above. The first, by Noolandi and Hong [82, 83], considers the case in which the two homopolymers are highly incompatible (i.e., there is practically no homopolymer A dissolved in the B-rich phase and vice versa). The second, by Leibler [84], deals with the opposite extreme in which the two homopolymers are relatively compatible (but are still demixed).

The treatment by Noolandi and Hong [82-83] is based on the functional representation of free energy density as developed in their earlier work, and the set of equations derived from it were solved numerically for the specific case of a symmetric system containing homopolymers A and B of infinite molecular weight, a diblock copolymer with A and B blocks of equal lengths, and a solvent which is equally good to polymers A and B. Some qualitative conclusions emerging from the analysis are as follows. With increasing copolymer concentration the interfacial tension is progressively reduced, and the reduction is approximately linear with the concentration of the copolymer and, at a fixed concentration, with the molecular weight of the copolymer. The calculated concentration profile across the phase boundary shows that the copolymer molecules accumulate in the boundary region, and this tendency is more pronounced with copolymers of higher molecular weight. Of the various physical factors contributing to the decrease in the interfacial tension, the main beneficial effect arises from the separation of homopolymers A and B at the boundary through the interposition of the copolymer between them, and the main counterbalancing effect from the loss of entropy due to the localization of copolymers at the boundary.

In the treatment by Leibler [84], the free energy density is expanded in a series in powers of concentration fluctuation, as in his earlier theory [55] of pure block copolymer. He also treats the symmetric case in which equal amounts of homopolymers A and B are mixed with a diblock copolymer AB having blocks of equal lengths, and the molecular weights of all three components are assumed equal. He finds also that the interfacial tension is reduced (very) approximately linearly with the amount of the copolymer added. He treats the concentration regime, near the critical point, in which the mixture is marginally incompatible, and finds that the copolymer is distributed about equally between the two phases, rich in A and rich in B. The reduction of interfacial tension arises mostly from the resulting reduction in the difference between the concentration of the monomer A and B between the two phases. The localization of copolymer at the interfacial region is found to be minor and therefore contributes only little to the reduction in the interfacial tension.

3.2.2 Experiment

There is a large body of experimental evidence supporting the interfacial activity of block (or graft) copolymers in mixtures with one or two homopolymers. For example, Gaines and Bender [85] have demonstrated a lowering of polymer melt surface tension on addition of styrene/dimethylsiloxane copolymer to polystyrene. Addition of only ~0.2% of the copolymer was shown to give a surface tension close to that of polydimethylsiloxane. More recently, the technique of X-ray photoelectron spectroscopy (ESCA) was employed by Dwight et al. [86, 87] to study the depth profile in the surface layer of blends of polycarbonate with siloxane/carbonate diblock copolymer. At a copolymer content of ca. 1% it was found that there was a large increase in the surface excess concentration of siloxane blocks, giving rise to an almost pure siloxane surface layer.

Evidence of the interfacial activity of diblock copolymers at the interface between two immiscible homopolymers, rather than between a single homopolymer and air, has been presented by Gaillard et al. [88], who used the spinning drop method to measure the interfacial tension in the system polystyrene/polybutadiene/styrene-monomer with addition of varying amounts of styrene/butadiene diblock copolymer. As expected, the interfacial tension decreases with increasing amount of copolymer, eventually levelling off beyond 5–10% of added copolymer. Noolandi and Hong [82] have compared these results with their theoretical prediction, the latter indicating that the interfacial tension should fall rapidly to zero for copolymer contents in excess of ca. 0.01%. Explanations offered for the discrepancy include non-equilibrium effects in the experimental system and possible shifts in the location of the block copolymer at the interface caused by rapid spinning of the drop during the experiment.

Other data illustrating the interfacial activity of diblock copolymers in homopolymer-homopolymer mixtures have involved measurements of the mechanical properties of blends and/or morphological investigations using the electron microscope. Early work of Riess et al. [48], who studied the effects of added graft and random copolymers as well as block copolymers, showed that transparent blends (implying either single phase or formation of droplets too small to scatter light) could be obtained, provided that homopolymer molecular weights were kept lower than the corresponding copolymer blocks. Kawai et al. [73, 89] have studied the morphology and mechanical properties of blends of polystyrene and polyisoprene to which (relatively large) amounts of styrene-isoprene diblock copolymer were added (as well as the properties of binary blends containing just one of the homopolymers). Both solubilization of the homopolymers and apparent macrophase separation were observed, depending on copolymer content.

A series of studies of the emulsifying effect of hydrogenated butadiene/styrene block copolymer on the morphology and mechanical properties of blends of high density polyethylene with polystyrene has been presented by Fayt et al. [90–93] (see also Ref. [94]). Through hydrogenation of butadiene-styrene diblock copolymers in which the butadiene block had either mixed 1,4/1,2 addition or only 1,4 addition, these workers obtained LDPE/PS and HDPE/PS block copolymers. They have demonstrated the utility of small amounts (1–9 %) of diblock copolymer in reducing the size of dispersed macrophases, and in some cases have shown that properties such as ultimate tensile strength can show synergistic improvement. The LDPE/PS copolymer was shown to act as a more efficient emulsifier when prepared in the form of a tapered diblock copolymer, possibly due to its lower tendency towards micelle formation.

Acknowledgement: This work was supported in part by the Office of Naval Research.

4 References

1. Krause, S., in "Polymer Blends", D. R. Paul and S. Newman, eds., Academic Press, NY, 1978, Vol. 1, Chapter 2
2. Paul, D. R., in "Polymer Blends", D. R. Paul and S. Newman, eds., Academic Press, NY, 1978, Vol. 2, Chapter 12
3. Schmitt, B. J.: Angew. Chem. Int. Ed. Engl. *18*, 273 (1979)
4. Bywater, S.: Polym. Eng. Sci. *24*, 104 (1984)
5. deGennes, P.-G.: "Scaling Concepts in Polymer Physics", Cornell University Press, Ithaca, NY, 1979
6. deGennes, P.-G.: J. Phys. (Paris) *31*, 235 (1970)
7. Roe, R.-J.: Adv. Chem. Ser. *176*, 599 (1979)
8. Roe, R.-J. and Zin, W.-C.: Macromolecules *13*, 1221 (1980)
9. Paul, D. R. and Barlow, J. W.: Polymer *25*, 487 (1984)
10. ten Brinke, G., Karasz, F. E., and MacKnight, W. J.: Macromolecules *16*, 1827 (1983)
11. Kambour, R. P., Bendler, J. T., and Bopp, R. C.: Macromolecules *16*, 753 (1983)
12. Molau, E.: J. Polymer Sci.: Polymer Lett. *3*, 1007 (1965)
13. Schmitt, B. J., Kirste, R. G., and Jelenic, J.: Makromol. Chem. *181*, 1655 (1980)
14. Fujioka, K., Noethiger, N., Beatty, C. L., Baba, Y., and Kagemoto, A.: ACS Adv. Chem. Ser. *206*, 149 (1984)
15. Kollinsky, F. and Markert, G.: Makromol. Chem. *121*, 117 (1969)
16. Kollinsky, F. and Markert, G.: ACS Adv. Chem. Ser. *99*, 175 (1971)
17. Rameau, A., Lingelser, J. P., and Gallot, Y.: Makromol. Chem. Rapid Commun. *3*, 413 (1982)
18. Scott, R. L.: J. Polymer Sci. *9*, 423 (1952)
19. a) Koningsveld, R. and Kleintjens, L. A.: J. Polymer Sci.: Polymer Symp. *61*, 225 (1977).
    b) Koningsveld, R., Kleintjens, L. A., and Markert, G.: Macromolecules *10*, 1105 (1977)
20. Shur, Y. J. and Rånby, B.: J. Appl. Polymer Sci. *20*, 3121 (1976)
21. Shaw, M. T.: J. Appl. Polymer Sci. *18*, 449 (1974)
22. McMaster, L. P.: ACS Adv. Chem. Ser. *142*, 43 (1975); Polymer Preprints *15(1)*, 254 (1974)
23. Stein, D. J., Jung, R. H., Illers, K. H., and Hendus, H.: Angew. Makromol. Chem. *36*, 89 (1974)
24. Chiou, J. S., Paul, D. R., and Barlow, J. W.: Polymer *23*, 1543 (1982)
25. Kruse, W. A., Kirste, R. G., Haas, J., Schmitt, B. J., and Stein, D. J.: Makromol. Chem. *177*, 1145 (1976)
26. Naito, K., Johnson, G. E., Allara, D. L., and Kwei, T. K.: Macromolecules *11*, 1260 (1978)
27. McBriety V. J., Douglas, D. C., and Kwei, T. K.: Macromolecules *11*, 1265 (1978)
28. Goh, S. H., Paul, D. R., and Barlow, J. W.: Polymer Eng. Sci. *22*, 34 (1982)
29. Hammer, C. F.: Macromolecules *4*, 69 (1971)
30. Shur, Y. J. and Rånby, B.: J. Appl. Polymer Sci. *19*, 1337 (1975)
31. Zabrzewski, G. A.: Polymer *14*, 347 (1973)
32. Shur, Y. J. and Rånby, B.: J. Appl. Polymer Sci. *19*, 2143 (1975)
33. Alexandrovich, P., Karasz, F. E., and MacKnight, W. J.: Polymer *18*, 1022 (1977)
34. Fried, J. R., Karasz, F. E., and MacKnight, W. J.: Macromolecules *11*, 150, 158 (1978)
35. Karasz, F. E. and MacKnight, W. J.: Pure Appl. Chem. *52*, 409 (1980)
36. Vukovic, R., Kuresevic, V., Karasz, F. E., and MacKnight, W. J.: Thermochim. Acta *54*, 349 (1982)

37. Vukovic, R., Karasz, F. E., and MacKnight, W. J.: Polymer 24, 529 (1983)
38. Vukovic, R., Karasz, F. E., and MacKnight, W. J.: J. Appl. Polymer Sci. 28, 219 (1983)
39. Zacharius, S. L., ten Brinke, G., MacKnight, W. J., and Karasz, F. E.: Macromolecules 16, 381 (1983)
40. Vukovic, R., Kuresevic, V., MacKnight, W. J., and Karasz, F. E.: J. Appl. Polymer Sci. 30, 317 (1985)
41. Vukovic, R., Kuresevic, V., Segudovic, N., Karasz, F. E., and MacKnight, W. J.: J. Appl. Polymer Sci. 28, 1379 (1983)
42. Cruz, C. A., Barlow, J. W., and Paul, D. R.: Macromolecules 12, 726 (1979)
43. Ziska, J. J., Barlow, J. W., and Paul, D. R.: Polymer 22, 918 (1981)
44. Harris, J. E., Goh, S. H., Paul, D. R., and Barlow, J. W.: J. Appl. Polymer Sci. 27, 839 (1982)
45. Leibler, L.: Makromol. Chem. Rapid Commun. 2, 393 (1981)
46. Koningsveld, R. and Kleintjens, L. A.: Macromolecules 18, 243 (1985)
47. Rigby, D., Lin, J. L., and Roe, R.-J.: Macromolecules 18, 2269 (1985)
48. Riess, von G., Kohler, J., Tournut, C., and Banderet, A.: Makromol. Chem. 101, 58 (1967)
49. Cohen, R. E. and Ramos, A. R.: Macromolecules 12, 131 (1979)
50. Boue, F. et al.: "Neutron Inelastic Scattering 1977", Vol. I, p. 563, International Atomic Energy Agency, Vienna, 1978; mentioned by deGennes, P.-G., Ref. 5
51. Chung, C. I. and Gale, J. C.: J. Polymer Sci., Polymer Phys. Ed. 14, 1149 (1976)
52. Chung, C. I. and Liu, M. I.: J. Polymer Sci., Polymer Phys. Ed. 16, 545 (1978)
53. Gouinlock, E. V. and Porter, R. S.: Polymer Eng. Sci. 17, 534 (1977)
54. Roe, R.-J., Fishkis, M., and Chang, J. C.: Macromolecules 14, 1091 (1981)
55. Leibler, L.: Macromolecules 13, 1602 (1980)
56. Benoit, H. and Benmouna, M.: Polymer 25, 1059 (1984)
57. Benoit, H. and Benmouna, M.: Macromolecules 17, 535 (1984)
58. Benoit, H., Wu, W., Benmouna, M., Mozer, B., Bauer, B., and Lapp, A.: Macromolecules 18, 986 (1985)
59. Nojima, S. and Roe, R.-J.: to be published
60. Krause, S.: in "Colloidal and Morphological Behavior of Block and Graft Copolymers", ed. by G. E. Molau, Plenum Press, NY, 1971, p. 223
61. Noolandi, J. and Hong, K. M.: Polymer Bulletin 7, 561 (1982)
62. Noolandi, J.: Polymer Eng. Sci. 24, 70 (1984)
63. Hong, K. M. and Noolandi, J.: Macromolecules 16, 1083 (1983)
64. Meier, D. J.: Polymer Preprints 18, 340 (1977)
65. Leibler, L., Orland, H., and Wheeler, J. C.: J. Chem. Phys. 79, 3550 (1983)
66. Whitmore, M. D. and Noolandi, J.: Macromolecules 18, 657 (1985)
67. Leibler, L. and Pincus, P.: Macromolecules 17, 2922 (1984)
68. Robeson, L. M., Meitzner, M., Fetters, L. J., and McGrath, J. E.: in "Recent Advances in Polymer Blends, Grafts and Blocks", ed. by L. H. Sperling, Plenum Press, NY, 1974, p. 281
69. Krause, S. and Biswas, A. M.: J. Poly. Sci.: Poly. Phys. Ed. 15, 2033 (1977)
70. Cohen, R. and Torradas, J. M.: Macromolecules 17, 1101 (1984)
71. Zin, W.-C. and Roe, R.-J.: Macromolecules 17, 183 (1984)
72. Roe, R.-J. and Zin, W. C.: Macromolecules 17, 189 (1984)
73. Inoue, T., Shoen, T., Hashimoto, T., and Kawai, H.: Macromolecules 3, 87 (1970)
74. Skoulios, A., Heltfer, P., Gallot, Y., and Selb, J.: Makromol. Chem. 148, 305 (1971)
75. Ptaszynski, B., Terisse, J., and Skoulios, A.: Makromol. Chem. 176, 3483 (1975)
76. Hashimoto, H., Fujimura, M., Hashimoto, T., and Kawai, H.: Macromolecules 14, 844 (1981)
77. Bates, F. S., Berney, C. V., and Cohen, R. E.: Macromolecules 16, 1101 (1983)
78. Rigby, D. and Roe, R.-J.: Macromolecules 17, 1778 (1984)
79. Rigby, D. and Roe, R.-J.: Macromolecules 19, 721 (1986)
80. Selb, J., Marie, P., Rameau, A., Duplessix, R., and Gallot, Y.: Polymer Bulletin 10, 444 (1983)
81. Roe, R.-J.: Macromolecules 19, 728 (1986)
82. Noolandi, J. and Hong, K. M.: Macromolecules 15, 482 (1982)
83. Noolandi, J. and Hong, K. M.: Macromolecules 17, 1531 (1984)
84. Leibler, L.: Macromolecules 15, 1283 (1982)
85. Gaines, G. L. and Bender, G. W.: Macromolecules 5, 82 (1972)

86. Dwight, D. W., McGrath, J. E., Beck, A. R., and Riffle, J. S.: Polymer Preprints *20(1)*, 702 (1979)
87. McGrath, J. E., Dwight, D. W., Riggle, J. S., Davidson, T. F., Webster, D. C., and Viswanathan, R.: Polymer Preprints *20(2)*, 528 (1979)
88. Gaillard, P., Ossenbach-Sauter, M., and Riess, G.: Makromol. Chem. Rapid Commun. *1*, 771 (1980)
89. Kawai, H., Hashimoto, K., Miyoshi, K., Uno, H., and Fujimura, M.: J. Macromol. Sci. — Phys. B *17*, 427 (1980)
90. Fayt, R., Jerome, R., and Teyssie, Ph.: J. Polymer Sci.: Polymer Letters Ed. *19*, 79 (1981)
91. Fayt, R., Jerome, R., and Teyssie, Ph.: J. Polymer Sci.: Polymer Phys. Ed. *19*, 1269 (1981)
92. Fayt, R., Jerome, R., and Teyssie, Ph.: J. Polymer Sci.: Polymer Phys. Ed. *20*, 2209 (1982)
93. Fayt, R., Hadjiandreou, P., and Teyssie, Ph.: J. Polymer Sci.: Polymer Chem. Ed. *23*, 337 (1985)
94. Heikens, D., Hoen, N., Barentsen, W., Piet, P., and Ladan, H.: J. Polymer Sci.: Polymer Symposia *62*, 309 (1978)
95. Gordon, P.: "Principles of Phase Diagrams in Materials Systems", McGraw-Hill, New York, 1968
96. Kraus, G., Rollman, K. W., and Gruver, J. T.: Macromolecules *3*, 92 (1970)
97. Yang, D., Zhang, B., Yang, Y., Fang, Z., Sun, G., and Feng, Z.: Polymer Eng. Sci. *24*, 612 (1984)
98. Rim, P. B. and Runt, J. P.: Macromolecules *16*, 762 (1983)
99. Runt, J. P. and Rim, P. B.: Macromolecules *15*, 1018 (1982)
100. Bernstein, R. E., Cruz, C. A., Paul, D. R., and Barlow, J. W.: Macromolecules *10*, 681 (1977)
101. Rim, P. B. and Runt, J. P.: J. Appl. Polymer Sci. *30*, 1545 (1985)
102. Nguyen, T. Q. and Kausch, H. H.: J. Appl. Polymer Sci. *29*, 455 (1984)
103. Massa, D. J.: ACS Adv. Chem. Ser. *176*, 433 (1979)
104. Barnum, R. S., Goh, S. H., Paul, D. R., and Barlow, J. W.: J. Appl. Polymer Sci. *26*, 3917 (1981)
105. Schultz, A. R. and Young, A. L.: J. Appl. Polymer Sci. *28*, 1677 (1983)
106. Welton, R. E., MacKnight, W. J., Karasz, F. E., and Fried., J. R.: Macromolecules *11*, 158 (1978)
107. Schultz, A. R. and Beach, B. M.: Macromolecules *7*, 902 (1974)
108. Aggarwal, S. L.: Polymer *17*, 938 (1976)
109. Amrani, F., Hung, J. M., and Morawetz, H.: Macromolecules *13*, 649 (1980)
110. Barnum, R. S., Barlow, J. W., and Paul, D. R.: J. Appl. Polymer Sci. *27*, 4065 (1982)
111. Davis, D. D., Taylor, G. N., and Kwei, T. K.: J. Appl. Polymer Sci. *26*, 2001 (1981)
112. Kimura, M. and Porter, R. S.: J. Polymer Sci.: Polymer Phys. Ed. *22*, 1697 (1984)
113. Clas, S. D. and Eisenberg, A.: J. Polymer Sci.: Polymer Phys. Ed. *22*, 1529 (1984)
114. Walsh, D. J., Higgins, J. S., and Rostami, S.: Macromolecules *16*, 388 (1983)
115. Walsh, D. J., Higgins, J. S., Rostami, S., and Weeraperuma, K.: Macromolecules *16*, 391 (1983)
116. Coleman, M. M., Moskala, E. J., Painter, P. C., Walsh, D. J., and Rostami, S.: Polymer *24*, 1410 (1983)
117. Nolley, E., Paul, D. R., and Barlow, J. W.: J. Appl. Polymer Sci. *23*, 623 (1979)
118. Akiyama, S. and Miasa, K.: Polymer J. *11*, 157 (1979)
119. Robeson, L. M. and McGrath, J. E.: Polymer Eng. Sci. *17*, 300 (1977)
120. Anderson, E. W., Bair, H. E., Johnston, G. E., Kwei, T. K., Padden, F. J., and Williams, D.: ACS Adv. Chem. Ser. *176*, 413 (1979)
121. Hickman, J. J. and Ikeda, R. M.: J. Polymer Sci.: Polymer Phys. Ed. *11*, 1713 (1973)
122. Matzner, M., Robeson, L. M., Wise, E. W., and McGrath, J. E.: Makromol. Chem. *183*, 2871 (1982)
123. Utracki, L. A. and Kamal, M. R.: Polymer Eng. Sci. *22*, 96 (1982)
124. Martuscelli, E., Silvestre, C., and Abate, G.: Polymer *23*, 229 (1982)
125. Ito, J., Mitani, K., and Mizutani, Y.: J. Appl. Polymer Sci. *30*, 497 (1985)
126. Ito, Y., Mitani, K., and Mizutani, Y.: J. Appl. Polymer Sci. *29*, 75 (1984)
127. Wang, C. B.: J. Polymer Sci.: Polymer Phys. Ed. *21*, 11 (1983)
128. Cheng, J. T. and Mantell, G. J.: J. Appl. Polymer Sci. *23*, 1733 (1979)
129. Varnell, D. F., Runt, J. P., and Coleman, M. M.: Polymer *24*, 37 (1983)

130. Aubin, M., Bedard, Y., Morrissette, M.-F., and Prud'homme, R. E.: J. Polymer Sci.: Polymer Phys. Ed. *21*, 233 (1983)
131. Tremblay, C. and Prud'homme, R. E.: J. Polymer Sci.: Polymer Phys. Ed. *22*, 1857 (1984)
132. Woo, E. M., Barlow, J. W., and Paul, D. R.: J. Appl. Polymer Sci. *28*, 1347 (1983)
133. Goh, S. H., Siow, K. S., and Yap, K. S.: J. Appl. Polymer Sci. *29*, 99 (1984)
134. Kenney, J. F.: in "Recent Advances in Polymer Blends, Grafts, and Blocks", L. H. Sperling, ed., Plenum Press, NY, 1974, p. 137
135. D'Orazio, L., Greco, R., Martuscelli, E., and Ragosta, G.: Polymer Eng. Sci. *23*, 489 (1983)
136. D'Orazio, L., Greco, R., Mancarella, C., Martuscelli, E., Rogosta, G., and Silvestre, C.: Polymer Eng. Sci. *22*, 536 (1982)
137. Wang, Y. Y., and Chen, S. A.: Polymer Eng. Sci. *21*, 47 (1981)
138. Nolley, E., Barlow, J. W., and Paul, D. R.: Polymer Eng. Sci. *20*, 364 (1980)
139. Chiu, W.-Y. and Fang., S.-J.: J. Appl. Polymer Sci. *30*, 1473 (1985)
140. Lu, Z.-H., Krause, S., and Iskandar, M.: Macromolecules *15*, 367 (1982)
141. Gaines, G. L.: Macromolecules *12*, 1011 (1979)
142. Watanabe, H. and Kotaka, T.: Macromolecules *16*, 769 (1983)
143. Schaefer, J., Sefcik, M. D., Stejskal, E. O., and McKay, R. A.: Macromolecules *14*, 188 (1981)
144. Hsiue, G. H. and Shih, S. W. F.: J. Appl. Polymer Sci. *30*, 1659 (1985)
145. Bajaj, P., Gupta, D. C., and Varshney, S. K.: Polymer Eng. Sci. *23*, 820 (1983)
146. Diamant, J., Soong, D., and Williams, M. C.: Polymer Eng. Sci. *22*, 673 (1982)
147. Choi, G., Kaya, A., and Shen, M.: Polymer Eng. Sci. *13*, 231 (1973)
148. Toy, L., Niimomi, M., and Shen, M.: J. Macromol. Sci. Phys. *B11*, 281 (1975)
149. Hsiue, G. H. and Ma, M.-Y. M.: Polymer *25*, 882 (1984)
150. David, C., Zabeau, F., and Jacobs, R. A.: Polymer Eng. Sci. *22*, 912 (1982)
151. Gupta, A. K. and Purwar, S. N.: J. Appl. Polymer Sci. *30*, 1777 (1985)
152. Gupta, A. K. and Purwar, S. N.: J. Appl. Polymer Sci. *30*, 1799 (1985)
153. Meyer, G. C. and Tritscher, G. E.: J. Appl. Polymer Sci. *22*, 719 (1978)
154. Hansen, D. R. and Shen, M.: Macromolecules *8*, 903 (1975)
155. Lindsey, C. R., Paul, D. R., and Barlow, J. W.: J. Appl. Polymer Sci. *26*, 1 (1981)
156. Ramos, A. R. and Cohen, R. E.: Polymer Eng. Sci. *17*, 639 (1977)
157. Ming, J., Xie, J.-V., and Yu, T.-Y.: Polymer *23*, 1557 (1982)
158. Jiang, M., Huang, X.-Y., and Yu, T.-Y.: Polymer *24*, 1259 (1983)
159. Seo, S. W., Jo, W. H., and Ha, W. S.: J. Appl. Polymer Sci. *29*, 567 (1984)
160. Heikens, D. and Banentsen, W.: Polymer *18*, 69 (1977)
161. McMaster, L. P.: Macromolecules *6*, 760 (1973)
162. Eisenberg, A., Smith, P., and Zhou, Z.-L.: Polymer Eng. Sci. *22*, 1117 (1982)
163. Nishi, T. and Kwei, T. K.: J. Appl. Polymer Sci. *20*, 1331 (1976)
164. Fujimura, T. and Iwakura, K.: Kobunshi Ronbunshu *31*, 617 (1974)
165. Markin, C. and Williams, H. L.: J. Appl. Polymer Sci. *25*, 2451 (1980)
166. Lebedev, Y. V.: J. Appl. Polymer Sci. *25*, 2493 (1980)
167. Kenney, J. F.: J. Polymer Sci.: Polymer Chem. Ed. *14*, 123 (1976)
168. Dumoulin, M. M. Farha, C., and Utracki, L. A.: Polymer Eng. Sci. *24*, 1319 (1984)
169. Inoue, T., Ougizawa, T., Yasuda, O., and Miyasaka, K.: Macromolecules *18*, 57 (1985)
170. Yasuda, O., Ougizawa, T., Inoue, T., and Miyasaka, K.: J. Polymer Sci.: Polymer Lett. Ed. *21*, 813 (1983)
171. Ougizawa, T., Inoue, T., and Kammer, H. W.: Macromolecules *18*, 2090 (1985)

Editor: K. Dušek
Received December 12, 1985

ESR Application to Polymer Physics
— Molecular Motion in Solid Matrix
in which Free Radicals are Trapped—

H. Kashiwabara, S. Shimada, Y. Hori and M. Sakaguchi*
Nagoya Institute of Technology, Nagoya 466, Japan

Examples of the studies on the relation between temperature dependences of ESR spectra of free radicals trapped in solid polymers and molecular motions of matrix-containing free radicals are shown. This method is less frequently used in "Molecular Motion Studies" compared with the so-called spin label or spin probe methods which have been widely applied. However, it safely can be said that the observation of spectra of the trapped radicals can yield more direct information on the matrix, since the trapped radicals themselves are used as "labels". Usually, a broad distribution of relaxation times has to be taken into consideration in studying polymeric systems. A few examples illustrate the arrangement of data of magnetic resonance based on this kind of approach. From these data it can be concluded that the simultaneous use of ESR and NMR measurements is advantageous. Though the details of so-called spin label or spin probe methods are not included in this article, some of the applications of the spin trapping method are shown, such as its useful application in studying materials containing unstable radicals.

---

* Ichimura College, Inuyama 484, Japan

Advances in Polymer Science 82
© Springer-Verlag Berlin Heidelberg 1987

1 Introduction

Rather soon after the nuclear magnetic reasonance (NMR) method was applied successfully to solid state physics, NMR was applied to the study of physical properties of polymers. To the best of our knowledge, a short paper by Wilson and Pake published in 1952 on the two-component structure of the proton NMR spectrum of polyethylene was the first paper in the field of NMR application to polymer science [1]. After that, many papers of NMR studies in polymer physics were published and many novel informations concerning molecular motions and structures of polymeric materials were presented. Recent advances in hardware of instrumentation of NMR measurement opened its application even to the medical and tomographic field [2].

Electron paramagnetic resonance (EPR or ESR), on the other hand, was applied to polymer science in the middle of the 1950s in the studies of polymerization [3] and irradiation effects [4], and its application to molecular motion study was established in the first half of the 1960s.

In the early stages of ESR application to polymer research, many studies on the identification of free radicals produced by irradiation with ionizing radiation, x-ray, and ultraviolet light were made. Some of the irradiation effects in polymeric materials were considered to originate from the radical processes and, therefore, clear identification of the radicals trapped in irradiated polymers was one of the most important problems at that stage. In this meaning, ESR application was considered to be a very convenient technique for this purpose, because detection and identification of the free radicals bearing unpaired electrons in principle can be done easily by the ESR method without any chemical modification of the materials.

However, successful identifications of the free radicals in polymers were not as easy as expected in early applications of this method. The reasons were as follows:
a) The irradiation effects in the materials are non-specific. Therefore, many kinds of free radicals are produced simultaneously. This means that a superposition of many kinds of patterns had to be observed originating from many kinds of free radicals produced.
b) Furthermore, conformation along the polymeric chain is not uniform; i.e., some sites are of *trans*, some are of *gauche* conformation and some sequences of monomeric units have helical structures. The variety of conformations of the chains lead to the various steric configurations between unpaired electron and interacting nuclei and this is also responsible for the different circumstances of hyperfine interaction resulting in different ESR-patterns even in chemically identical free radicals. Therefore, we have to observe the superposition of the different patterns even if the inconvenience caused by simultaneous trapping of the many kinds of free radicals can be removed.

These problems, which make the interpretation of observed ESR spectra of polymer radicals difficult, required several methods to be solved. One was the successive heat treatment of materials so that the fast decaying radicals vanish while preserving the stable radicals. Heat treatment is a rather simple technique, which gives interesting information concerning the relation between radical decay and relaxation processes in solid high polymers. This was one of the motives for the study presented in Sect. 4.

In the course of ESR studies of oxidation processes in irradiated polyethylene, an interesting property of peroxy radicals in polyethylene was found and this was an incentive for studying the molecular motion in polymeric materials by use of peroxy radicals as a label. This topic will be treated in Sect. 6.

ESR is a magnetic resonance method using microwave frequencies of the order of 10 GHz. This means that ESR has a different time scale compared to NMR (electromagnetic waves of the order of 100 MHz are common in most NMR spectrometers). A few examples of motional narrowing of line width, which is connected with the time scales will be discussed in Sect. 8.

Since the expression "ESR application to the study of molecular motion" easily reminds of the so-called spin label and spin probe methods, these topics are included in the present article; the related problems will be considered in Sect. 9 and 10. However, any trapped free radical must be a label for the study of molecular motion when the temperature dependences of the spectral parameters are adequately measured. The discussion described in Sect. 6 is an example of the study using peroxy radicals as labels. The discussion in Sect. 4 is another example of ESR application to molecular motion study using the spectral intensity. Examples of molecular motion based on the changes of spectral parameters of trapped free radicals are presented in Sects. 5 and 8. All this should show how the trapped radicals can be used as "direct labels" in order to apply ESR without substantially modifying the investigated materials.

Concerning the observation of ESR spectra, an important technical point must be added: Since spectra of the free radicals in polymeric chains are sensitive to power saturation, the spectra must be observed at very low microwave power. A most remarkable example in the authors' experience was the case of polyisobutylene [103], where a power of less than 0.39 μW had to be applied in order to obtain a good spectrum. Usually, a power of less than 50 μW is used for observing carbon radicals like alkyl or allylic radicals. The saturation behavior is less pronounced when the spectrum of peroxy radicals is observed. This character of peroxy radicals is helpful for enhancing signals of peroxy radicals when the spectra of peroxy and carbon radicals are superposed (Sect. 6). Of course, the saturation of the spectrum is closely related to the spin-lattice relaxation and constitutes an essential problem of polymer physics. However, this point will be excluded in the present article and only a few studies on the power saturation in the case of polymer radicals will be listed in the reference section [101, 103, 104, 105].

2 Brief Explanation of ESR Spectral Parameters

The principles and theoretical problems of ESR measurement are described in several excellent textbooks on this topic [5]. In the following, the terminology used in the succeeding section will be briefly explained to facilitate the further reading.

ESR spectra of free radicals trapped in polymer matrices reflect the physical properties of the matrix polymer, usually indicated by g-values, hyperfine coupling constants, and line widths of the components of the whole spectra. Effects of the crystalline field and quadrupole interaction are usually very small in the case of polymer radicals. The g-value is defined by the resonance condition:

$$h\nu = g\beta H \tag{2.1}$$

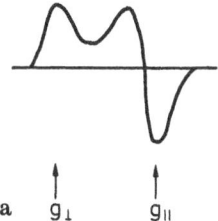 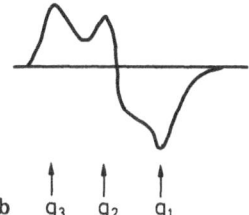

a $g_\perp$ $g_\parallel$ b g_3 g_2 g_1

Fig. 2.1a and b. Schematic illustrations of first derivatives of ESR spectra with g-anisotropy. **a** case of axial symmetry; **b** case of $g_1 < g_2 < g_3$

where ν is the frequency of the microwave used, β the Bohr magneton (unit of magnetic moment), and H the applied static magnetic field. If spin-orbit interaction is negligible, the g-value is very close to 2.0023, the value of the isolated electron, However, the electrons in the molecules are in orbital motion and, hence, g-values derived from ESR spectra of some radicals show values different from that of the isolated electron. It is obvious that an anisotropic character can be attributed to the g-factor, which can be written in a tensor form usually expressed by the principal values along the principal axes in the molecules. We shall call these principal axes g_1-, g_2-, and g_3-axes. Sometimes, an axially symmetric case can be discussed when two values, $g_\parallel$-value and $g_\perp$-value, are used instead of three principal values, g_1, g_2, and g_3. Typical shapes of the spectra with g-anisotropy are shown in Fig. 2.1. The most remarkable spectrum showing g-anisotropy in the case of polymer application is the ESR spectrum of peroxy radicals. When molecular motions of a certain mode occur, the directions of the principal axes can be different than those in the rigid state and the respective principal values of the g-tensor change. When the motion is completely random and three dimensionally, the three principal values are averaged out. In case of rotation around one principal axis, e.g., the g_1-axis, the other (g_2 and g_3) values shall be averaged out, but the g_1 values does not change. In this way, variation in g-ansiotropy can be used as a measure of molecular motion.

The hyperfine structure of the ESR spectrum is used to identify trapped radicals. This is due to the interaction of unpaired electrons with magnetic moments of the nuclei surrounding unpaired electrons localized at the radical site. Hyperfine interaction is described by two terms. One is the isotropic interaction, a, originating from the contact interaction of the wave functions of electron- and nuclear spins, and the other is the anisotropic interaction, D, originating from the magnetic dipole-dipole interaction. The total hyperfine interaction, a + D, is therefore essentially anisotropic and can also be expressed in the form of a tensor $\tilde{A}$. As in the case of $\tilde{g}$-tensor, $\tilde{A}$ can be described by three principal values, A_1, A_2, and A_3, corresponding to the principal axes in a proper molecular coordinate system. The magnitude of the isotropic term of hyperfine interaction depends on an overlap of the wave packets of the unpaired electrons and nuclei; in the case of the hyperfine interaction of an unpaired electron with a β-proton (see Fig. 2.2), McConnell's relation [Eq. (2.3)] has been derived [6].

$$a = a_0 + B \cos^2 \theta \qquad (2.2)$$

In Eq. (2.2) a_0 is a very small constant, B is also a constant, and θ is the angle shown in Fig. 2.2. in the plane perpendicular to the direction of the C—C bond. The magnitude of a is also depending on the spin density ϱ at the unpaired electron.

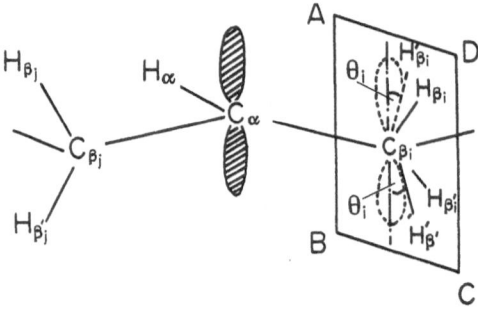

Fig. 2.2. Schematic illustration of steric configuration of methylene protons around an unpaired electron. Plane ABCD is perpendicular to the direction of the C_α-C_β bond. Dotted lines in plane ABCD mean projections of the corresponding items on the plane ABCD

$$a = Q\varrho \qquad\qquad (2.3)$$

Equations (2.2) and (2.3) show the possibilities of temperature dependence of a in addition to that of the anisotropic term which can easily be temperature dependent. The reader may be confused with the temperature dependence of a found in Eq. (2.3), but this has been observed experimentally for oxygen-containing radicals, such as $\sim CH_2-\dot{C}H-O-CH_3$ trapped in irradiated polyethyleneoxide. In these cases, molecular motion results in the delocalization of the unpaired electron at the oxygen atom and leads to smaller hyperfine interaction with γ-protons (methyl protons). Examples of temperature dependence of hyperfine splitting in the case of polymer radicals will be shown in Sect. 5.

3 Relation Between the Variation of ESR Spectra of Free Radicals and their Locations [7]

As mentioned in Sect. 1, most ESR spectra of irradiated polymers consist of superpositions of several patterns corresponding to different radicals or to chemically identical radicals trapped at different sites. The separation of the superposed patterns is an important problem in elucidating the relation between the variation of the ESR spectrum and the physical properties of trapping sites. In order to illustrate this point, a simple example of solution-grown polyethylene will be shown first. High density polyethylene, Sholex 6050 (Product of Showa Electric Co. Ltd.), was purified and recrystallized in dilute xylene solution at 85 °C for a week. The recrystallized materials had a density of 0.9810 g · cm^{-3} and a lamellar thickness, determined by small-angle X-ray scattering, of 14.7 nm. Even though the free radicals in those materials should be mainly the alkyl radicals trapped in the crystalline part, it was found that the spectrum observed was a superposition of at least two kinds of patterns originating from similar alkyl radicals which were trapped in different regions of the crystallites. This was confirmed by the following experiment.

One sample of the solution-grown material (called here M-85) was subjected to *f*uming *n*itric *a*cid *t*reatment (abbreviated FNAT) at 60 °C for 10 hr, and another sample for 196 hr. The former is called N-10, the latter N-196. After FNAT, N-10 and N-196 were washed thoroughly to remove nitric acid and homopolymer. M-85, N-10, and N-196 are powdered materials. Another sample of solution-grown material

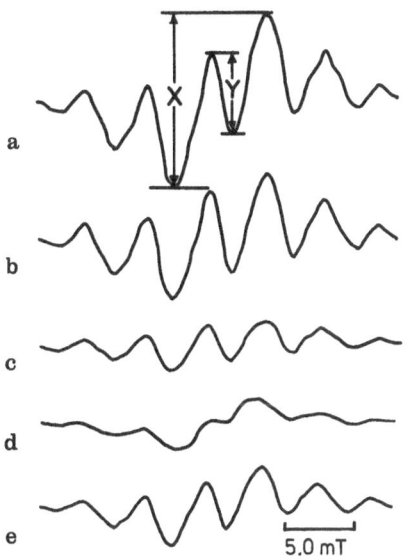

Fig. 3.1a–e. ESR spectra of irradiated polyethylene M-85. Observation was made at liquid nitrogen temperature after irradiation at the same temperature, sharpness = Y/X. **a**, spectrum after heat treatment at 245 K for 10 min; **b**, spectrum after heat treatment at 277 K for 10 min; **c**, spectrum after heat treatment at 317 K for 10 min; **d**, spectrum obtained by subtracting **b** from **a**; **e**, spectrum obtained by subtracting **c** from **b** (Ref. [7])

was prepared from Sholex 6050 by recrystallization in dilute xylene solution in a similar way as M-85, but the plate-shaped samples were prepared on a glass filter at last. This sample is called M-85P. All materials were irradiated up to about 1 Mrad. *In vacuo* at about 1.3×10^{-2} Pa at liquid nitrogen temperature.

ESR spectra of irradiated M-85 were observed at liquid nitrogen temperature and after post-irradiation heat treatment at various temperatures higher than 77 K. Several examples of spectra are shown in Fig. 3.1. Curve a is the spectrum observed at 77 K after heating to 245 K in order to remove the radicals trapped in the amorphous part. Spectrum b was observed after heating to 277 K, and spectrum c after heating to 317 K. Subtracting b from a gives d. e was obtained in a similar manner by subtracting c from b. A remarkable change in the ESR spectra is observed during heat treatment. The reason for this change must be that many free radical species or structural isomers exist before heat treatment. The unstable free radicals disappear at lower temperatures and much more stable radicals could be observed after heat treatment. The patterns in Fig. 3.1. d and e show the spectra of the free radicals which disappeared during heat treatment at temperatures ranging from 245 K to 277 K and from 277 K to 317 K, respectively. According to the spectral character of the patterns, both d and e in Fig. 3.1. can be attributed to the alkyl radicals; d is a very broad spectrum whereas e is a sharp one. Since the former corresponds to the radicals having disappeared at lower and the latter at higher temperatures, the apparent difference in sharpness of d and e corresponds to the differences in apparent line widths of the radicals disappearing at low temperature and at high temperature. The line width of the spectrum is usually due to the relaxation times T_1 and T_2, and it also depends on the mobility of the trapping sites and on radical concentration. The other reason for the difference in apparent line width is the broad distribution of θ appearing in Eq. (2.2). At any rate, it can be said that ESR spectra of alkyl radicals depend on the trapping site. In the present

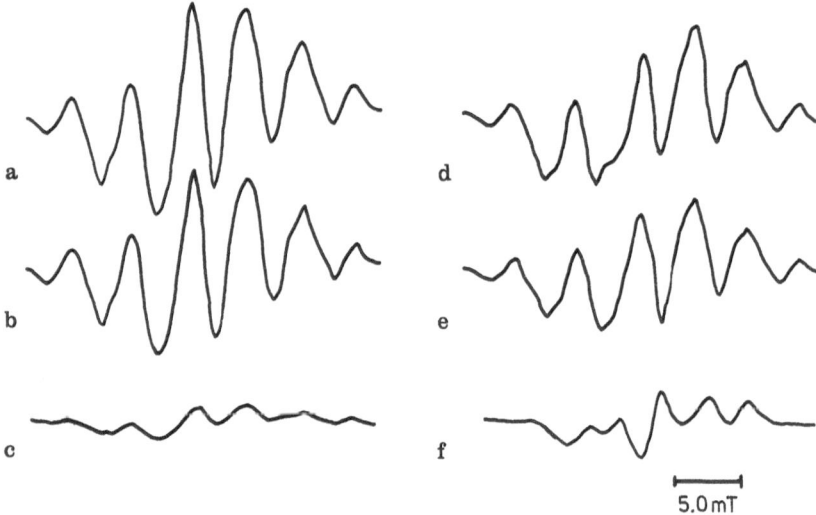

Fig. 3.2a–f. ESR spectra of irradiated polyethylene N-10 and N-196. Observation was made at liquid nitrogen temperature after irradiation at the same temperature. **a**, spectrum of N-10 after heat treatment at 244 K for 10 min; **b**, spectrum of N-10 after heat treatment at 276 K for 10 min; **c**, spectrum obtained by subtracting **b** from **a**; **d**, spectrum of N-196 after heat treatment at 245 K for 10 min; **e**, spectrum of N-196 after heat treatment at 277 K for 10 min; **f**, spectrum obtained by subtracting **e** from **d** (Ref. [7])

case, we assume that the free radicals corresponding to the broad spectrum are trapped at the surface of the crystallites and that those corresponding to the sharp spectrum are trapped inside the crystallites. Other experimental facts support the above interpretation. Using N-10 and N-196, similar heat treatment experiments were made. In the case of N-10, the spectrum of Fig. 3.2c., which is corresponding to Fig. 3.1d., is very broad but no broad spectrum was observed in the experiment using N-196. FNAT is a well-known treatment for removing the surface area of the polymer crystallite selectively (etching effect). Based on the results shown in Fig. 3.2., it can be said that the etching effect in N-10 is less complete compared to that in N-196. Therefore, it is quite reasonable that the broad spectra shown in Fig. 3.1d. or Fig. 3.2c. correspond to the free radicals trapped at the surface of the crystallites.

On the other hand, a spectral sharpness, Y/X, is defined as indicated in Fig. 3.1a. The sharpness and spectral intensity of M-85 were measured at various stages after storage at 265 K. Before storage at 265 K, the sample was heated to 196 K for 3 hr. This was in order to remove the unstable radicals trapped in the amorphous part, just similar to the heat treatment at 245 K in Fig. 3.1. The sharpness of the spectrum and the concentration of the remaining radicals were plotted against the storage time (Fig. 3.3.). This indicates that the sharpness increases and the concentration decreases with time and that both quantities approach a constant value after the same time. This shows that the free radicals corresponding to the broad spectrum are unstable and that the free radicals corresponding to the sharp spectrum hardly decay at 265 K. A separation of the superposed spectra can be achieved by this way. Related problems will be discussed further below.

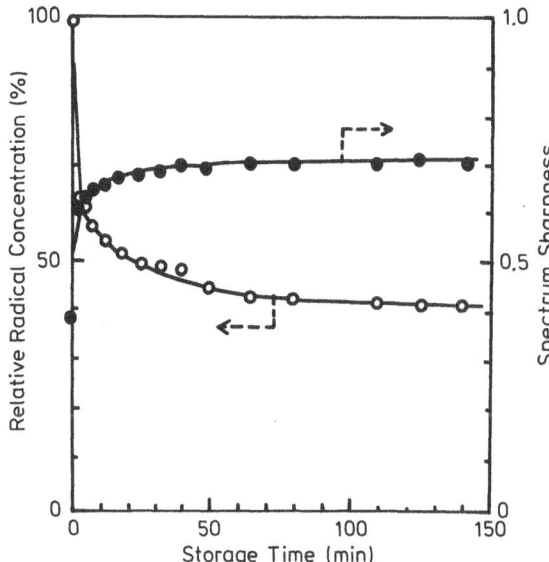

Fig. 3.3. The decay curve (○) of alkyl radicals produced in irradiated polyethylene A-85 at 265 K and the sharpness (●) of ESR spectra of the correspondence (Ref. [7])

Figure 3.3. also shows the constant value of the remaining relative concentration of stable free radicals to be 0.4 (Sect. 4). The approximate thickness of the interior of the crystallite, l, and that of the lamellar surface, d, can be estimated by the following equation:

$$\varrho = \varrho_c - (\varrho_c - \varrho_a)\frac{d}{1 + d} \tag{3.1}$$

where ϱ, ϱ_c, and ϱ_a are the densities of the sample, inner crystalline part and amorphous part, respectively. ϱ is 0.9810 g · cm^{-3} and l + d is 14.7 nm as mentioned above. Since the values of ϱ_c and ϱ_a are established values, d and l can be estimated. In the present case, d and l were found to be 2.7 and 12.0 nm, respectively. From the values of d and l, the local concentrations in the surface area, R_a, and interior of the crystallite, R_c, can be estimated; R_a/R_c was 7 in the present case. According to the paper by Kusumoto et al. [8], R_a/R_c was ca. 15, though the materials used were not identical. At any rate, we can say that a considerable amount of free radicals migrate into the surface area where high local concentrations appear, since no specific site of trapping at the initial stage of irradiation can be considered. This fact is rather interesting from the radiation chemical viewpoint; related discussions have been reported[9]. An experiment employing a similar heat treatment as in the case of Fig. 3.1. was made for a sample of M-85P, in order to investigate the angle dependence of the ESR spectrum. It was shown that the subtracted spectra corresponding to the radicals disappearing at low temperature, Figs. 3.4c. and 3.4f. do not show any angle dependence. This indicates that the arrangement of the molecular chain in the trapping site of the vanished radicals is at random, reflecting that the site must be the surface area. According to these results, it should be possible to label the spins on the inner part and the surface area of the crystallite separately by γ-irradiation and discuss the molecular motions in the respective parts. An example of this kind of study will be shown in Sect. 4.

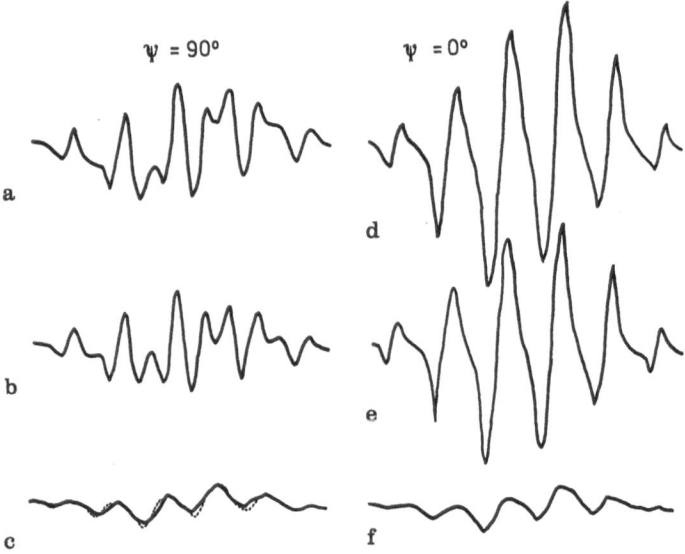

Fig. 3.4a–f. ESR spectra of the alkyl free radicals produced in plate-shaped crystals observed with different angles between direction of c-axis and magnetic field, ψ. Observation was made at 122 K after irradiation at liquid nitrogen temperature. **a, b, c** spectra at $\psi = 90°$; **d, e, f**, spectra at $\psi = 0°$; ψ represents the angle between the applied magnetic field and the direction of the c-axis of the crystals. ---------, corresponds to **f. a, d**, spectra after heat treatment at 243 K for 10 min; **b, e**, spectra after heat treatment at 279 K for 10 min; **c**, spectra obtained by subtracting **b** from **a**; **f**, spectrum obtained by subtracting **e** from **d** (Ref. [7])

4 Decay Reaction of Free Radicals and its Relation to the Molecular Motion in Matrix Polymer

As mentioned in the previous section, heat treatment after irradiation is a convenient method for identifying ESR spectra of free radicals trapped in irradiated polymers. In the course of heat-treatment studies, it was found that the decay of free radicals in irradiated polymers, especially in polyethylene and polypropylene, is closely related to the molecular motion in matrix polymers[10,11]. In order to describe the decay quantitatively, some authors used a first-order reaction scheme [12,13] and some a second-order reaction scheme [8,14,15]. However, only few explanations of the experimental data were obtained, though analysis with two first-order rate constants gave good results [13]. A schematic description of the nature of the decay reaction of the free radicals in irradiated polymer is shown in Fig. 4.1.; the initial stage of the process deviates from the linear plot for both the logarithmic concentration plot (first-order scheme) and the inverse concentration plot (second-order scheme). Though a first-order scheme with two reaction rates, fast and slow, was seen to be successful in explaining the experiment [13], a "bi-molecular" reaction would be expected, since many ESR studies on decay reaction were made at temperatures higher than 77 K and at a time scale of several hours. In the case of polyethylene, which was investigated by many authors, the "fate" of the decayed radical must be

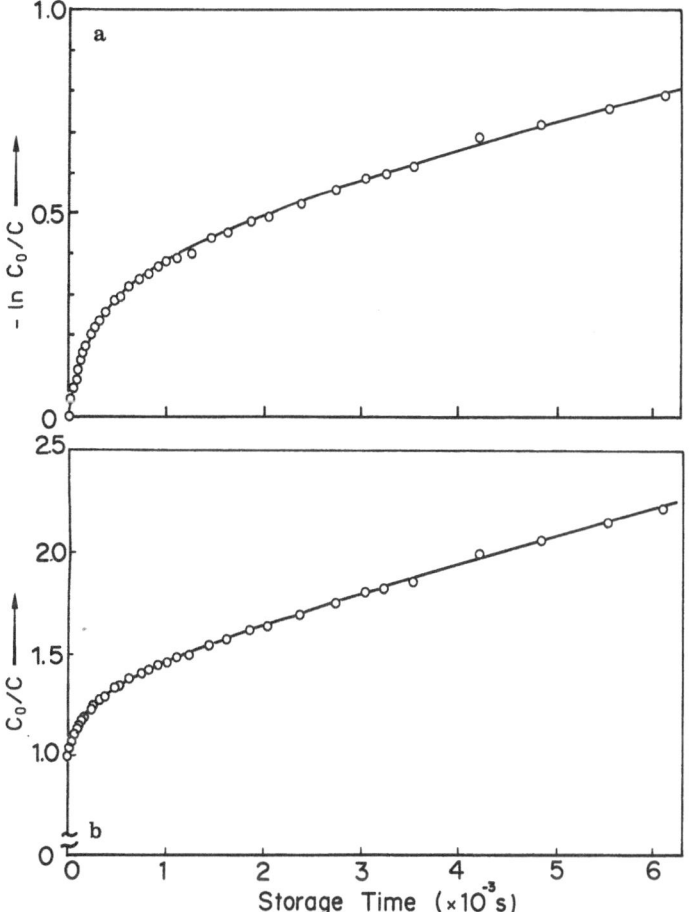

Fig. 4.1 a and b. Examples of the fact that actual decay reactions of the alkyl free radicals trapped in polyethylene can not be interpreted by simple first-order or second-order reaction schemes: **a** deviation from first-order reaction scheme, **b** deviation from second-order reaction scheme (Ref. [16b])

the formation of interchain cross-linking. In view of these circumstances, Shimada et al. [16] tried to analyze the decay data of the free radicals in irradiated polyethylene with a scheme involving a diffusion-controlled bi-molecular reaction. These authors modified the equations in the paper by Waite [17] and the fundamental diffusion equations used for the decay reaction of the polymer radicals were:

$$\frac{C_0}{C} = 1 + 8\pi\gamma_0 DC_0 \left[1 + \frac{\sqrt{2\gamma_0}}{\pi Dt} \right] t \tag{4.1}$$

$$= 1 + A\sqrt{t} + Bt = 1 + B \left[1 + \frac{A'}{\sqrt{t}} \right] t \tag{4.2}$$

$$(A'B = A)$$

$$A = 8\sqrt{2}\,\gamma_0^2 C_0(\pi D)^{1/2} \tag{4.3}$$

$$B = 8\pi\gamma_0 DC_0 \tag{4.4}$$

$$D = (2/\pi)\,(B/A)^2\gamma_0^2 \tag{4.5}$$

where C indicates the concentration of the radicals studied at a certain time t, C_0 the initial value of C, D is the diffusion constant, and A, A', and B are the constants satisfying Eqs. (4.3), (4.4), and (4.5). The constant γ_0 is the capture radius which is defined as the distance at which two radicals can recombine with a minor activation energy. Equations (4.1) to (4.5) can be used when only one kind of radical takes part in the decay process, all being reactive. However, some polymer radicals are stable and do not participate in the decay process as in the case discussed in the previous section (Fig. 3.3.). This requires a further modification of the equations. In a simple example, radical 1 is assumed to react, whereas radical 2 is stable. Then C and C_0 in Eq. (4.1) must be replaced by the following quantities:

$$C = C_1 + C_2 \qquad C_0 = C_1^0 + C_2^0 \tag{4.6}$$

where C_1 and C_2 are the concentrations of radical 1 and radical 2, respectively. Superscript ^{0} means the initial concentration. Using the above expression (4.6), Eq. (4.7) can be derived in place of (4.2):

$$\frac{C_0 - C_2^0}{C - C_2^0} = \frac{1-a}{x-a} = 1 + B\left[1 + \frac{A'}{\sqrt{t}}\right]t \qquad \left(x = \frac{C}{C_0}\right) \tag{4.7}$$

where a equals C_2^0/C_0 and must be constant because radical 2 does not react within a certain temperature range used in the study. Equation (4.7) can be rewritten as:

$$x = \frac{1}{B[1 + A'/\sqrt{t}]} \cdot \frac{1-x}{t} + a \tag{4.8}$$

Therefore, if the quantity x is plotted against $(1 - x)/t$ and $(1 - x)/t$ is extrapolated to 0, we can estimate the value of a experimentally. Since the quantity $x - a$ is the concentration of the reacting radicals, the best fitted constant A, A', or B can be determined from a plot of $x - a$ vs t by computer simulation. If a is not estimated by use of Eq. (4.8), the constants A, A', and B cannot be determined. An example of determining a and related simulations is shown in Figs. 4.2. and 4.3. These indicate the decay reaction of the alkyl radicals trapped in the crystalline region at temperatures ranging from $-23\,°C$ to $1.5\,°C$. The sample used is almost the same material as that used for Fig. 2.1. As shown in Fig. 4.2., the constant a can be determined as 0.4. Figure 4.3. shows that this constant is necessary for determining the best fitted constants involving the diffusion constant. Black circles in Fig. 4.3. are the "raw" concentrations obtained from the integral of the ESR spectra observed, but simulations with these values where not satisfying. However, when the value for a is subtracted from the values indicated within the black circles, a simulation for determining the

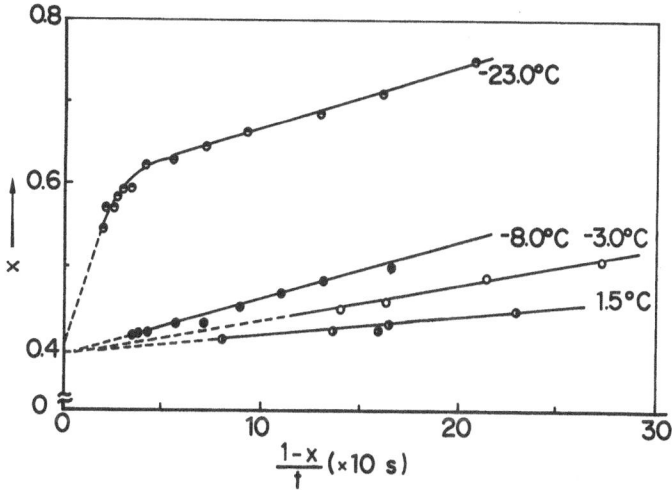

Fig. 4.2. The relative concentration of alkyl radicals vs value of $(1 - x)/t$ at various temperatures [plots according to Eq. (4.8)]: ESR measurements were made at 77 K (Ref. [16b])

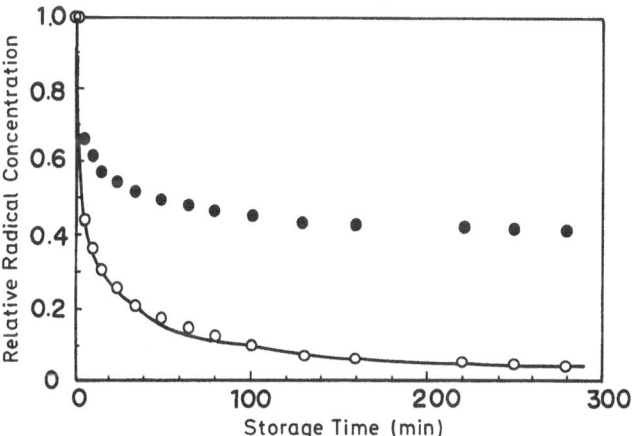

Fig. 4.3. The decay curve of alkyl radicals trapped in irradiated single crystals of polyethylene A-85 at 269 K after heating at 195 K for a long time. ESR observations were made at 77 K: (●) Plot of raw data without any reduction, (○) plot of reduced amount after subtracting the amount of non-vanishing radical species. Solid line shows calculated decay curve with $A = 4 \times 10^{-2} \ sec^{-1/2}$ and $B = 1 \times 10^{-3} \ sec^{-1}$ (Ref. [16b])

constants A and B can be made as shown in Fig. 4.3. Thus we can analyze the decay reaction of the "unstable" radicals within the investigated temperature region separately. These radicals are considered to be trapped on the surface of the crystallite and their decay reaction should be closely related to the so-called α'relaxation in polyethylene, as discussed later. The constant value $a = 0.4$ is the same as the saturation value in Fig. 3.3., indicating that the interpretations of Figs. 3.1. and 4.2. are quite consistent and thus reflect the validity of the analysis using Eq. (4.8). When the decay

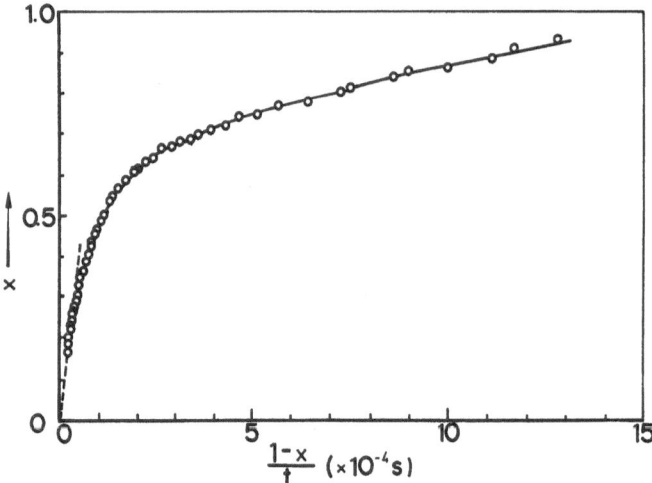

Fig. 4.4. The relative concentration of alkyl radicals vs value of $(1 - x)/t$ at 307 K (Ref. [16b])

reaction is studied at high temperature, the constant a needs not to be taken into account since all radicals can be considered to be reactive. In this case, the radical concentration as observed can be plotted against the duration of the decay reaction and the best fitted constant A, A′, or B can be estimated. At high temperature, the value of a is 0 which can be obtained by the plot based on Eq. (4.8) as shown in Fig. 4.4.

By means of the described procedures, it is possible to determine the constants A and B at various temperatures and thus to observe a temperature dependence of diffusion constants following Eqs. (4.3) to (4.5). The expression "diffusion" has been used several times in this section, there by referring to diffusion of reactants and therefore diffusion of free radicals participating in a bi-molecular decay reaction. How does the radical diffuse in a rather rigid crystalline phase? (Most of the observed radicals are trapped in the crystalline section.) The "diffusion" in the present case can best be considered as the "diffusion" of the radical site rather than actual "mass transfer". Therefore, "diffusion" of the radical requires an exchange of unpaired electrons and protons at neighboring sites. This process calls for a slight twisting of the molecular chains. In other words, a twisted or distorted region must move with a certain diffusion constant within the "sea" of polymer. Formally, we can use the Stokes-Einstein equation, Eq. (4.9), which describes the motion of a sphere with a radius b moving in the medium with viscosity η:

$$D = RT/3\pi b\eta \qquad\qquad (4.9)$$

where R is the gas constant and T the temperature in K. Usually, viscosity of the materials is related with Young's modulus E and relaxation time τ by the following equation:

$$\eta = \tau \cdot E \qquad\qquad (4.10)$$

Therefore, if distorted region is assumed to be a sphere with similar dimensions as the lattice constants of polyethylene crystals, the relaxation time τ can be estimated from the value of D determined from the data of the decay reaction. The relaxation time is considered to be a time constant of the molecular motion causing a slight distortion associated with the "diffusion" of the free radical. Thus, the relaxation time of the molecular motion associated with the decay reaction can be estimated. In order to validate this procedure, the diffusion constant was estimated from the known relaxation time obtained in dynamic mechanical studies of polyethylene within the temperature region of the so-called relaxation process in a crystalline phase and

Table 4.1. Values of A, B, and D for the decay reaction of alkyl radicals in polyethylene*
a) Reaction in the lamellar surface

Temperature (K)	A $(s^{-1/2})$	B (s^{-1})	D $(cm^2 \, s^{-1})$
265	4×10^{-2}	1.0×10^{-3}	9.5×10^{-20}
270	4×10^{-2}	1.7×10^{-3}	2.7×10^{-19}
274.5	4×10^{-2}	3.4×10^{-3}	1.1×10^{-18}

b) Reaction in the interior of the crystallite

Temperature (K)	A $(s^{-1/2})$	B (s^{-1})	D $(cm^2 \, s^{-1})$
307	1.1×10^{-2}	6.0×10^{-5}	3.0×10^{-21}
314	1.2×10^{-2}	1.3×10^{-4}	1.2×10^{-20}
319	1.0×10^{-2}	1.7×10^{-4}	3.6×10^{-20}
327	1.4×10^{-2}	6.0×10^{-4}	1.8×10^{-19}

* The capture radius, r_0, was assumed as 1.54 Å (length of the C—C bond)

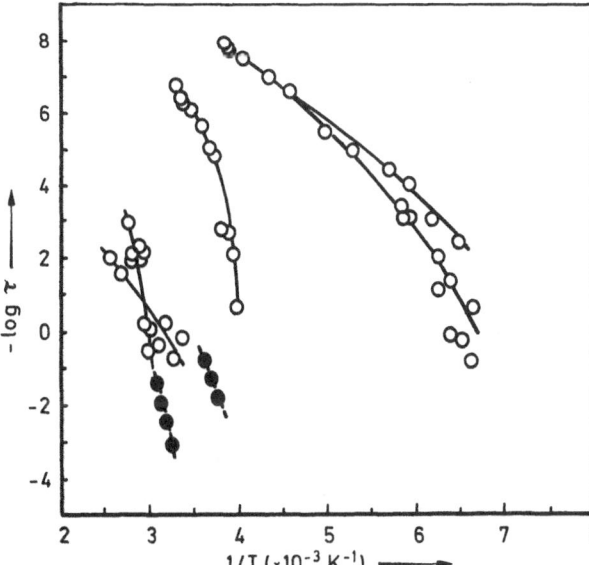

Fig. 4.5. A relaxation map of polyethylene: ○, plot arranged by Wad [16]; ●, plots based on the decay reaction of free radicals (Ref. [20])

its value was compared with the diffusion constant obtained from the decay reaction of the radicals in the same temperature region. The values obtained were in good agreement. Table 4.1a shows the values of A, B, and D obtained from the decay reaction of the free radicals in polyethylene crystals within the low temperature region, meaning the decay reaction of the "unstable" radicals described in connection with the constant a in Eq. (4.7) and Fig. 4.2. Table 4.1b shows similar values for the decay reaction at high temperatures. According to the discussions concerning the diffusion constant and relaxation time, τ was estimated for the various temperatures studied, and $-\log \tau$ was plotted in so-called relaxation maps arranged by Wada [18]; Fig. 4.5. was obtained [20]. It can be said that the relaxation times obtained from the decay reaction of the free radicals at low temperatures (the decay reaction of the radicals corresponding to the broad spectra described in Sect. 3) are within the extrapolated range of the plots of the relaxation time obtained from mechanical studies of α'-relaxation; the relaxation times obtained from the decay of the radicals at high temperature are within the same range as the data obtained from mechanical studies of α-relaxation. From the plots of the relaxation time in the relaxation map of Fig. 4.5., the activation energies of the respective molecular motion associated with the respective decay reactions can be obtained. The activation energy obtained from the data of the reaction at higher temperatures was found to be 180 kJ/mol and this value is very close to the activation energy of the molecular motion associated with the α-relaxation process in the crystalline phase of 192 kJ/mol reported by Takayanagi et al. [19]. The above-mentioned proposes the possibility of separately studying the molecular motions on the surface and inside the crystallites by use of data from decay reactions of the free radicals trapped at the respective locations.

A similar application of Eq. (4.8) was also used for the decay reaction of free radicals in polyoxymethylene. It is well known that scission-type free radicals:

$$\sim O-\overset{\overset{\displaystyle H}{|}}{\underset{\underset{\displaystyle H}{|}}{C}}\cdot$$

and chain radicals:

$$\sim O-\overset{}{\underset{\underset{\displaystyle H}{|}}{\dot{C}}}-O\sim$$

are trapped in irradiated polyoxymethylene [21]. The ESR spectrum of the chain radical is doublet due to the hyperfine interaction with one α proton. The spectrum of the scission radical is a triplet, but it is unstable and, therefore, at room temperature, the doublet spectrum corresponding to chain radicals can be observed only. The decay reaction of the stable radicals, which were trapped when polyoxymethylene was heated to room temperature after γ-irradiation at the temperature of liquid nitrogen, was studied at various temperatures ranging from 343 K to 364 K. However, the ESR spectrum observed was not a pure doublet. A plot based on Eq. (4.8) was constructed resulting in Fig. 4.6. and a value of a = 0.2 was determined; i.e., 20% of the initial intensity is due to a radical whose stable pattern is superposed. This stable pattern may result from an anti-oxidant reagent. Figure 4.7. shows that the simulated curve made from the plot of the spectral intensities can be drawn simply by use of the reduced intensity (open circles), as in the case of Fig. 4.3., instead of the intensities as observed

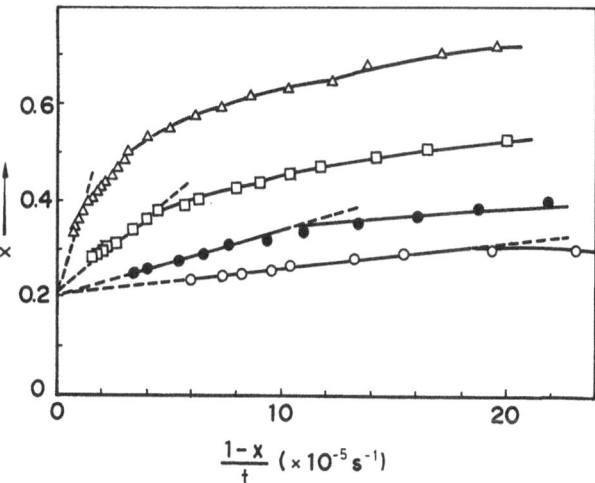

Fig. 4.6. Relative concentration of free radicals of polyoxymethylene, x, vs. value of $(1 - x)/t$ at temperatures: $\triangle$, 343.5 K; $\square$, 349.0 K; $\bullet$, 354.0 K; $\bigcirc$, 364.0 K (Ref. [20])

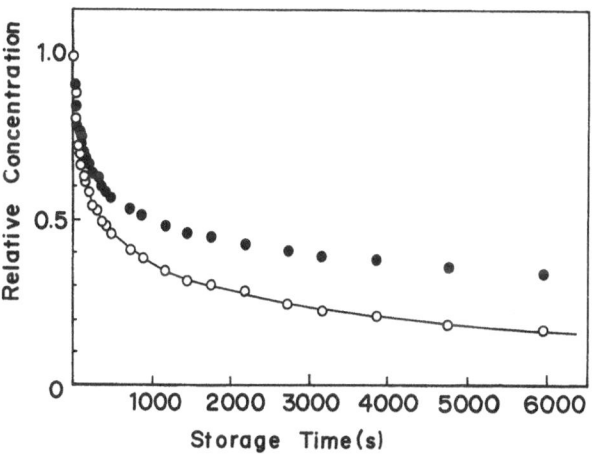

Fig. 4.7. Decay curve of free radicals trapped in irradiated polyoxymethylene at 354 K after heating at room temperature for an extended time period: $\bullet$, plot of raw data without any reduction; $\bigcirc$, plot after subtracting the amount of non-vanishing radicals obtained from Fig. 4–6; solid line indicates calculated decay curve with $A = 5.0 \times 10^{-2}$ s$^{-1/2}$ and $B = 1.6 \times 10^{-4}$ s (Ref. [20])

(black circles). As in the case of the above-mentioned polyethylene, constants A and B, at various temperatures, can be estimated from the data of the decay reactions at various temperatures, and temperature dependence of D (hence of τ) also can be discussed [20]. Figure 4.8. plots the relaxation time in the relaxation map of polyoxymethylene. Here, the free radicals correspond to the doublet spectrum, which is rather stable after heating the sample to room temperature, and are caused by the molecular motion associated with the relaxation process in the crystalline phase of polyoxy-

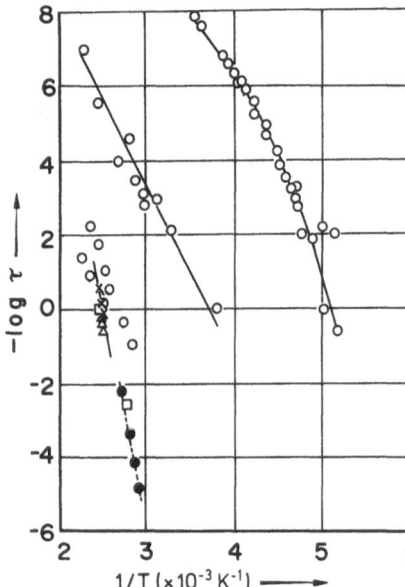

Fig. 4.8. Relaxation map of polyoxymethylene: ●, data obtained from decay reaction of free radicals; ○, ×, □, △, data of mechanical or dielectric studies (Ref. [20])

methylene, since relaxation times obtained from the decay reaction are on the line corresponding to α-relaxation studied by the mechanical or dielectric method.

5 Temperature Dependence of Hyperfine Coupling Constants and Relations to Molecular Motion

5.1 Hindered Oscillation of Alkyl Radicals in Polyethylene [22]

As mentioned in Sect. 2, hyperfine splitting due to β-protons depends on the steric configuration of the local site of the unpaired electron and is dependent on temperature. Using this property of hyperfine splitting, information on the molecular motion of the polymeric chain can be obtained. As an example, the case of alkyl radicals in polyethylene will be discussed here.

Experiments on two types of materials will be presented, namely (1) a powder of solution-grown polyethylene, Sholex 6050, and (2) a urea-polyethylene inclusion complex, in which polyethylene is of the same origin. The inclusion complexes of urea and low-molecular-weight hydrocarbon molecules have been studied extensively for more than thirty years, but the urea-polyethylene complex (UPEC) was successfully prepared rather recently. We prepared the complex according to the method described by Monobe et al. [23].

The materials were γ-irradiated up to ca. 1 Mrad at 77 K at about 10^{-2} Pa, and the ESR spectra of the alkyl radicals trapped in the polyethylene chain of the respective materials were recorded. The temperature dependence of hyperfine splitting due to β protons will mainly be discussed based on McConnell's relation (Eq. (2.2)). After

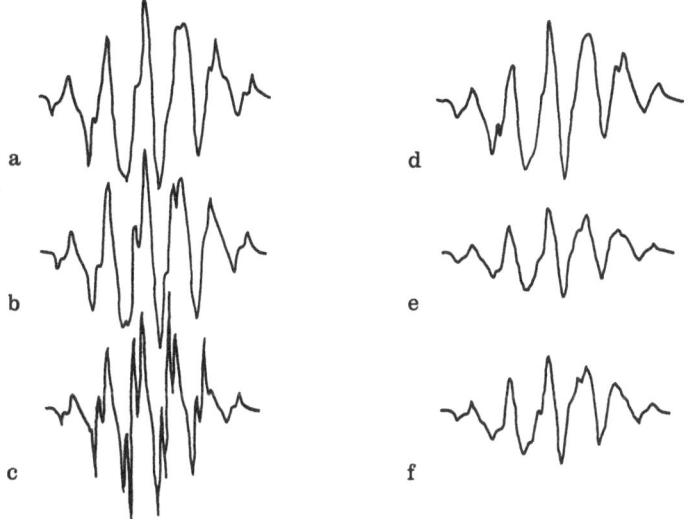

Fig. 5.1a–f. ESR spectra of alkyl radicals in polyethylene in UPEC (**a, b, c**) and solution grown polyethylene (**d, e, f**): Observation temperatures; **a**) 108 K, **b**) 243 K, **c**) 321 K, **d**) 114 K, **e**) 242 K, and **f**) 319 K (Ref. [22])

proper heat treatment, by which only the alkyl radicals were trapped in the materials, the ESR spectrum was observed, first at 77 K, then the temperature was gradually raised. Computer simulations were made in order to obtain the best-fitted patterns and the values of line width and hyperfine splittings of the protons at various temperatures were determined. Several examples of ESR spectra of alkyl radicals in UPEC and solution-grown polyethylene are shown in Fig. 5.1. It can be seen that the spectra of UPEC are very sharp compared to those of solution-grown polyethylene. This is a reflection of the greater mobility of polyethylene in UPEC compared to that in bulk systems. The shape of the wing peaks shows the effect of anisotropy of hyperfine coupling due to α protons, the pattern being typical for the amorphous pattern in both materials. The hyperfine coupling constant due to β-protons obtained by the simulation method indicates that θ in Eq. (2.2) deviates from $30°$ even at a rigid state and, hence it was concluded that the angles corresponding to θ in Eq. (2.2) are $25.4°$ and $34.6°$, respectively (see Fig. 5.2). Furthermore, the values of splitting due to β protons at various temperatures were obtained as shown in Fig. 5.3. The following interpretation of the temperature dependence of the values in Fig. 5.3. is given: Since the chemical structure of the alkyl radicals is:

$$
\begin{array}{cccccc}
\text{H} & \text{H} & & \text{H} & \text{H} \\
{\sim}C_\gamma{-}C_\beta{-}\dot{C}_\alpha{-}C_\beta{-}C_\gamma{\sim} & , \\
\text{H} & \text{H} & \text{H} & \text{H} & \text{H}
\end{array}
$$

four β protons have to be considered whose steric configurations are shown in Fig. 5.2. in the rigid state. Considering the hyperfine splittings due to the β-protons $\Delta H_{\beta1}$, $\Delta H_{\beta2}$, $\Delta H_{\beta1'}$, and $\Delta H_{\beta2'}$, a symmetric structure of the alkyl radicals implies the rela-

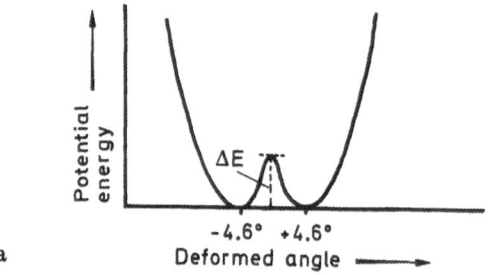

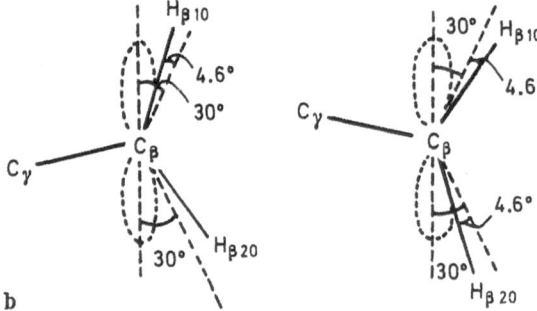

Fig. 5.2a. Schematic illustration of the potential curve of the alkyl free radical for hindered oscillation around the C_α-C_β bond. **b** Two stable conformations which are the mirror images of each other (Ref. [22])

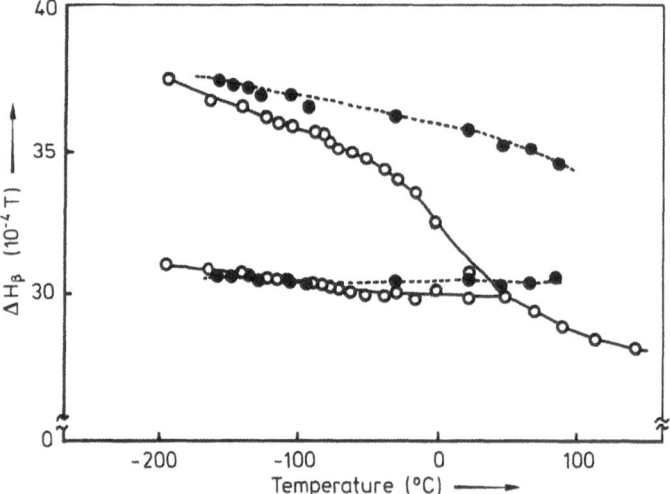

Fig. 5.3. Variation of the value of ΔH_β with the observation temperature: ○, UPEC; ●, solution grown polyethylene (Ref. [22])

tions $\Delta H_{\beta1} = \Delta H_{\beta1'}$ and $\Delta H_{\beta2} = \Delta H_{\beta2'}$. It can be assumed that the methylene group is in a hindered oscillation around the stable configuration shown in Fig. 5.2. at higher temperatures than 77 K. This leads to the vibration of the θ values around ($30°$ $\pm\ 4.6°$), the values of the rigid state. Since a double minimum potential is assumed as shown in Fig. 5.2., the two θ values corresponding to two methylene protons are different at low temperatures, $RT < \Delta E$, becoming identical at high temperatures, $RT > \Delta E$. This is a reason why the hyperfine splittings due to β-protons change as shown in Fig. 5.3. If a harmonic oscillation is assumed, the amplitude of the oscillation of θ below the "barrier" temperature ($RT < \Delta E$), $\Delta\theta$, will be less than 4.6° and that above the "barrier" temperature ($RT > \Delta E$), $\Delta\theta'$, larger than 9.2°. Therefore, the hyperfine splitting above the barrier temperature, $\langle \Delta H_a \rangle$, can be estimated as follows:

$$\langle \Delta H_\alpha \rangle = B[\cos^2 30° \langle \cos^2 \Delta\theta' \rangle + \sin^2 30°(1 - \langle \cos^2 \Delta\theta' \rangle)] \qquad (5.1)$$

This equation indicates that $\langle \Delta H_a \rangle$ is less than $\beta \cos^2 30°$ if $\Delta\theta$ is not zero and decreasing with increasing $\Delta\theta'$. Since $\Delta\theta'$ must increase with raising temperature, the experimental fact that hyperfine splitting due to β-protons is decreasing beyond a certain temperature, as shown in Fig. 5.3., is reasonable.

The sharper spectrum of the radicals in UPEC shown in Fig. 5.1 c. reflects a much more mobile character of the radical site in UPEC compared to the radical site in the bulk polyethylene. A similar phenomenon is also shown in Fig. 5.3.; i.e., in the case of UPEC, two kinds of hyperfine splitting due to β-protons at low temperature become merged into one kind at a certain temperature (barrier temperature). The barrier temperature for bulk materials, on the other hand, is much higher than that of UPEC.

In this section, it has been shown that the temperature dependence of the hyperfine splitting can be valuable information in studying molecular motion.

5.2 Separate Observation of Motions of Methyl and Methylene Groups in Scission Radicals of Polyethyleneoxide [24]

The urea-polyethyleneoxide complex (UPEOC) was prepared in a similar way as in the case of UPEC. The DSC curve of the UPEOC was obtained in order to confirm the successful preparation of UPEOC; the temperature of decomposition of UPEOC has been determined as 146 °C.

UPEOC was γ-irradiated up to 2.8 Mrad and ESR spectra of the produced radicals were observed under various conditions. Though a detailed discussion of the observed spectra must be omitted here. a stable spectrum observed after the heat treatment at room temperature was identified as coressponding to the free radicals $\sim CH_2-\dot{C}H-O-CH_3$. The stable spectrum observed at 293 K is shown in Fig. 5.4. in which a stick diagram explains the afore-mentioned identification. This stable radical is a secondarily produced radical originating from a scission-type radical[24], $\sim CH_2-CH_2-O-\dot{C}H_2$.

Based on the values of hyperfine coupling constants due to β-protons, the steric configuration of the atoms in the stable radical must be as illustrated in Fig. 5.5.: It is well known that the conformation of the polyethyleneoxide chain in its normal crystalline phase is of trans-trans-gauche type; i.e., a 7_2-helical structure as illustrated by the left structure in Fig. 5.5. Since this conformation is in good agreement with that of the stable free radical represented by the right structure in Fig. 5.5., it can be assumed that the 7_2-helical structure of the polyethyleneoxide chain still exists in irradiated UPEOC. The values of the line width ΔH_{ms1}, and hyperfine coupling constants due

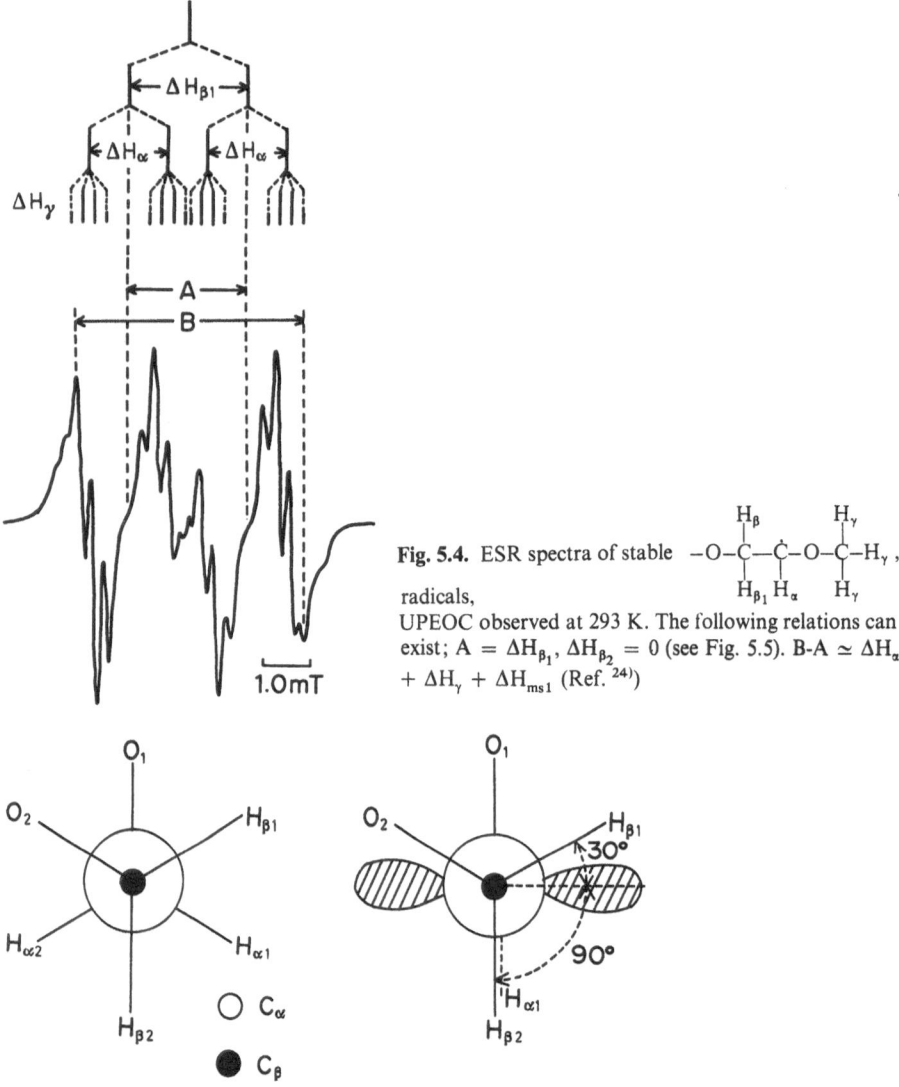

Fig. 5.4. ESR spectra of stable $-O-\overset{H_\beta}{\underset{H_{\beta_1}}{C}}-\overset{H_\gamma}{\underset{H_\alpha}{\dot{C}}}-O-\overset{H_\gamma}{\underset{H_\gamma}{C}}-H_\gamma$, radicals, UPEOC observed at 293 K. The following relations can exist; $A = \Delta H_{\beta_1}$, $\Delta H_{\beta_2} = 0$ (see Fig. 5.5). $B-A \simeq \Delta H_\alpha + \Delta H_\gamma + \Delta H_{ms1}$ (Ref. [24])

Fig. 5.5. Schematic illustrations of steric configurations of the atoms in usual PEO (left) and at the site of stable radical (right) (Ref. [24])

to β-protons, ΔH_β, and γ-protons, ΔH_γ, were determined for the spectra observed at various temperatures by means of computer simulation, as shown in Fig. 5.6. The value of the hyperfine coupling constant due to the α proton, ΔH_α, was also determined, but it did not show any detectable change throughout the temperature range used in this study. Figure 5.6. shows that ΔH_β does not change up to 150 K where a rather abrupt decrease of ΔH_β appears. On the other hand, the value of ΔH_γ begins to decrease at a very low temperature compared with ΔH_β. This indicates that γ-protons are quite mobile even at the temperature of liquid nitrogen. This problem will be discussed later.

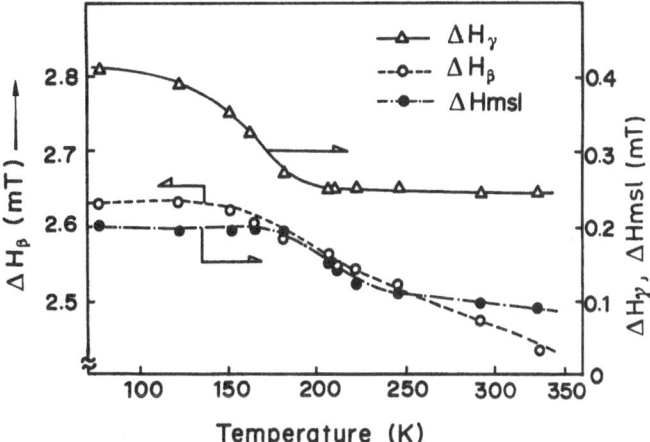

Fig. 5.6. Hyperfine splitting width due to β- and γ-protons and line widths at various temperatures (Ref. [24])

Recently, the effect of hindered rotation of the molecular site relative to the localization of the unpaired electron was discussed by Bullock and Howard in the case of the cation of 1,2,4,5-tetrahydroxybenzene and related compounds [25]. They discussed the temperature dependence of hyperfine splitting and proposed that the value of the hyperfine coupling constant due to the proton of the hydroxyl group depends on the spin density of the carbon atom bonded to the oxygen and the spin polarization. The latter is correlated either with a twisted angle of the plane of the benzene ring or the value of the overlap integral of p-orbitals on oxygen and carbon atoms. According to the above interpretation proposed by Bullock and Howard, the variation in ΔH_γ in the present case was considered: Rotation around the axis of the C_α—O bond in the stable radical (scission radical), $-C_{(\beta)}H_2-\dot{C}_{(\alpha)}H-O-C_{(\gamma)}H_3$, can easily be considered. This rotation results in the delocalization of the unpaired electron at the oxygen atom and leads to a smaller hyperfine coupling constant due to γ-protons. This interpretation fits well with experimental results shown in Fig. 5.6.

The variation in ΔH_β can be interpreted in terms of hindered oscillation about the axis of the C_α—C_β bond, like the case of alkyl radicals in polyethylene [22]. An average value for ΔH_β can be expressed by Eq. (5.2) connecting McConnell's relation [Eq. (2.1)] and the equilibrium angle illustrated in Fig. 5.5.

$$\langle\Delta H_\beta\rangle = K[1/4 + (1/2)\cos^2\theta] \tag{5.2}$$

K is a constant and $\Delta\theta$ means the displacement due to torsion. This relation is a decreasing function of $\Delta\theta$, which is quite consistent with experimental data as $\Delta\theta$ must increase with raising temperature.

According to the above discussion concerning the stable radicals in UPEOC, it can be said that the temperature dependences of ΔH_β and ΔH_γ can be investigated separately. Therefore, the local mobilities of methylene and methyl groups can be studied separately.

6 Study of the Molecular Motion of Polymeric Chains Labelled with Peroxy Radicals [26, 27]

As mentioned already in Sect. 2, changes of the principal value of the g-tensor also give important information on molecular motion in polymeric chains. The spectrum of peroxy radicals shows a typical g-anisotropy and the study on molecular motion by observing peroxy radicals is, therefore, a kind of spin label study of the materials containing peroxy radicals. In the course of studying the oxidation of irradiated polyethylene, an interesting feature of the ESR spectrum of peroxy radicals was found.

Generally, the oxidation process is one of the important problems in the industrial application of radiation processing of polymers and many investigations have been made [28, 29, 30]. However, only a few ESR applications in this field were reported and precise knowledge on the oxidation process based on the nature of peroxy radicals was needed. In the case of irradiated polyethylene, especially high-density polyethylene, the observation of high-intensity ESR spectra originating from peroxy radicals is not easy. However, when rather high microwave power, more than 1 mW, is used, the spectra of the carbon radicals, allylic or alkyl radicals, mixed with peroxy radicals disappear due to the power saturation, but the spectrum of the peroxy radicals is relatively enhanced [31], since peroxy radicals are less susceptible to power saturation compared to carbon radicals [32]. Though this is also an interesting problem concerning spin-lattice relaxation in polymer radicals, it will be omitted here and the character of the observed spectrum of peroxy radicals will mainly be considered.

Usually, oxidation in the crystalline region is regarded as unimportant according to the diffusion studies of oxygen by pressure measurements [32]. However, using the

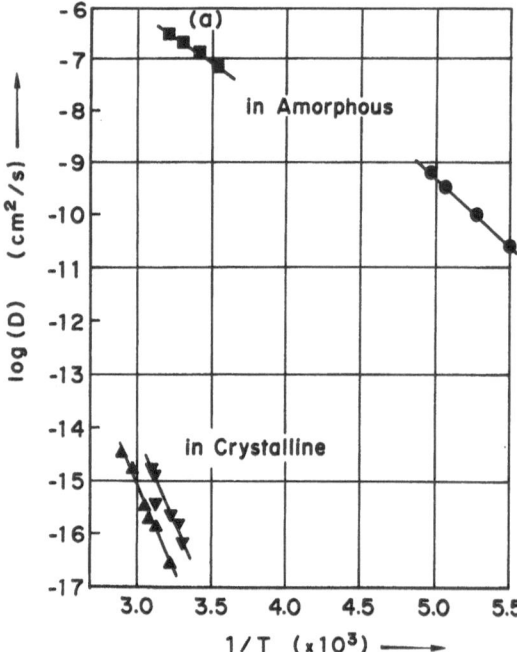

Fig. 6.1. Arrhenius plots of the diffusion constant of oxygen into polyethylene (amorphous and crystalline regions). Several points located at upper left, ■, marked by a) are based on the results in the paper by Michaels et al. [30] (Ref. [35])

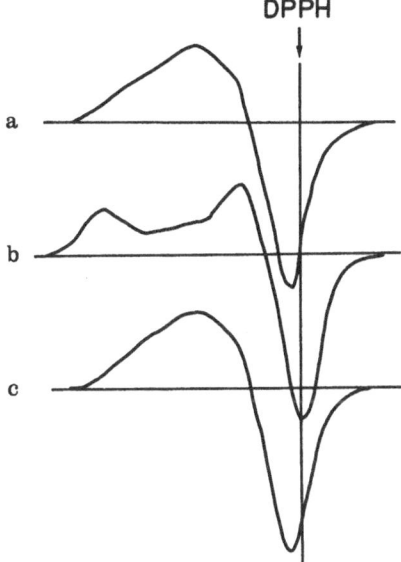

Fig. 6.2a–c. Reversible change of ESR spectra of peroxy radicals in polyethylene at the following temperatures. **a** 241 K; **b** 118 K; **c** 245 K (Ref. [26])

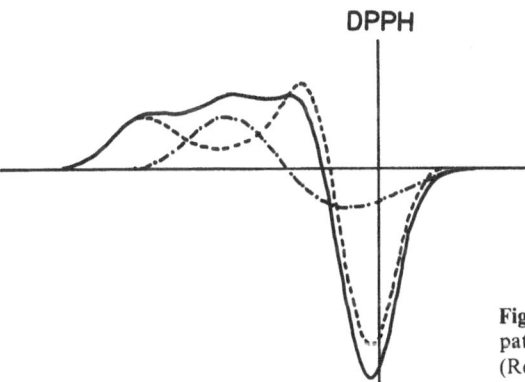

Fig. 6.3. An example of the superposition of pattern A (------------) and pattern B (—·—·—·—) (Ref. [26])

ESR method, it was concluded by Hori et al. [33] that oxidation in the crystalline region is possible. As a result of studying the oxidation process, the diffusion constant of oxygen in polyethylene was obtained at various temperatures both for crystalline and amorphous regions [31, 34, 35]; these are shown in Fig. 6.1. showing that the difference in the values of diffusion constants in amorphous and crystalline regions amounts to about 10 orders of magnitude.

Figure 6.2. shows a few examples of the spectra of peroxy radicals in irradiated polyethylene contacted with oxygen gas after irradiation. This same figure also shows a temperature dependence of the spectrum; i.e., Fig. 6.2a. is the spectrum observed at 241 K, b at 118 K, and c at 245 K; the integral intensities of these three patterns are the same indicating that no radical decay occurs throughout the temperature range studied, but the spectral shape varies reversibly between a high temperature pattern and low temperature pattern. According to the analysis by Hori et al. [26], the ESR spectrum of peroxy radicals in irradiated polyethylene is a superposition of at least two kinds of patterns with different sets of principal values of the g-tensor. An example is shown in Fig. 6.3.

Since the intensities of the spectra shown in Fig. 6.2. are the same, only the superposition of different patterns changed with the temperature. This must be a reflection of the fact that trapping sites of the peroxy radicals are not the same and, therefore, the motional character of each of the sites is different. Of course many radical sites can be considered in polymers. Two different site can be assumed, the one being more mobile than the other. The spectral simulations of Fig. 6.3. show a case observed at 196 K. The principal values of g-tensor of pattern A were assumed to be $g_1 = 2.0042$, $g_2 = 2.0099$, and $g_3 = 2.0331$ and those of pattern B were assumed to be 2.0042, 2.0149, and 2.0236, respectively. The mixing ratio of patterns A and B was assumed to

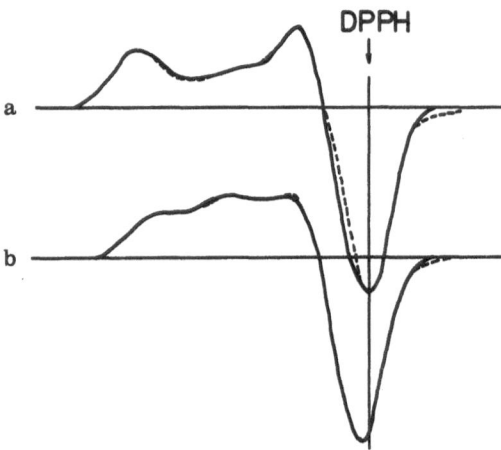

Fig. 6.4a and b. Comparison of the simulated spectra (————) and observed spectra (------------) of peroxy radicals at 118 K (**a**) and 195 K (**b**) (Ref. [26])

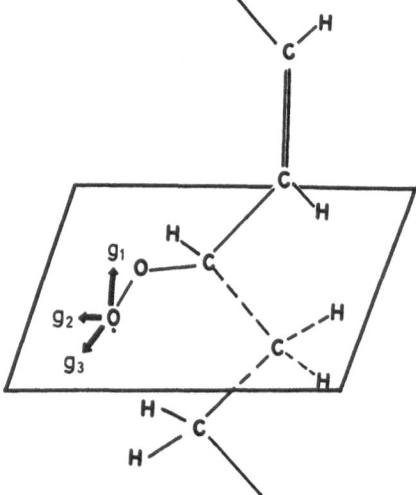

Fig. 6.5. Schematic illustration of the structure of peroxy radicals from allylic radicals (Ref. [26])

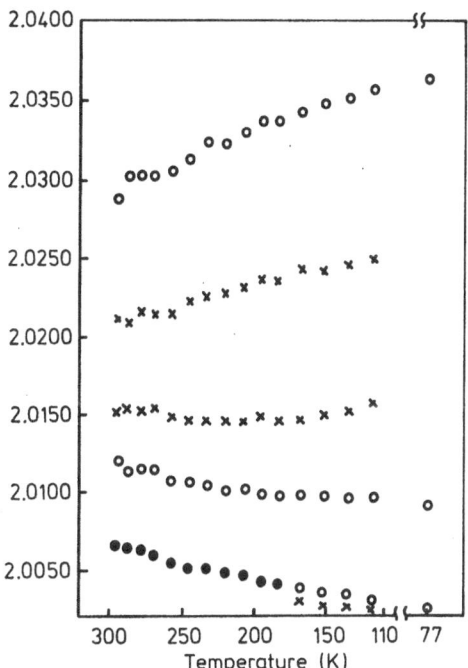

Fig. 6.6. Change of g-values for normal poly-ethylene with temperature: ○, A radical; ×, B-radical; ●, means complete coincidence of ○ and × (Ref. [26])

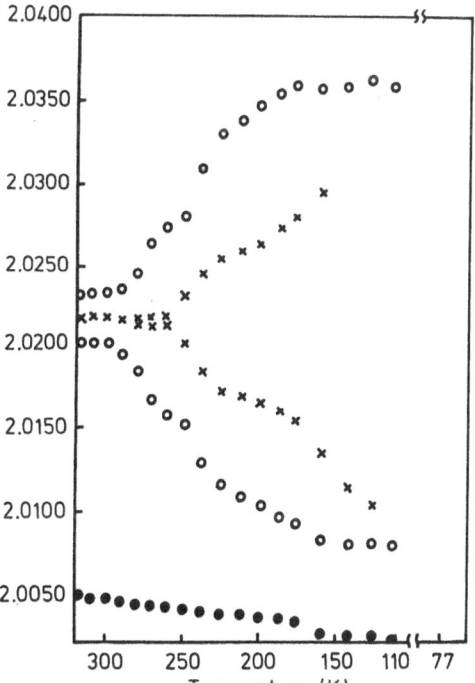

Fig. 6.7. Change of g-values for UPEC with temperature: ○, A radical; ×, B radical; ●, means complete coincidence of ○ and × (Ref. [27])

be 0.4. The observed and simulated spectra for low and high temperatures are nearly identical as shown in Fig. 6.4. As indicated above for the principal values of the g-tensor in the simulation of Fig. 6.3., the difference in values of g_2 and g_3 of pattern B is smaller than that of pattern A and the values of g_1 are almost the same for A and B. this means that a rotational motion around the g_1 axis of the radicals corresponding to the B pattern is much more mobile than the radicals of the A pattern (in the following, the terms A-radical and B-radical will be used). Therefore, conditions of the radical site of peroxy radicals can be illustrated as shown in Fig. 6.5. The A-radical is less mobile than the B-radical, and the less mobile site of the A-radical is considered to be changed to the mobile site with raising temperature. In other words, the ratio of the mixture of pattern A and B and principal values of the g-tensor vary with temperature. Thus, the ratio and principal values of the g-tensor can be the parameters in computer simulations of the spectra observed at various temperatures; the principal values varied with temperature as shown in Figs. 6.6. and 6.7. The variations of principal values of the g-tensor of the spectra of peroxy radicals shown in Figs. 6.6. and 6.7. were investigated with two kinds of materials, powdered materials of washed Sholex 6050[26] and UPEC [27] (urea-polyethylene inclusion complex). In Figs. 6.6. and 6.7. little variation is seen for g_1 throughout the temperature range, and g_2 and g_3 are averaged out with raising temperature. Averaging of g_2 and g_3 in the case of UPEC (Fig. 6.7.) is much more remarkable than that for bulk materials (Fig. 6.6.). Since the directions of g_1, g_2, and g_3 are defined as shown in Fig. 6.5., it can be said that vibration or rotation around the c-axis of the polymeric chain in UPEC is much greater than in bulk polyethylene. The motional character of the polyethylene chain in UPEC will be discussed again in a later section. The change of the spetrum with temperature is drastic in the case of UPEC; some spectra are shown in Fig. 6.8.

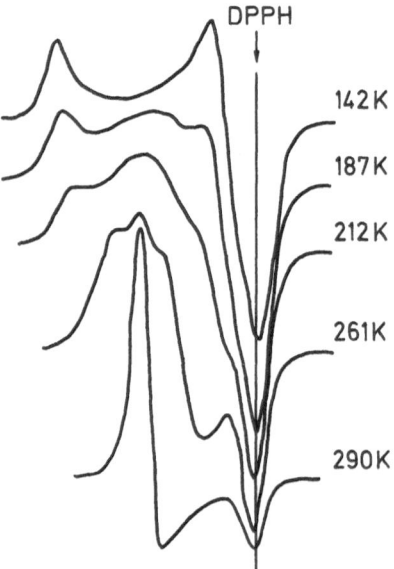

Fig. 6.8. Change of ESR spectra of peroxy radicals in UPEC with temperature (Ref. [27])

The above discusses the idea of two kinds of radical sites, a mobile and a less mobile one. A similar interpretation can also be made in the case of peroxy radicals in polypropylene. Isotactic polypropylene, Noblen MA-4 (Mitsubishi Petrochemical Co. Ltd.), was studied. As in the case of polyethylene, washed polypropylene materials were γ-irradiated and then contacted with oxygen molecules in order to trap the peroxy radicals. In a temperature range where no radical decay occurs, a reversible change of the ESR spectrum with temperature was also observed. The spectrum was analyzed at various temperatures with the assumption of two kinds of patterns as in the case of peroxy radicals in polyethylene. And the variations of principal values of the g-tensor with temperature were obtained by spectrum simulations. Also in this case, rotation around the chain-axis is plausible, but a characteristic for the peroxy radicals in polypropylene is that the different variation of g-values of the patterns from mobile and less mobile radicals is very distinct compared with polyethylene. The pattern of the less mobile radicals is seen to be very rigid throughout the temperatures of the study and the g_1-value of the pattern of the mobile radical varies more distinctly than for polyethylene. This is due to the fact that isotactic polypropylene chains have a 3_1-helical conformation and the g_1-axis is not parallel to the chain axis. Therefore, when rotation around the chain axis occurs, g_1-values also must be changed [36]. According to the studies by Shimada et al. [37], the angle between the direction of the g_1-axis and that of the chain axis was found to be about 40°. A similar molecular motion of the peroxy radicals in polytetrafluoroethylene was also discussed with the model of chain axis rotation [38, 39, 40]. Concerning the temperature dependence of the ESR spectrum of peroxy radicals, a different interpretation was proposed by the group of Kevan for polyethylene [41] and polypropylene [42], while they discussed analogous problems in polytetrafluoroethylene along the same model as in the present article [40] (see also Sect. 7).

Molecular motion of polymeric chains was discussed on the basis of temperature dependence data of ESR spectra of peroxy radicals. However, peroxy radicals also can be used as a probe to study orientation problems. Since the three principal axes of the g-tensor of peroxy radicals are orthogonal to each other and a change of direction can be sensitively determined by the ESR spectrum, proper analysis of the data can provide helpful information on chain orientation [37].

7 Motional Character of the Polyethylene Chain in the Urea-Polyethylene Inclusion Complex

In Sects. 5 and 6, a few investigations of urea-polyethylene complexes (UPEC) were discussed. The UPEC is an interesting material because a single polyethylene chain is located in an hexagonal canal of urea molecules and it must be expected that the polyethylene chain can behave differently from the bulk systems like solution-grown crystals or materials recrystallized from the melt. The inclusion complex system composed of short hydrocarbon molecules and urea molecules was studied more than 30 years ago. The crystalline structures of urea-hydrocarbon complexes are known [43]. The urea-polyethylene complex system was prepared rather recently by Monobe et al. [23], replacing the hydrocarbon molecules in the urea-hydrocarbon complex by

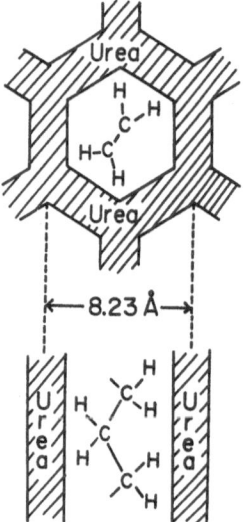

Fig. 7.1. Schematic illustrations of UPEC (Ref. [27])

polyethylene. Considering the similarity of linear saturated hydrocarbon molecules and polyethylene, the crystalline structure of UPEC, at least its section view perpendicular to the chain axis, is considered to be the same as that of the urea-hydrocarbon complex studied by Smith [43]. Therefore, a schematic illustration of UPEC can be given by Fig. 7.1. According to Ref. [43], the distance between the centers of pores made by urea molecules is 8.23 Å, and a similar value would also be expected for UPEC.

In the previous sections, a much more mobile character of the polyethylene chain in UPEC than in bulk systems was shown. In this section, this problem will be summarized more systematically.

7.1 Results of Broad-Line NMR Studies [44]

One method of investigating molecular motion in polymer physics is the observation of the temperature dependence of the line width of broad-line NMR spectra. However, since UPEC is composed of polyethylene and urea molecules, the protons in urea molecules must be replaced by deuterons in order to observe the behavior of the polyethylene chain by proton magnetic resonance. For this purpose, deuterated urea molecules were used in the preparation of UPEC (d-UPEC). In the preparation of d-UPEC, deuterated methanol has been used as a solvent in order to prevent proton exchange. In order to compare the new data with the data of bulk polymers, solution-grown polyethylene and extended-chain crystals of polyethylene were also used in the NMR study.

The spectra at various temperatures obtained by a broad-line NMR spectrometer, JEOL-JES-BE-1, operating at 40 MHz were processed by computer in order to calculate second moments. Variations of the second moment with observation temperature are shown in Fig. 7.2. for three kinds of materials. All parts of bulk polyethylene

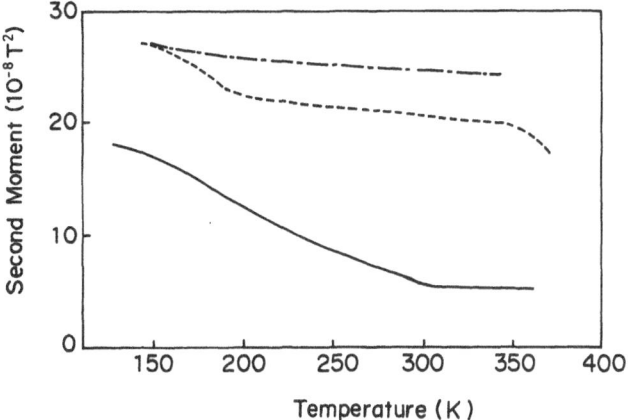

Fig. 7.2. Second moments of broadline NMR spectra of polyethylene in d-UPEC (————), solution grown polyethylene (------------) and extended chain crystals of polyethylene (—·—·—) at various temperatures (Ref. [44])

Table 7.1. Calculated (intramolecular) and observed second moments of polyethylene (10^{-8} T)

Calculated for		Observed at	
rigid lattice	19.17		
rotational oscillation		125 K	17.7
amplitude[a] 30	17.72		
60	14.17		
		216 K	11.3
90	10.32		
free rotation	5.94	295 K	5.5

[a] The amplitude means 2α in the equation of angular oscillation,
$\varphi = \varphi_1 + \alpha \sin(\omega t)$ in Ref. [47]

are rigid below 180 K, while the amorphous part is mobile and the crystalline part is immobile between 200 K and 300 K. The crystalline part becomes mobile above 330 K. For extended chain crystals of polyethylene, the situation seems to be similar to that for bulk polyethylene, though the mobile fraction is smaller than that of bulk polyethylene. On the other hand, the second moment of d-UPEC is completely different from normal polyethylene. The second moment of d-UPEC is smaller than that of solution-grown materials or extended chain crystals throughout the temperature range of observation. This means that the polyethylene chain in d-UPEC is much more mobile than in the other two materials which are rigid below 150 K. However, the intramolecular contribution to the second moments of the polyethylene chain was calculated by use of Van Vleck's equation [45] under the conditions given by Gutowsky and Pake [46] or Andrew [47], and compared with the values estimated from the observed spectra (Table 7.1.). The observed second moment at 125 K, 17.7×10^{-8} T^2, is slightly smaller than the calculated value for the rigid state, 19.17×10^{-8} T^2, but very close to the calculated value for vibration with an amplitude of 30°. Furthermore, the observed second moment reaches 5.5×10^{-8} T^2 at 295 K and then remains constant at higher temperatures than 295 K. This constant value is very close to the value calculated with the assumption of free rotation, 5.94×10^{-8} T^2. Therefore, the second moment at 125 K can reasonably be interpreted as reflecting the vibration around the chain axis with an amplitude of 30°; i.e., these two features, vibration at low temperature and free rotation at high temperature, are quite consistent.

7.2 Hyperfine Splitting due to β-Protons in Alkyl Radicals

The temperature dependence of the hyperfine splitting width due to β-protons of alkyl radicals shows much of a mobile character in the case of the polyethylene chain in UPEC than in the case of bulk systems, as shown in Sect. 5 and vibration appears to occur around the chain axis since data analysis suggested vibrational motion of methylene groups.

7.3 Mobility of Peroxy Radicals

Based on the temperature dependence of the principal values of the g-tensor obtained from the ESR spectrum of the peroxy radicals trapped in polyethylene, the rotational motion around the chain axis of polyethylene in UPEC is much more mobile than that in bulk systems, as shown in Sect. 6.

7.4 Detailed Discussion of Chain Axis Rotation of Guest Molecules in Their Complex with Urea Molecules

From Sect. 7.1, 7.2, and 7.3 it follows that there is a more mobile rotating motion around the chain axis of polyethylene in UPEC compared to the bulk systems, and a remarkable difference is shown in the study of temperature dependence of g-anisotropy of peroxy radicals. On the other hand, however, Schlick and Kevan [41] proposed a different interpretation concerning the temperature dependence of ESR spectra of peroxy radicals. They assumed a jumping rotation of the O—O group around the axis of the C—O bond and made spectral simulations under various conditions of rotation angle and jumping time. Based on these simulations, they concluded that the variation of the spectrum with temperature was caused by the 180° jumping rotation of the O—O group around the axis of the C—O bond with the angle of C—O—O being 135° and not by the rotation around the chain axis, as presented by Hori et al. [26, 27]. In order to provide a clear evidence supporting the rotation around the chain axis, an experiment using a model compound was made by Hori et al. [48], preparing a single crystal of the urea-n-tetracosane complex. n-Tetracosane can be considered to be a model compound of linear polyethylene, though there are 24 carbon atoms in one n-tetracosane. These materials will be called UP$_{24}$C. The formation of an inclusion complex of n-tetracosane with urea molecules was confirmed by DSC measurement as in the case of UPEC. A single crystal of UP$_{24}$C was prepared by very slow cooling of the mixture of a solution of urea in isopropyl alcohol and a solution of n-tetracosane

Table 7.2. Principal values of g-tensor for peroxy radicals in $UP_{24}C$

Powder pattern (77 K)	Single crystals (room temperature)
$g_1 = 2.0025$	$g_{//} = 2.0044$ ($g_{//}$ // c-axis)
$g_2 = 2.0082$	$g_\perp = 2.0210$ ($g_\perp \perp$ c-axis)
$g_3 = 2.0355$	
$(g_2 + g_3)/2 = 2.0218$	

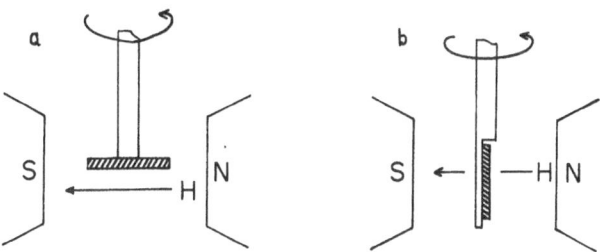

Fig. 7.3a and b. Placings of $UP_{24}C$ sample in static magnetic field (Ref.[48])

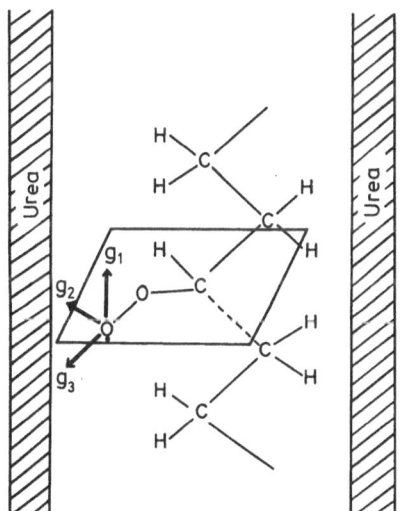

Fig. 7.4. Schematic illustration of peroxy radicals in hydrocarbon molecule in its inclusion complex with urea (Ref. [48])

in *isooctane*. Powdered samples of $UP_{24}C$ were also prepared. Peroxy radicals were trapped in *n*-tetracosane of $UP_{24}C$ by γ-irradiation in air and the ESR spectra of peroxy radicals in *n*-tetracosane were observed by an X-band spectrometer with 100 KHz field modulation. From the ESR spectra of the powdered materials, principal values of the g-tensor at 77 K were determined as shown in Table 7.2. ESR spectra of the single crystal materials were observed at room temperature with two kinds of placings of the materials in the static magnetic field, as shown in Fig. 7.3.; i.e., in one case the chain axis was parallel to the direction of the magnetic field (H // C-axis) and in the other the chain axis was perpendicular to the direction of the magnetic field (H ⊥ C-axis). In both cases, the sample was rotated around the axis perpendicular to the direction of the magnetic field (see Fig. 7.3.) in order to observe the spectra at various angles of rotation. If the definition of principal axes of the g-tensor is also correct in the present case, the model of peroxy radicals in $UP_{24}C$ must be as shown in Fig. 7.4. Since all molecular chains in the single crystal materials are aligned in the same direction, singlet-like patterns must be observed in both cases at room temperature, in which case g_2 and g_3 axes are expected to average out. But as observed with the sample placing shown in Fig. 7.3a., the position of the pattern must change with the angle of rotation of the sample. On the other hand, the position of the pattern does not change with the rotation of the sample for the sample shown in Fig. 7.3b. The unchanging position in the latter must be the same as the position corresponding to the averaging out of the g_2 and g_3 values. The observed results are shown in Fig. 7.5., which indicates the conditions as expected. In Fig. 7.5., the positions of the patterns are expressed in terms of g-values, and the theoretical curve indicating the positions with various angles of rotation of the sample show a good agreement with observed results. According to the results of Fig. 7.5., it can be said that the variation of the observed spectrum with the temperature of ESR observation is reflecting the rotational motion of the chain, which traps the peroxy radicals, around the chain axis.

On the other hand, the order-disorder transition in the urea-linear-hydrocarbon [49, 50, 51, 52] complexes has been observed in thermal [50], NMR [49, 51, 52], and X-ray diffraction [49] measurements, and free rotation of the guest molecules above the transi-

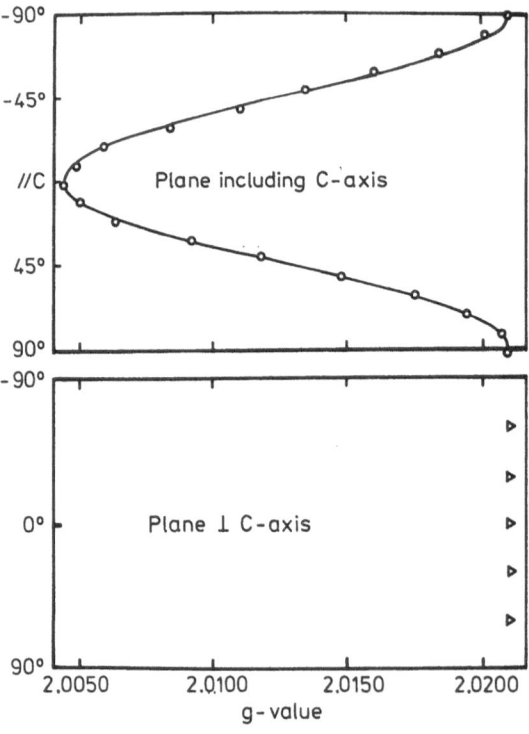

Fig. 7.5. The angular dependence of g-values of the peroxy radical in $UP_{24}C$ at room temperature. (upper) Rotating in a–c plane; the solid line is the theoretical curve. (lower) Rotating in a–b plane (Ref. [48])

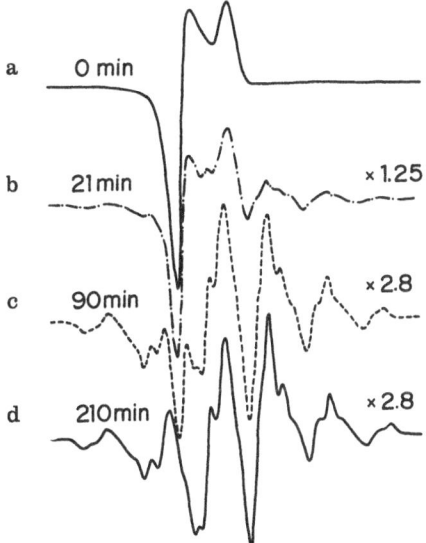

a   0 min

b   21 min   × 1.25

c   90 min   × 2.8

d   210 min   × 2.8

Fig. 7.6a–c. Conversion of ESR spectrum of peroxy radicals into that of alkyl radicals in UPEC by heat treatment *in vacuo* at 361 K (Ref. [27])

tion temperatures were concluded [49, 51, 52]. Since the transition temperature of UP$_{24}$C was found to be 176 K, it can safely be said that the *n*-tetracosane molecule as a guest is freely rotating at room temperature in the canal of urea molecules. Therefore, though the molecules containing the peroxy radicals are different from pure *n*-tetracosane, the interpretation of the rotation of the peroxy radicals around the chain axis discussed in this section can be accepted, since it was reported that lauric acid in urea-lauric acid complexes is freely rotating at room temperature [49].

According to the facts mentioned above concerning the molecular motion of the compounds in their complex with urea molecules, the interpretations in Sect. 6 and in this section should be reasonable. Recently, Suryanarayana et al. [53] reported a similar interpretation of the temperature dependence of ESR spectra of peroxy radicals in urea-*n*-alkane complexes. Chain axis rotation was also reported in the study on the peroxy radicals in polytetrafluoroethylene by the same group [40]. Of course, a rotation of the O—O group around a C—O bond appears possible at very high temperatures. Actually, it was found at 361 K that the peroxy radicals trapped in UPEC abstracted the hydrogen atom at a neighboring site and converted to alkyl radicals without any change of radical concentration [54]. This change of the spectrum is shown in Fig. 7.6. and it indicates that the intensity of the spectrum of peroxy radicals decays and that of alkyl radicals increase, but total intensity does not change remarkably. This is a reflection of the rotation of the O—O bond around the C—O bond, and this kind of motion may occur a little even at lower temperature than 300 K, since a slight change in the g_1-value can be seen in Fig. 6.7.

The nature of the molecular motion has been discussed in the first part of this section, but the value of the activation energy must also be obtained in order to describe the molecular motion quantitatively. For this purpose, the data of the simultaneous observations of NMR and ESR spectra considering the distribution of the relaxation time has been successfully analyzed, as described in the following section.

8 Simultaneous Observations of ESR and NMR Spectra
— Consideration of Distribution of Correlation Time —

8.1 Motional Narrowing of Line Width and Distribution of Correlation Time

Correlation times are usually calculated based on data of broad-line NMR spectra in NMR application to polymer physics, and the temperature dependence of the correlation time is usually the basis for estimating the activation energy associated with the molecular motion contributing to the narrowing of the line width of the NMR spectrum. The most conventional method for estimating the correlation time from the line width is to apply the so-called BPP (Bloembergen, Purcell, and Pound) equation [55]. However, in the case of motional narrowing in polymers, the activation energy obtained from the temperature dependence of correlation time calculated by use of the BPP equation is very small compared with the energy obtained, for example, in dynamic mechanical studies. This phenomenon was pointed out by Miyaka [56] more than twenty years ago, and he suggested that the distribution of the correlation time in solid polymers must be taken into consideration, since the BPP equation is valid for the case of random Brownian motion with a single correlation time. According to his paper, correlation time and line width which is proportional to the inverse of the spin-spin relaxation time, T_2, must be related by Eq. (8.2) instead of the usual BPP equation, Eq. (8.1). Equation (8.1) is not the original BPP equation, but the equation modified by Gutowsky and Meyer [57] in order to apply the BPP relation to the case of two-step narrowing which is common in rubbery or polymer materials.

$$\langle \Delta\omega^2 \rangle - \langle \Delta\omega^2 \rangle_F = \frac{2}{\pi\alpha} (\langle \Delta\omega^2 \rangle - \langle \Delta\omega^2 \rangle_F) \tan^{-1} (\tau/_{T_2}) \qquad (8.1)$$

$$\langle \Delta\omega^2 \rangle - \langle \Delta\omega^2 \rangle_F = \frac{2}{\pi\alpha} (\langle \Delta\omega^2 \rangle_R - \langle \Delta\omega^2 \rangle_F) \int_{-\infty}^{\infty} \tan^{-1}(\tau/T_2) \, I(\tau) \, d \log \tau \qquad (8.2)$$

where $\langle \Delta\omega^2 \rangle_R$ and $\langle \Delta\omega^2 \rangle_F$ are the second moments at rigid and free states, respectively; $\langle \Delta\omega^2 \rangle$ without suffix is a second moment observed at a certain temperature and T_2 is the spin-spin relaxation time which is inversely proportional to the line width observed. The line shape parameter α can be assumed to be unity. Equation (8.2) contains a distribution of the correlation time, $I(\tau)$, and this function can be expressed by Eq. (8.3) assuming a symmetric distribution:

$$I(\sqrt{\langle \Delta\omega^2 \rangle}^{-1} \cdot a(T)) \simeq 2[\langle \Delta\omega^2 \rangle/(\langle \Delta\omega^2 \rangle_R - \langle \Delta\omega^2 \rangle_F)]$$
$$\times \left[1 + \frac{d \log a(T)}{d \log \sqrt{\langle \Delta\omega^2 \rangle}} \right]^{-1} \qquad (8.3)$$

where a(T) means a shift factor familiar in the theory of relaxation phenomena in polymeric materials. In other words, Eq. (8.3) also implies the temperature-time superposition principle:

$$\tau(T) = a(T) \, \tau(T_r) \tag{8.4}$$

where $\tau(T_r)$ is the correlation time at a reference temperature T_r, which is also a familiar quantity in relaxation problems of polymers. Since the Arrhenius relation can be generally adopted for the temperature dependence of the correlation time, the following equations can be accepted in the present case:

$$\tau(T) = A \exp (E/RT) \tag{8.5}$$

$$a(T) = \exp \{(E/R) \, (1/T - 1/T_r)\} \tag{8.6}$$

Thus, since the relaxation spectrum $I(\tau)$ contains a(T) as shown in Eq. (8.3) and a(T) can be determined with E as a parameter, $I(\tau)$ can also be determined experimentally by use of Eq. (8.3) and parameter E from the data of both ESR and NMR observations, data of line widths, or second moments. Therefore, if both values of $I(\tau)$ obtained from ESR and NMR data are identical with a certain value of E, this value E must be the true activation energy of the molecular motion, provided that ESR and NMR observe the same molecular motions. However, further evidence concerning this problem is needed. Due to the nature of $I(\tau)$, the symmetric function, and characteristic of the arctangent function appearing in Eq. (8.1) and Eq. (8.2), the following relation can be obtained:

$$\langle \Delta \omega^2 \rangle_m - \langle \Delta \omega^2 \rangle_F \simeq (\langle \Delta \omega^2 \rangle_R - \langle \Delta \omega^2 \rangle_F) \int_{\tau_m}^{\infty} I(\tau) \, d \log \tau$$

$$= \frac{1}{2} (\langle \Delta \omega^2 \rangle_R - \langle \Delta \omega^2 \rangle_F) \tag{8.7}$$

where τ_m is the correlation time at maximum $I(\tau)$, and $\langle \Delta \omega^2 \rangle_m$ is the second moment from which τ_m is estimated by simple application of the BPP equation. In other words, τ_m can be a representative correlation time in the temperature range in which the motional narrowing occurs. τ_m can be estimated from ESR data and from NMR. Since the time constants of the observations of ESR and NMR are different, estimations of two τ_m values obtained from ESR and NMR measurements must correspond to the estimations of relaxation times from dynamic mechanical measurements with different time constants of observations (different frequencies). τ_m values obtained this way correspond to different temperatures; i.e., the temperature, to which $\langle \Delta \omega^2 \rangle_m$ in ESR measurements corresponds, is different from the temperature that corresponds to $\langle \Delta \omega^2 \rangle_m$ in NMR. In this way, we can obtain the temperature dependence of representative correlation time just like in the case of temperature dependence of relaxation time in dynamic mechanical studies. Of course, the above interpretation must be valid when ESR and NMR both detect motional narrowing attributed to the same molecular motion. If this is the case, the temperature dependence of τ_m can yield an activation energy which must be the same as the activation energy obtained as a parameter which causes the $I(\tau)$ values obtained from ESR and NMR data to become the same. An example based on the present discussion will be shown below.

8.2 Example: UPEC [58]

The motional narrowing of the line width obtained from the spectrum of broad-line proton NMR of d-UPEC and the motional narrowing of the line width from the ESR spectrum of alkyl radicals trapped in polyethylene chains in UPEC were analyzed based on the method described in the preceding section.

The NMR data from Sect. 7.1 were used and the alkyl radicals were trapped by γ-irradiation of UPEC *in vacuo* $(1.3 \times 10^{-2}$ Pa$)$ up to 3 Mrad. Irradiated UPEC was heat treated so that only the alkyl radicals were trapped. Line widths of ESR spectra observed at various temperatures were determined by comparing the observed spectra and the best-fitted simulated spectra at various temperatures. The alkyl radicals show a sextet ESR spectrum as repeatedly shown (e.g., Fig. 5.1.). Since ESR spectra of alkyl radicals in powdered materials show a so-called amorphous pattern caused by the anisotropy of the hyperfine splitting due to α-protons (refer to Sect. 2), computer simulation is indispensable in order to determine the true line widths. Variations of line widths of NMR and ESR spectra with temperature are shown in Fig. 8.1., in which the temperature dependence of hyperfine splitting of the ESR spectrum is also shown by open rectangulars to be described at a later point.

$I(\tau)$ can be obtained by use of Eq. (8.3) both from NMR and ESR data. Figure 8.2. shows $I(\tau)$ at the reference temperature 183.5 K with various values of E as a parameter

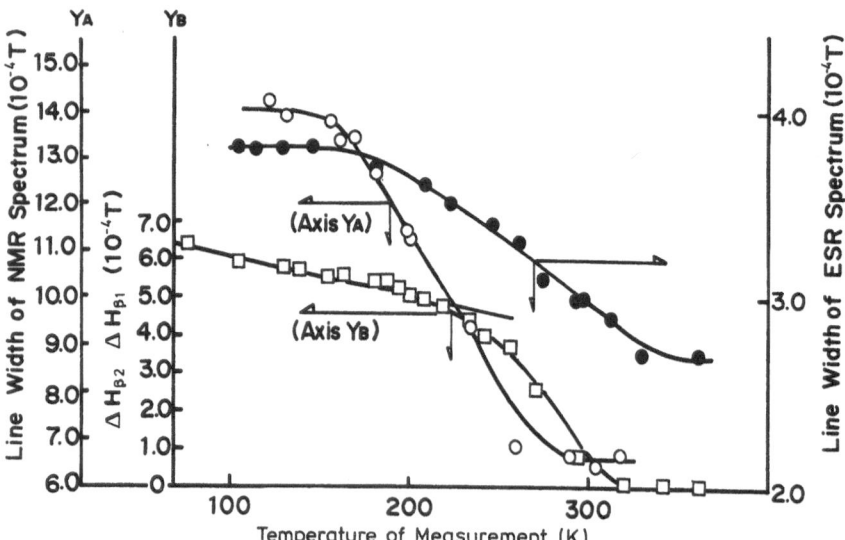

Fig. 8.1. Variation of line width and difference in hyperfine splittings due to two methylene protons of alkyl radicals with observation temperature: ○, broad line NMR line width; ●, ESR line width; □, difference in hyperfine splittings (Ref. [58])

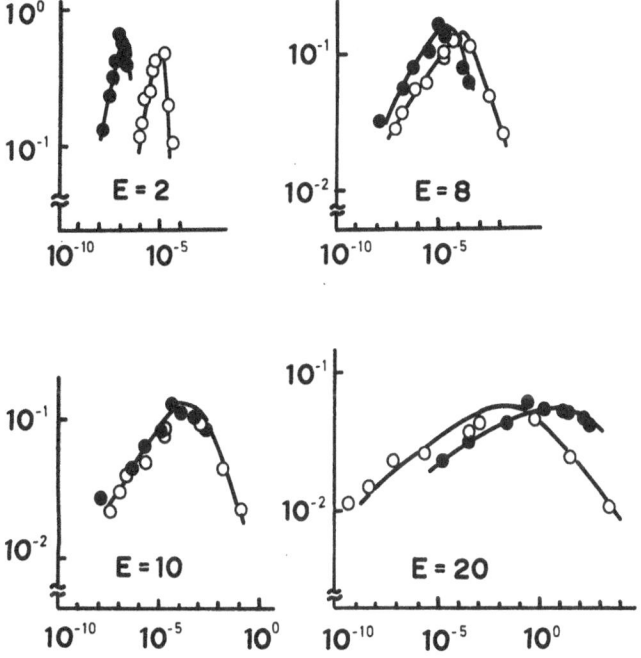

Fig. 8.2. Distributions of correlation time obtained from NMR (○) and ESR (●) data. Calculations were made for $T_r = 183.5$ K with various activation energies in kcal/mol as shown in the figure: horizontal axis, relaxation time (s) (Ref. [58])

for obtaining $I(\tau)$. As shown in Fig. 8.2., the relaxation spectra $I(\tau)$ obtained from NMR and ESR data are not identical, except of $E = 42$ kcal/mol. Therefore, it can be said that the activation energy associated with the molecular motion should be 42 kcal/mol if ESR and NMR measurements in this case are the reflections of the same molecular motion of polyethylene chains in UPEC.

As described above, Fig. 8.1. shows the temperature dependence of ESR-hyperfine splitting of alkyl radicals. Rectangulars in Fig. 8.1. mean the differences in hyperfine splitting due to β-protons, as discussed for Fig. 5.3. The varying quantity plotted by rectangulars in Fig. 8.1. reflects a temperature dependence of the amplitude of the hindered oscillation of methylene protons, as discussed in Sect. 5, and this phenomenon must resemble the phenomenon of spin exchange in the case of high-resolution NMR discussed by Gutowsky and Holm [59] by solving the modified Bloch equation. They derived a relation between peak separations observed at various temperatures and correlation time, τ, of molecular motion causing the change in peak separation. This was expressed by:

$$\frac{1}{\tau} = \sqrt{2\pi} \{(\Delta v_0)^2 - (\Delta v)^2\}^{1/2} \tag{8.8}$$

where Δv_0 and Δv are the peak separations expressed by frequencies at slow and intermediate rates of exchange, respectively. Equation (8.8) can be applied to ESR spectra and then a difference in the hyperfine splitting due to two methylene protons will correspond to Δv_0 or Δv in Eq. (8.8). According to a similar application of Eq. (8.8) to ESR studies by Ohnishi et al. [60)], in the case of allylic free radicals in polyethylene, Eq. (8.8) can be modified as follows:

$$\Delta v = \Delta v_0 \sqrt{1 - \frac{1}{2\pi^2\tau^2 \, \Delta v_0^2}} = \Delta v_0 \sqrt{1 - \frac{1}{2\pi^2\tau^2 \, \Delta v^2 + 1}} \qquad (8.9)$$

$$\omega = \omega_0 \sqrt{1 - \frac{2}{\tau^2\omega^2 + 2}}, \qquad \begin{pmatrix} \omega = 2\pi \, \Delta v \\ \omega_0 = 2\pi \, \Delta v_2 \end{pmatrix} \qquad (8.10)$$

As discussed earlier, the distribution of the correlation time also must be considered in this case. Therefore, Eq. (8.10) can be rewritten as including $I(\tau)$ and the shift factor $a(T)$ [as in Eq. (8.2)]:

$$\omega = \omega_0 \int_{-\infty}^{\infty} \sqrt{1 - \frac{2}{\omega^2 a^2(T) \, \tau^2 + 2}} \cdot I(\tau) \, d \log \tau \simeq \omega_0 \int_{\sqrt{6}/3a(T)\cdot\omega}^{\infty} \frac{I(\tau)}{\tau} \, d\tau \, .$$

$$(8.11)$$

Based on a similar procedure by Miyake [56)], an equation similar to Eq. (8.3) can be obtained:

$$I\left(\frac{\sqrt{6}}{3a(T) \cdot \omega}\right) \simeq \frac{\omega}{\omega_0}\left(1 + \frac{d \log a(T)}{d \log \omega}\right)^{-1} \qquad (8.12)$$

Based on Eq. (8.12), $I(\tau)$ can be obtained experimentally with a parameter E like in using Eq. (8.3), assuming that the temperature dependence of τ is also Arrhenius type. When E is assumed to be 42 kcal/mol, $I(\tau)$ obtained from temperature dependence data of different hyperfine splittings (indicated by □ in Fig. 8.1.) shows good agreement with $I(\tau)$ values obtained from line width data of ESR and NMR studies, as shown in Fig. 8.3. This can be a reflection of the fact that hindered oscillation of methylene groups causing the temperature dependence of different hyperfine splittings due to methylene protons is attributed to the same molecular motion causing narrowing of line widths of ESR and NMR spectra, and the activation energy of the motion should be 42 kcal/mol.

The arrangement of various data for obtaining $I(\tau)$ shows that these data are consistent, as described above. However, data analysis according to the considerations made for Eq. (8.7) is necessary. The values of τ calculated by simple application of the BPP equation [Eq. (8.1)] at various temperatures are plotted in Fig. 8.4. for ESR and NMR measurements. Open triangles in the same figure are the values of correlation times calculated by use of Eq. (8.10), which are corresponding to the values shown in Fig. 5.3. at high temperatures. Three groups of plotted values for τ indicate similar slopes in Arrhenius plots corresponding to $9.6 \sim 10.0$ kcal/mol. Since two

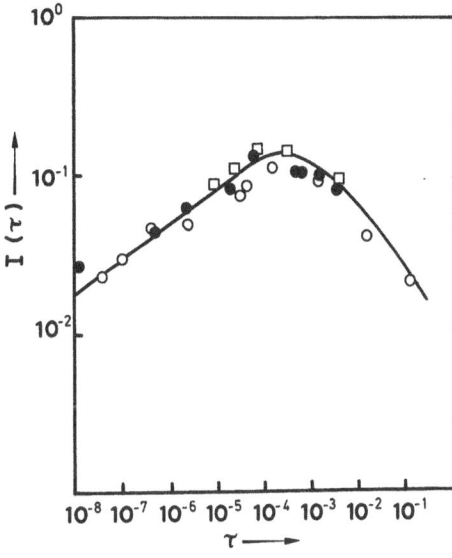

Fig. 8.3. Comparison of the distributions of correlation time obtained from NMR line width, ○, ESR line width ●, and difference in hyperfine splitting width □, assuming 10 kcal/mol (42 kJ/mol) of the activation energy (Ref.[58])

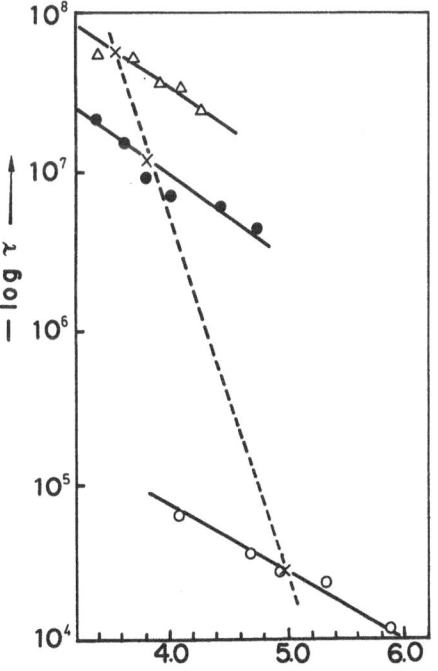

Fig. 8.4. Correlation times as a function of inverse temperature: ○, NMR; ●, ESR; □, difference in hyperfine splitting; ×, representative correlation time in the respective observations (see text) (Ref. [58])

values for τ_m in Eq. (8.7) can be estimated from ESR and NMR measurements, they are plotted in Fig. 8.4. by cross-marks. The dotted line connecting these two cross-marks reaches to the cross-mark located in the group of triangles. This latter cross-mark means half of two values of τ estimated for the rigid state and for the highest temperature of observation. The slope of the dotted line in Fig. 8.4. corresponds to an activation energy of 44 kcal/mol. This is consistent with the discussions concerning the value of parameter E, for obtaining the same curve of I(τ) obtained from ESR and NMR data; the true activation energy of the observed molecular motion must be 42 kcal/mol, since 42 kcal/mol mentioned for Fig. 8.2. and 44 kcal/mol for Fig. 8.4. must be the same within the experimental error. Thus, the molecular motion of the polyethylene chain in UPEC, being a rotation around the chain axis, has been confirmed by analyzing the ESR and NMR data which consider the distribution of the correlation time in polymeric materials and its activation energy was found to be 42 kcal/mol. This value seems appropriate, since the motion is not necessarily along the whole chain.

8.3 Example: Polytetrafluoroethylene [61]

A similar method of analysis as for UPEC (above section) was applied to a transition phenomenon in polytetrafluoroethylene around room temperature, and the analysis was confirmed by comparing the results with the established knowledge in dynamic mechanical studies of the same materials.

A sample of polytetrafluoroethylene, Aflon G80 (Product of Asahi Glass Company), was γ-irradiated up to 40 Mrad at room temperature under vacuum of 10^{-2} Pa. The irradiated sample was heated at 500 K for 40 min in order to observe only the trapped stable fluorinated alkyl radicals. The sample for the NMR study was irradiated in the same way as the sample for the ESR study. Line widths of ESR spectra of alkyl radicals at various temperatures were measured by a X-band spectrometer, and the line width and second moment of broad-line NMR spectra at various temperatures were obtained at 40 MHz. The temperature dependences of line widths (second moment) of ESR and NMR spectra are shown in Fig. 8.5., from which relaxation spectra I(τ) can be obtained both from ESR and NMR data based on Eq. (8.3). τ_m values can be estimated by use of Eq. (8.7) both for ESR and NMR observations.

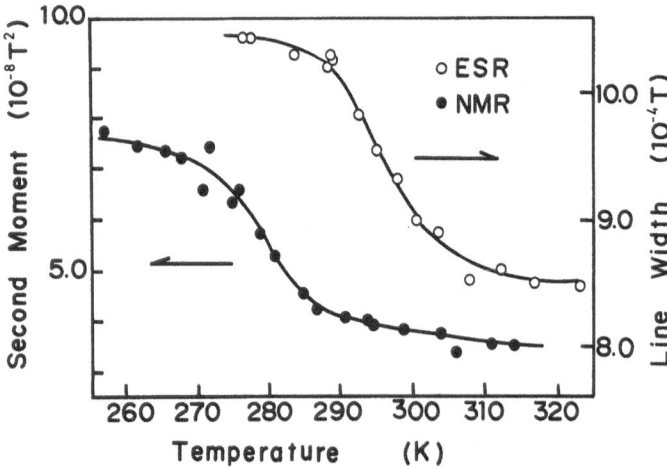

Fig. 8.5. Line widths of ESR spectrum of fluoroalkyl radical in PTFE ($\bigcirc$) and second moments of BL-NMR of irradiated PTFE ($\bullet$) at various temperatures (Ref. [61])

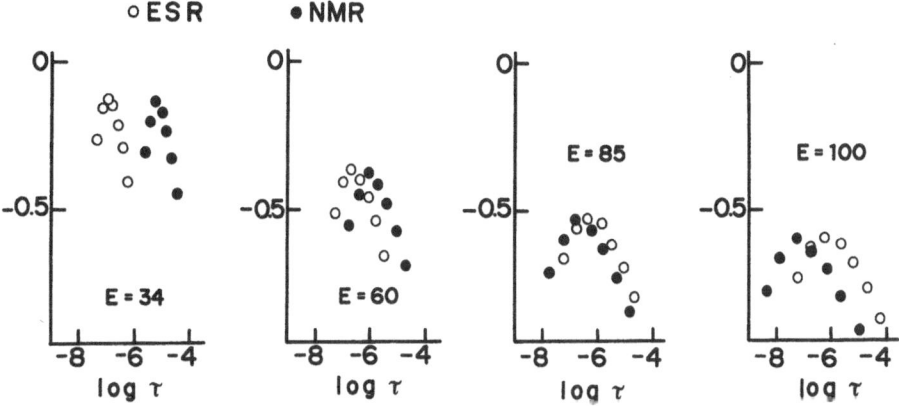

Fig. 8.6. Comparison of relaxation spectra at reference temperature (T_r = 290 K) obtained from ESR data (○) and NMR data (●) assuming various activation energies in kcal/mol (Ref. [61])

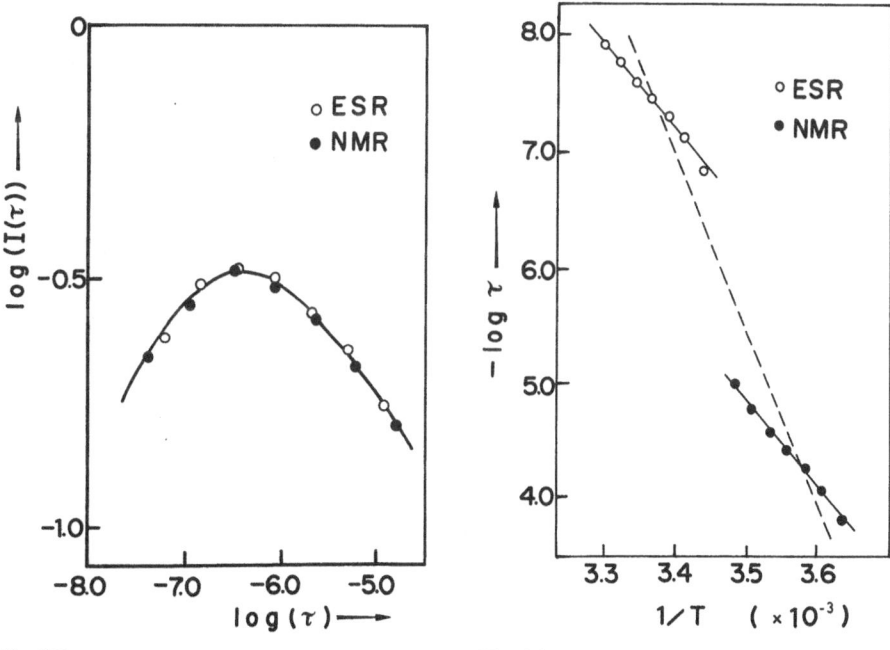

Fig. 8.7.   **Fig. 8.8.**

Fig. 8.7. Relaxation spectrum of PTFE at reference temperature (T = 290 K) when activation energy is assumed to be 75 kcal/mol (315 kJ/mol); ○, plots based on ESR data; ●, plots based on NMR data (Ref.[61])

Fig. 8.8. Arrhenius plot of correlation times estimated from ESR line width (○) and BL-NMR line width (●) of PTFE: Dotted line is connecting the representative correlation times in both measurements (Ref. [61])

I(τ) values obtained at the reference temperature $T_r = 290$ K with various activation energies are shown in Fig. 8.6. as in the case of Fig. 8.2. As shown in Fig. 8.6., relaxation spectra obtained from ESR and NMR data are not in good agreement with the activation energies described in the same figure. But when the activation energy is assumed to be 315 kcal/mol, I(τ) values obtained from ESR and NMR data coincide well, as shown in Fig. 8.7. On the other hand, τ_m discussed in connection with Eq. (8.7) was also obtained both for ESR and NMR measurements: τ_m in ESR was 0.4 $\times 10^{-7}$ s, the temperature corresponding to τ_m in ESR was 297 K, τ_m in NMR was 0.7×10^{-4} s and the temperature corresponding to τ_m in NMR was 280 K. In Fig. 8.8. (comparable to Fig. 8.4. for UPEC) a dotted line connecting two τ_m values is shown. The slope of the dotted line in Fig. 8.8. corresponds to an activation energy of 310 kcal/mol. The motional narrowing in the temperature region shown in Fig. 8.5. is attributed to the crystalline phase change in polytetrafluoroethylene and also has been studied by dynamic mechanical measurement; the activation energy was found to be 290 kcal/mol[19]. The activation energy indicated in Fig. 8.8., the slope of the dotted line connecting two τ_m values, is very close to the value of 290 kcal/mol obtained in the dynamic mechanical study. It also can be considered to be the same as the value of E, a parameter for obtaining the same relaxation spectrum both from ESR and NMR, as shown in Fig. 8.7. According to the above results, the motional narrowing phenomena observed both in ESR and NMR at room temperature for polytetrafluoroethylene must reflect the crystalline transition in the same materials established in dynamic mechanical studies. Also, the coinciding activation energies obtained by the presented analysis and in dynamical studies reflect the validity of the analysis and interpretations made in Sect. 8.1. This also means that the interpretations in Sect. 8.2 are valid and, therefore, the mode of the molecular motion discussed in Sect. 7 is also acceptable.

The example of ESR and NMR data analysis for polytetrafluoroethylene is a reexamination of the study made by Tamura [62] more than fifteen years ago. However, in the present paper the analysis is consistent in that it considers the distribution of the correlation time in polymeric materials. In this sense, the case of polytetrafluoroethylene is a good example to which the method described in Sect. 8.1 has been applied.

9 Application of Spin Label and Spin Probe Methods

9.1 General Remarks

More than twenty years ago, in Kivelson's famous paper [63] a relation between hyperfine splitting due to nitrogen nuclei and correlation time of the motion of molecules containing a nitroxide radical was proposed. A few years later, stable free radicals containing nitroxide radicals, such as:

were used as a spin label reagent and the molecular motion of spin labelled molecules with nitroxide radicals were widely studied [64, 65]. In earlier applications of the spin label technique, bio-related materials were mainly investigated. In the late 1960, this method was applied to molecular motion studies of synthetic high polymers [66]. Since the spin label technique requires covalent bonding of the stable nitroxide radical to the polymer chains, the "spin probe" method was also deviced to keep pace with the spin label method [67, 68, 69]. In the spin probe method, nitroxide radicals, like:

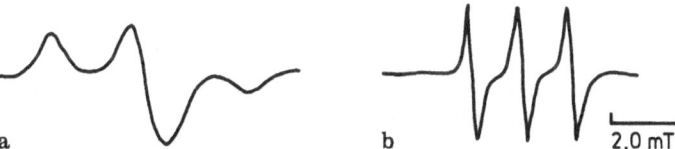

are mixed with polymer powders in an appropriate solvent. In both spin probe and spin label methods, a temperature dependence of the spectral shape including hyperfine splitting due to nitrogen nuclei with spin 1 is usually observed and the correlation time and other quantities specifying the molecular motion are thereby derived.

A similar technique which stabilizes the unstable radicals by converting them to nitroxide radical adducts, called the spin trapping method [70], is widely used. A few examples of the spin trapping method as they are applied to the study of molecular motion will be shown in the next section. The applications of the spin label and spin probe methods are described in various excellent textbooks [71], review articles [72], and proceedings of related symposia [74]. Therefore, in the present article, only a few examples will be considered.

Generally, the ESR spectrum of the nitroxide radical shows a triplet pattern as shown in Fig. 9.1. depending on the temperature of observation. The patterns in Fig. 9.1. are schematic illustrations. Figure 9.1a. appears when the nitroxide radical is trapped in a rigid matrix. In this case, the motion of the label or probe molecules are frozen in and random orientation of rigid probe molecules causes a pattern indicating an anisotropic hyperfine interaction in rigid state; i.e., the separation of the three components due to hyperfine interaction with the nitrogen nucleus is very broad resulting in the asymmetric pattern with a broad line width of each of the components. However, when the motion of the probe molecules and trapping matrix increases, anisotropies of the hyperfine tensor or $\tilde{g}$-tensor can be averaged out and three sharp components with narrow separations appear as shown in Fig. 9.1b.; i.e., the symmetry of entire spectrum is increased.

a   b   2.0 mT

Fig. 9.1. Typical triplet spectra of nitroxide radicals in most of spin-label or spin-probe studies (upper): A, rigid state; B, rapid motion.

The molecular axis of the probe is usually determined as follows: the p_z direction of the unpaired electron is taken as the z axis, the x axis is in the N—O direction.

According to the theory by Kivelson, a short correlation time in the order of $10^{-9} \sim 10^{-12}$ s can be calculated by the following equation:

$$\tau = \omega_0 [(h_0/h_1)^{1/2} - (h_0/h_{-1})^{1/2}] \frac{-15\pi \sqrt{3}}{8b \, \Delta\gamma \, H} \tag{9.1}$$

$$= \omega_0 [(h_0/h_1)^{1/2} + (h_0/h_{-1})^{1/2} - 2] \frac{4\pi \sqrt{3}}{b^2} \tag{9.2}$$

where h_1, h_0, and h_{-1} are the respective peak heights of the three components as shown in Fig. 9.1 b., W_0 is the line width of the central component, H is the static magnetic field applied. b and $\Delta\gamma$ in Eqs. (9.1) and (9.2) are the quantities from the anisotropies of the $\tilde{g}$- and $\tilde{A}$-tensors, and are expressed as follows:

$$b = \left(\frac{4}{3}\right) \pi(A_{zz} - A_{xx})$$

$$\Delta\gamma = -(\beta/\hbar) \left\{ g_1 - \frac{1}{2} (g_2 + g_3) \right\} \tag{9.3}$$

Equations (9.1) and (9.2) can be derived from Eq. (9.4), originally proposed by Kivelson [63] assuming that hyperfine anisotropy has an axial symmetry, and that the motion of the labelled site is isotropic with a very short correlation time; the line shape is Lorentzian.

$$[T_2(M)]^{-1} = A + BM + CM^2 \tag{9.4}$$

The above indicates that the line width of the component corresponding to the quantum number M can be expressed by a quantity proportional to M and a term proportional to M^2. Constants A, B, and C in the same equation include also the quantities defined by Eq. (9.3).

In studying the molecular motion of synthetic polymers, many cases have been observed, however, in which the time constant of the motion is slow and direct application of Eq. (9.1) is not appropriate. For slow molecular motion, $10^{-7} > \tau > 10^{-9}$, Freed et al. [74] proposed an equation for calculating correlation times based on the simulation of the spectra; and this equation is widely applied:

$$\tau = a(1 - S)^b \tag{9.5}$$

where S is the ratio of A_z (or $A_{//}$ in the case of axial symmetry), the principal value of the A-tensor, at a certain temperature and that for the rigid state, but, practically, outermost separation (extrema separation) can correspond to A_z for usually amorphous systems[5d]. The constant quantities a and b in Eq. (9.5) depend on the motional model. According to Freed et al.[74], the following values are proposed:

$$a = 5.4 \times 10^{-10} \; ; \quad b = -1.36 \quad \text{for "Brownian" diffusion} \qquad (9.6\,a)$$

$$a = 6.10 \times 10^{-9} \; ; \quad b = -1.01 \quad \text{for "Moderate jump" diffusion} \quad (9.6\,b)$$

$$a = 2.55 \times 10^{-9} \; ; \quad b = -0.615 \; \text{for "Large jump" diffusion} \qquad (9.6\,c)$$

Equation (9.5) appears to be easily applicable, but the selection of one model from three models, Eq. (9.6), also requires a spectral simulation in order to make an exact calculation. However, the "moderate jump" model has been used widely as of convenience.

9.2 Properties of the Crystalline Surface

Kusumoto et al. [75] studied the mobility of molecules on the surface of crystallites in solution-grown crystals of polyethylene by using the spin probe method. They observed a temperature dependence of the outermost separation of the ESR spectrum [the quantity corresponding to S in Eq. (9.6)] originating from the spin probe reagent dispersed in the solution-grown crystals of polyethylene. Since the size of the spin probe reagent is rather large, the reagent can only behave as a probe for the surface area of the crystallite. Correlation times at various temperatures were estimated as illustrated in Fig. 9.2., and it was found that the correlation time decreased with raising temperature (solid line); the slope of the decrease became steep at the temperature of crystallization of the materials, as seen from the Arrhenius plot of the correlation time. When the temperature of observation was lowered from the highest temperature, the correlation time increased again but values were larger than that observed at raising temperature (dotted line). The steep slope in the Arrhenius plot in the case of raising temperature can be a reflection of the fact that the molecular motion becomes active due to the lamellar thickening during heating at higher temperatures than the temperature of crystallization. When lamellar thickening proceeds at a suffi-

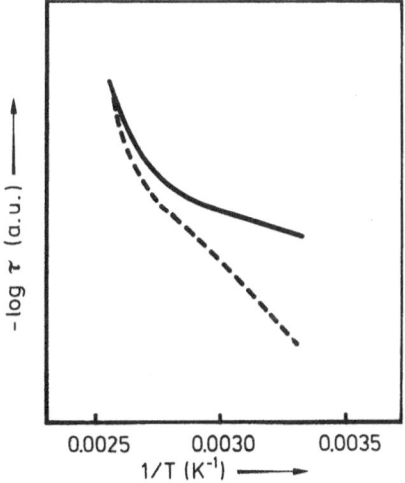

Fig. 9.2. Correlation time estimated from the spectrum of the spin-probe molecule at surface area of solution-grown crystals of polyethylene (based on Ref. [75]).

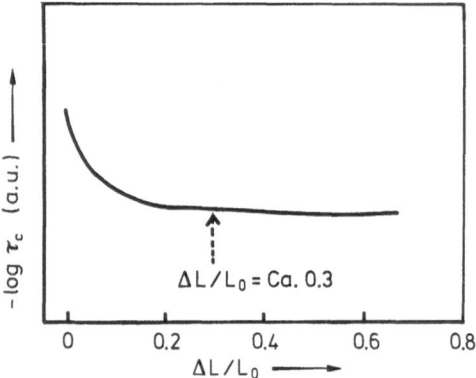

Fig. 9.3. Relation between correlation time estimated from spin-probe experiment and $\Delta L/L_0$ (see text) (based on Ref. [75])

ciently rate, folding of the molecular chain must be more tight than before thickening and the correlation time observed with decreasing temperature after the first heating must be longer than during the first raising of the temperature. Since lamellar thickening is essentially an irreversible process, it is quite resonable that the correlation times obtained for decreasing temperature after the first heating do not take the same values as in the case of the first raising of the temperature. Concerning the formation of tight loops due to lamellar thickening, the following facts were also found by Kusumoto et al. [75]: Several portions of high density polyethylene were crystallized in dilute xylene solution at various temperatures and crystallites with various lamellar thickness were prepared. They were heated up to various temperatures higher than the respective crystallization temperatures in order to allow lamellar thickening; the correlation times associated with the molecular motion at the crystalline surface were estimated using the spin probe method. It was also found that when the estimated correlation times are plotted against the ratio of the excess thickness due to the thickening, ΔL, to the thickness of lamellars as crystallized, L_0, $\Delta L/L_0$, most of the plotted points can be located on one curve. The correlation time increases with increasing values of $\Delta L/L_0$ tending to become saturated at values of $\Delta L/L_0$ larger than 30%, as illustrated in Fig. 9.3. This means that large values of $\Delta L/L_0$ generally mean a more tight loop structure, since increasing values of the correlation time reflect a less mobile character.

9.3 Relation Between Outermost Separation Width and Transition ind the Polymer

As described for Eq. (9.5), the width of the outermost separation of the ESR spectrum of the nitroxide radical has a significant meaning associated with the mobility of the labelled site. Therefore, the temperature dependence of the outermost separation width can be a measure of the mobility of a system; i.e., narrowing of the outermost separation occurs with the beginning of motion at the labelled site. Usually, it is said that the temperature at which the outermost separation takes a value of 50 Gauss has a significant meaning (this temperature is called T_{50G}). Actually, the correlation between T_{50G} and the glass rubber transition temperature T_g has been discussed by many authors [73]. The problem of the correlation between T_{50G} and T_g or the melting point T_m was first approached by Rabold [67], but a rather clear correlation between T_{50G} and T_g was discussed by Boyer [76]. Discussions on this problem were continued by Boyer and Kumler [77] leading to the following relation:

$$T_{50G} = T_g/(1 - 0.03\, T_g/\Delta H_a) \tag{9.7}$$

where ΔH_a is the activation enthalpy for the glass rubber transition. In analyzing the relation between T_{50G} and T_g, the probe size has to be taken into account. Kusumoto et al. [78] considered the molar volume of the probe and obtained:

$$T_{50G} - T_g = 52 \left[2.9\, f \left(\ln \frac{1}{f} + 1 \right) - 1 \right] \tag{9.8}$$

where f is the ratio of the volume of the probe to that of the segment. According to Eq. (9.8), T_{50G} and T_g can be identical when the value of f is ca. 0.1.

In Memoriam Professor N. Kusumoto

We wish to recall the excellent achievments of Professor N. Kusumoto in the field of ESR application to polymer science. Prof. Kusumoto, who died of leukemia in April 1981 at the early age of fourty six, was a person of gentle but strong will. He continued his research until late in 1980 inspiring us during many fruitful discussions with new ideas and insights.

It is said that Prof. Kusumoto observed ESR signals from his blood and compared them with normal blood signals ever since 1978; ligand conditions in blood are claimed to change due to leukemia.

Sections 9.2 and 9.3 in this paper contain material from studies of Prof. Kusumoto and his group.

10 Application of the Spin Trapping Method to Molecular Motion Studies

10.1 Stabilization of Unstable Radicals

Considering the limitations associated with the spin-probe method caused by the large size of probe molecules, it can be said that one of the main advantages of using the nitroxide radical as a probe or label is the conversion of the very unstable radical site to a stable nitroxide radical adduct. Then we can observe a temperature dependence of the spectrum reflecting the physical properties of the site where the unstable radical was trapped at first, even though we can hardly observe the spectra of an original radical at high temperatures. Nitrosobenzene and its derivatives are widely used as a spin trap reagent. Though there are many spin trap reagents used for various investigations for identifying the unstable free radicals produced [70], precise description of spin trap reagents or examples of their use will be omitted in the present article since this article is mainly focussed on molecular motion studies. Since the application of the spin trapping method is based on the observation of nitroxide

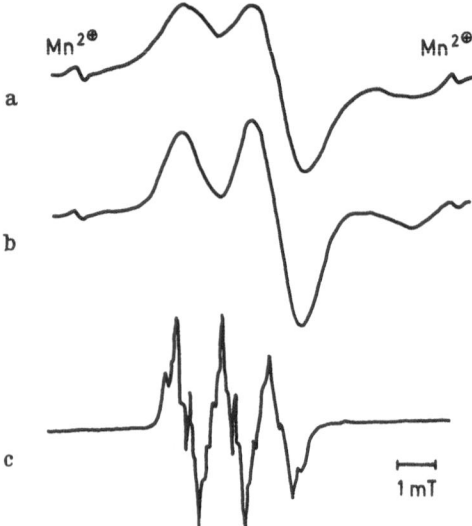

Fig. 10.1 a–c. ESR spectra from polypropylene fractured in the presence of nitrosobenzene *in vacuo* at 77 K: (**a**) observed at 77 K before heat treatment; (**b**) observed at 77 K after heat treatment at 302 K for 5 min; (**c**) observed at 458 K after heat treatment. The peaks marked with Mn^{2+} are the peaks due to Mn^{2+} inserted in the cavity in order to calibrate the width of the magnetic field (Ref. [79])

radicals, various quantities and relations mentioned in the preceding section are also useful here.

One of the effective applications of the spin trapping method is the study of molecular motion at chain ends. The chain end can usually be considered to be much more mobile than the inside part of the molecular chain. The mobility of the chain end at surfaces is also an interesting subject of study. As is well known, mechanical degradation of polymeric substances produces a radical trapped at the chain end due to chain scission. However, the free radicals produced by the chain scission due to mechanical degradation are very unstable. Therefore, when the polymer is mechanically fractured together with the spin trapping reagent, like nitrosobenzene mixed with polymer materials, the free radicals produced at the chain end shall be converted to the stable nitroxide radical. This was experimentally confirmed by Sakaguchi et al [79] as shown in Fig. 10.1. They observed a typical ESR spectrum of a nitroxide radical originating from a spin adduct trapped in polypropylene destroyed mechanically together with mixed nitrosobenzene. The mechanical destruction was conducted by use of the apparatus originally designed by Sakaguchi [80] in a glass ampoule sealed under a vacuum of 10^{-2} Pa. The spin adduct produced in the case of mechanically destroyed polyethylene is also a similar nitroxide radical, but nitrosobenzene combines with the polyethylene chain to produce the spin adduct $\sim\underset{\underset{\varnothing-\dot{N}O}{|}}{CH}-CH_3$. Identification

of the spin adduct was confirmed by computer simulation of the spectrum [82]. The combination with the spin trapping reagent is understandable, because a rapid conversion of the end radical $\sim\dot{C}H_2$ to $\sim\dot{C}HCH_3$ was also observed even at low temperature [83]. At any rate, very unstable radical sites at the chain end can be stabilized (spin labelled) with nitroxide radicals.

10.2 Molecular Motion of the Chain End of Fractured Polymers

ESR spectra of spin-labelled polyethylene and polypropylene mentioned above were observed at various temperatures and the molecular motions at chain ends were discussed.

As mentioned in the preceding section, the spin adducts in polyethylene and polypropylene fractured under the presence of nitrosobenzene at 77 K are nitroxide radicals at or near chain ends. Since it is reported that the mechano radicals (the free radicals produced by mechanical destruction) are trapped on the fractured surface [80, 81], the application of the spin trapping method is useful for examining the molecular motion of the molecules on the surface.

The temperature dependence of ESR spectra of fractured polyethylene labelled by nitroxide radicals is shown in Fig. 10.2., which shows the case of high density polyethylene. W indicated in Fig. 10.2. means extrema separation (outermost separation) and is the same quantity as S in Eq. (9.5). W is a measure of the mobility of the materials and the value of W is shown to decrease with raising temperature reflecting a much more mobile character of the observed molecules at elevated temperatures. As mentioned in the previous section, the varying extrema separation with temperature has been widely discussed in connection with molecular motion.

Concerning the mechanical destruction of polymers, an interesting fact was reported. This is a phase transformation in the crystal structure from the normal orthorhombic to a monoclinic phase due to mechanical degradation. Kurokawa et al. [85] studied the increase of the monoclinic phase after ball-milling of polyethylene. Yemni and McCullough [86] concluded that the monoclinic phase is more stable energetically

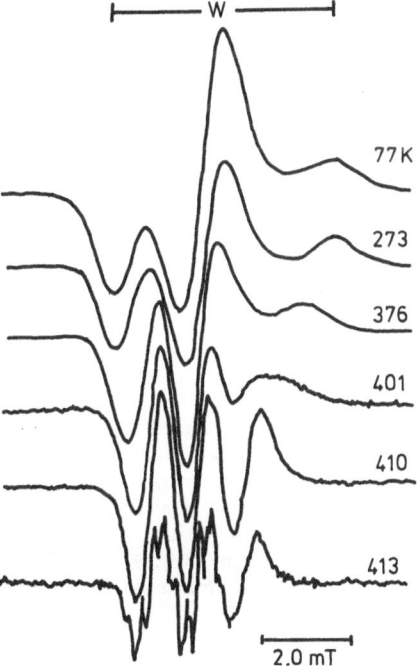

Fig. 10.2. ESR spectra of high density polyethylene fractured in the presence of nitrosobenzene. Observation temperatures are shown on respective spectra. W is extrema separation (outermost separation width) (Ref. [84])

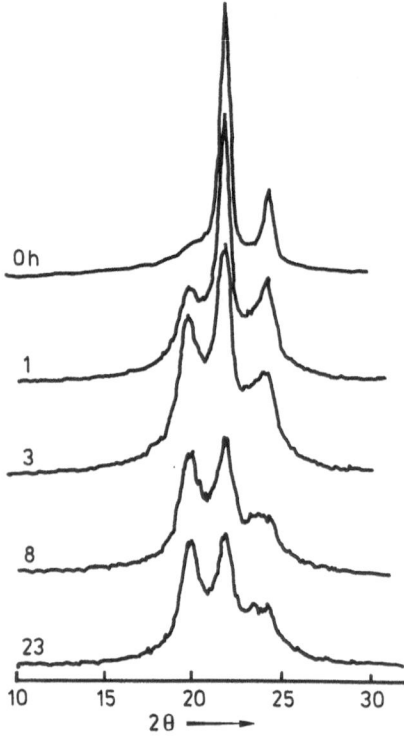

Fig. 10.3. X-ray diffraction patterns of high density polyethylene fractured for various time periods shown on respective patterns (Ref. [84])

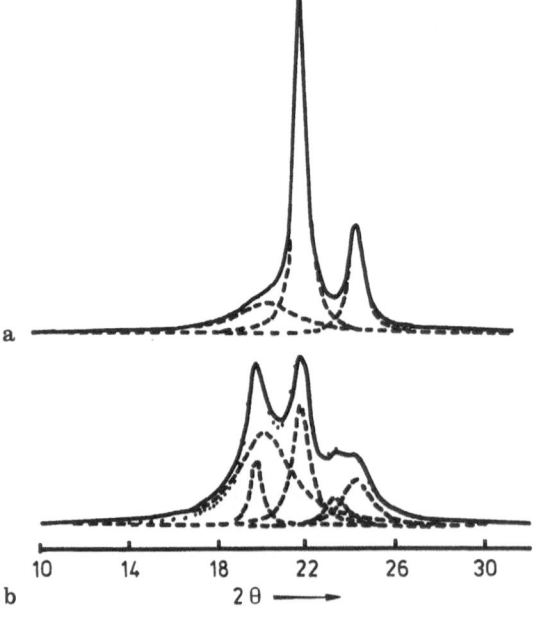

Fig. 10.4a and b. Decomposition of the X-ray diffraction spectra into components.
(a) high density polyethylene before fracture;
(b) high density polyethylene after fracture for 23 h (Ref. [84])

than the orthorhombic phase and the same conclusion was also reported by Tadokoro et al. [87]. Vivatpanachart et al. [84] performed a much more precise study on the phase transformation in ball-milled polyethylene and obtained a change in the X-ray diffraction pattern of high-density polyethylene as shown in Fig. 10.3., which indicates a variation of the pattern due to the milling time. Figure 10.4. shows the effect of ball-milling on the X-ray diffraction pattern more clearly. Figure 10.4a., the pattern obtained before milling, is a superposition of three peaks centered at 2θ equal to 19.92°, 21.66°, and 24.08° which correspond to the reflections from the amorphous region, and the (110) and (200) planes of the orthorhombic unit cell, respectively. However, the pattern of the milled materials, Fig. 10.4b., contains the other two peaks centered at 2θ of 19.75° and 23.19° which are the reflections from the (001) and (200) planes of the monoclinic unit cell, respectively. Though the positions of new peaks in Fig. 10.4b. deviate slightly from the values corresponding to the (001) and (200) planes of the monoclinic unit cell reported in the literature [88], 19.50° and 23.12°, it can be said within an experimental error that the monoclinic phase was certainly produced by the mechanical degradation. Figure 10.5. shows this more clearly; i.e., changes of the degree of crystallinity, and the contents of orthorhombic and monoclinic phases with milling time are indicated. Numerical values of the related quantities are also shown in Table 10.1., in which the correlation times calculated from the extrema separations observed at 30 °C are included. A moderate jump diffusion model was applied in using Eq. (9.5).

Figure 10.6. shows temperature dependences of the extrema separations observed for the materials subjected to ball-milling for various durations. This indicates that the narrowing temperature shifts to higher temperatures by the longer milling time, but no effect of milling time appears when the duration is longer than 10 h. A high temperature shift generally must reflect a less mobile character and the correlation time tends to be longer at longer milling times, also as indicated in Table 10.1. Reduc-

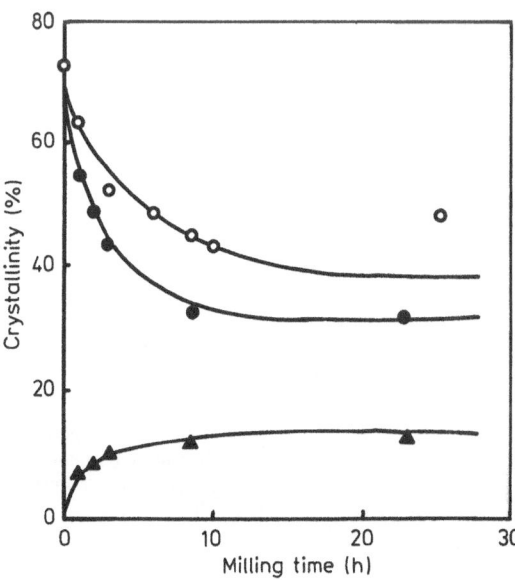

Fig. 10.5. Variation of crystallinity (O), orthorhombic (●) and monoclinic (▲) contents as a function of milling time (Ref. [84])

Table 10.1. Milling time dependences of mobility and crystalline phases in high-density polyethylene, Hizex Million

Milling time (h)	Orthorhombic content (%)	Monoclinic content (%)	Amorphous part (%)	Correlation time at 30 °C (s)
0	73.4	—	26.4	—
1	54.8	7.65	37.6	2.4×10^{-8}
4	40.0	10.0	50.0	2.8×10^{-8}
10	32.0	12.4	56.0	3.4×10^{-8}
23	31.01	12.4	56.5	3.4×10^{-8}

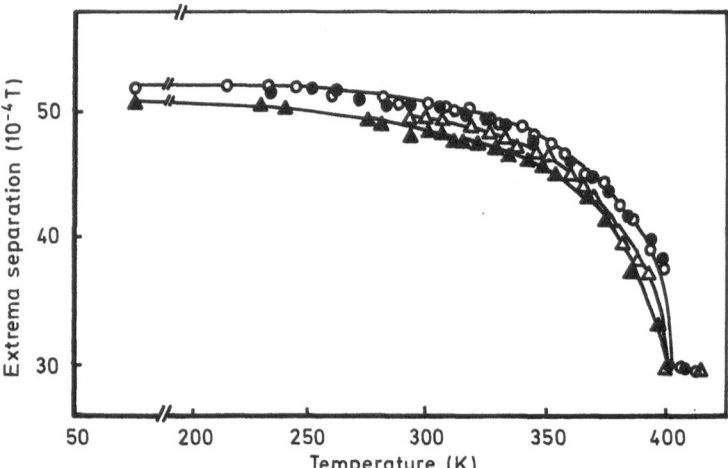

Fig. 10.6. Variation of extrema separation with temperature of observation for high density polyethylene fractured for 1 h (▲), 4 h (△), 10 h (●), and 25 h (○) (Ref. [84])

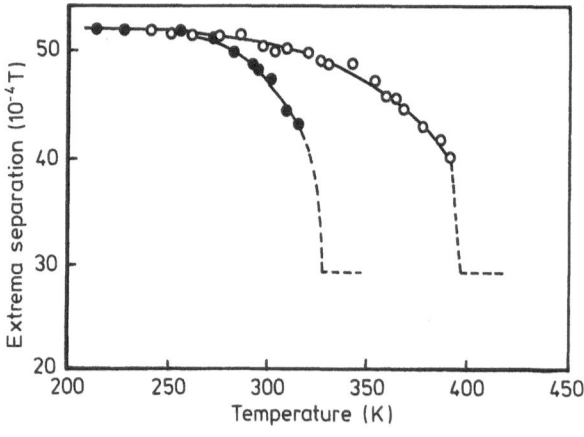

Fig. 10.7. Variation of extrema separation with temperature for high density (○) and low density (●) polyethylenes (Ref. [84])

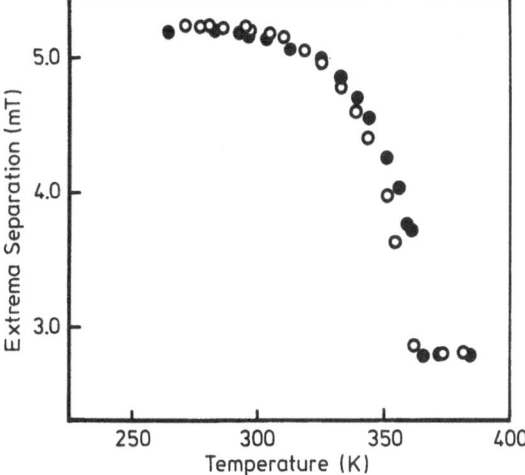

Fig. 10.8. Extrema separation vs. temperature of fractured atactic polypropylene labeled with nitroxide radical: ○, before melt; ●, after melt (Ref. [89])

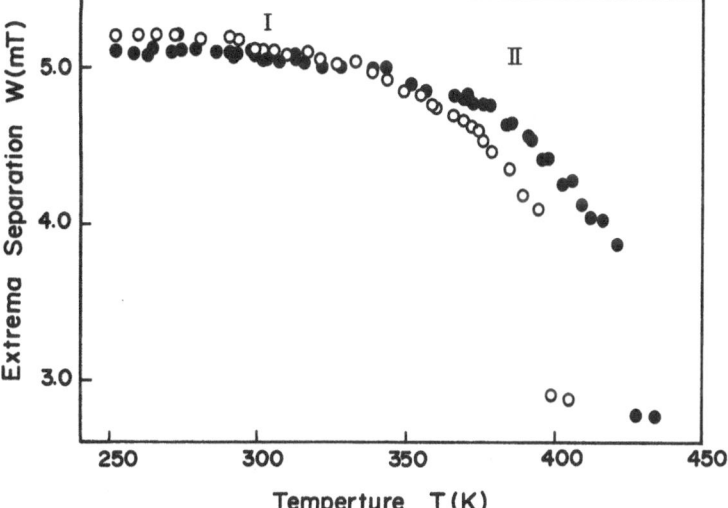

Fig. 10.9. Extrema separation vs. temperature of fractured isotactic polypropylene labeled with nitroxide radical: ○, before annealing; ●, after annealing at 423 K for 51 h (Ref. [90])

tion of the mobility and increase of the content of the monoclinic phase are also parallel to the facts mentioned above. However, a possible effect of the amorphous content to the narrowing temperature must be examined. Figure 10.7. shows the results of comparing the narrowing curves for low-density polyethylene and high-density polyethylene; i.e., the narrowing temperature of the materials with a greater amorphous part is lower than that for materials with a smaller amorphous part. Based on the above results, it can be said that the less mobile character in the materials subjected to milling for longer time periods is associated with an increase in the content of the monoclinic phase.

Fractured polypropylene labelled with nitroxide radicals were also studied in a similar manner as in the case of polyethylene. In the case of atactic polypropylene [89], the transition temperature observed for the first heating appears to be slightly lower than the transition temperature observed for the second heating made after the materials are heated to the melting point (Fig. 10.8.). This is quite reasonable since the labelled nitroxide radicals are located at the chain ends on the fresh surface of the polymer [81] and the radical site must be located in the bulk after the first heating. However, for very large label molecules, melting of the materials first may cause an occlusion of the label molecules into a bulky space and may shift the transition temperature to lower values. The example of 2,4,6-tri-*tert*-butyl nitrosobenzene, described in the next section, will illustrate such a case.

Figure 10.9. shows the case of isotactic polypropylene. There are two temperature regions of narrowing, regions I and II. Many experiments were made under various conditions and it was concluded that the narrowing in region I reflects a very small-scale motion of just the site of the nitroxide radical and that the narrowing in region II reflects the motion of rather long "tie-like" parts attached to the crystalline surface of isotactic polypropylene [90].

It is interesting to note the annealing in the case of isotactic polypropylene. As shown in Fig. 10.9., heat treatment results in an up shift of the narrowing temperature in region II [90]. On the other hand, however, an effect of annealing appeared in the down shift of the transition (narrowing) temperature in the case of fractured polyethylene [91]. A similar phenomenon was also reported by Bullock et al. for polyethylene by the spin label method [92].

The effect of annealing of the crystalline polymer on the chain ends located of the surface can be explained in two different ways. One [93] is that the heat treatment, leading to a thickening of the lamella, would move the surface chain end into the inside of the crystallite. The other explanation [94, 95] would be that the heat treatment results in the exclusion of chain ends from the crystallite.

Vivatpanachart et al. [91] concluded that exclusion of the chain end from the crystalline region should be a cause of the down shift of the transition temperature after annealing in the case of fractured spin-labelled polyethylene; this was based on the comparison of annealing effects in low- and high-density polyethylenes. However, the crystalline structure of isotactic polypropylene should be more bulky than polyethylene since the former has a 3_1 helix and the latter a planar zig-zag conformation. Therefore, it is reasonable that the spin-labelled chain end of isotactic polypropylene can be included in the thickened crystals after annealing; this is a cause of the annealing effect mentioned above, the high temperature shift of the transition temperature.

10.3 Comparison of Molecular Motions at the Chain End and Inside the Chain [96]

The spin trapping method is also useful for comparing molecular motions at chain ends and inside of the molecular chain.

High-density polyethylene, Sholex 6050, may serve as an example. The materials were subjected both to thermal degradation and γ-irradiation together with 2,4,6-tri-*tert*-butyl nitrosobenzene. As in the case of mechanical degradation of polymers in the presence of nitrosobenzene, ESR spectra from nitroxide radicals were observed and variations of extrema separations were investigated. The spin adducts produced by the above procedures were identified as:

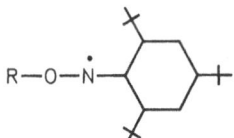

$$\text{(II)}$$

Spin adducts I and II are different from the spin adducts identified in the preceding section $\left(R\!-\!\underset{\underset{O}{|}}{N}\!-\!\bigcirc \right)$. This is due to the very "bulky" structure of 2,4,6-tri-*tert*-butyl nitrosobenzene used instead of nitrosobenzene. The spectra observed at very high temperatures are illustrated in Fig. 10.10. which shows reasonable h.f.s. corresponding to the spin adducts mentioned above; i.e., the spin adduct trapped in thermally degraded materials show triple-triplets and that in the irradiated materials shows quadro-triplets. The values of hyperfine coupling constants due to nitrogen ($a_N = 1.02$ mT) and protons ($a_H = 0.2$ mT) are comparable to the values reported in the literature [97,98]. Values of $a_N = 1.$ 15 mT and $a_H < 0.2$ mT have been reported for the case of:

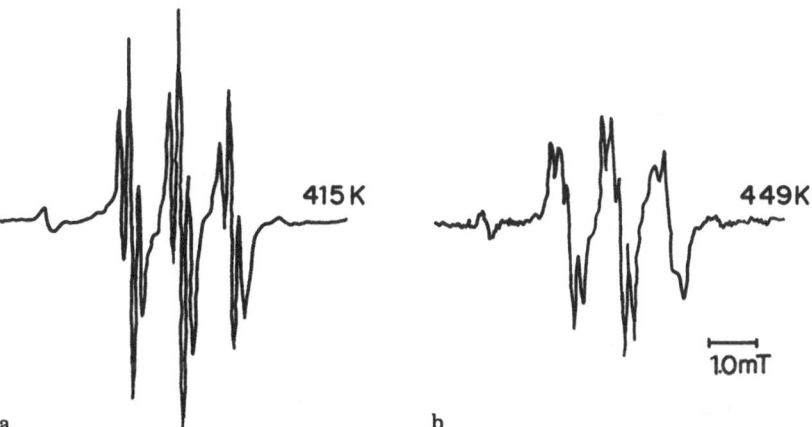

Fig. 10.10a and b. ESR spectra of nitroxide radicals trapped in (a) thermally degraded and (b) γ-irradiated materials (△) first heating and (▲) second heating (Ref. [97])

while $a_N = 1.35$ mT and $a_H = 1.79$ mT in the case [98] of:

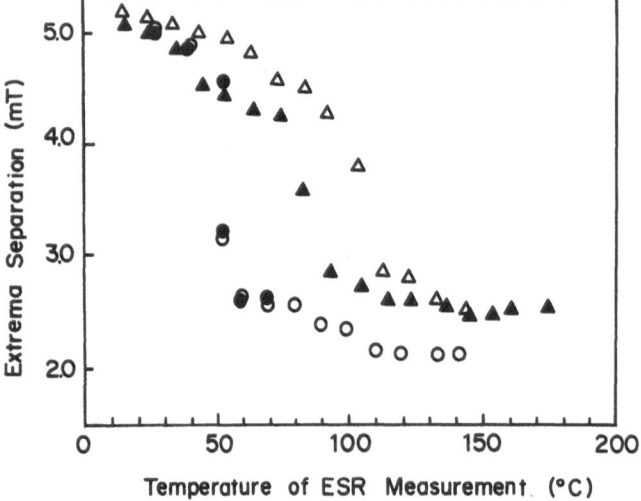

Since the values of a_N and a_H from the spectra in Fig. 10.10. are quite similar to those of the former case, the spectra in Fig. 10.10. can be easily interpreted. The low wing peaks in the same figure may correspond to the minor radicals of the latter case.

Figure 10.11. shows varying extrema separations with temperature for the respective materials. It shows that the narrowing of the extrema separation of the spectrum of irradiated materials occurs at higher temperatures than that of thermally degraded materials. The difference in the narrowing temperature indicates a different mobility. Figure 10.11. reflects the much more mobile character of the chain end compared to the inner part of molecular chains. Since the less mobile character attributed to the irradiated materials can be considered to result from the effect of radiation-induced cross-linking, the dose dependence of the narrowing temperature was examined. However, no dose dependence was found. Therefore, it safely can be said that the difference in the mobilities of the chain end and inner part of the molecular chain was reflected in the temperature dependences of ESR spectra of both materials spin labelled by the spin trapping method.

Figure 10.11. also shows a downshift of the narrowing temperature for irradiated materials to ca. 80 °C in measuring with decreasing temperature in contrast to the first heating (ca. 110 °C). No change of the narrowing temperature is seen for the

Fig. 10.11. Variation of extrema separation of ESR spectra with temperature: thermally degraded materials, (○) first heating and (●) second heating; γ-irradiated materials (△) first heating and (▲) second heating (Ref.[97])

second heating. This can be interpreted as a rearrangement of molecular chains after the melting due to the first heating; i.e., there is possibly a "bulky" arrangement of the molecular chains occluding the large spin trapping reagent and the more mobile feature can be attributed to such a bulky arrangement. This kind of arrangement is established at the first heating and no change is seen for the second heating. On the other hand, in the case of thermally degraded materials, the mobility of the chain end is not affected by the heating because the molecular chain is already heated in "thermal" degradation before the measurement.

10.4 The Nature of Molecular Motion [99]

As mentioned above, the spin trapping method is useful for studying molecular motion by ESR when the radical site is very unstable or for comparing such molecular motion with other molecular motions in different parts of the polymer.

Furthermore, it is desirable to make a more detailed investigation of the nature of the molecular motion reflected in the narrowing of the extrema separation of ESR spectra. Here, a rather detailed analysis of the temperature dependence of the ESR spectrum of irradiated materials labelled by the spin trapping method will be described.

As for the molecular motion in the polyethylene chain in UPEC (Sect. 8), the temperature dependence of the line width of the spectrum coming from nitroxide radicals was studied by means of computer simulation. In order to simplify the procedure, line widths of the central component of the main triplet at various temperatures were obtained and are plotted against the observation temperature together with the extrema separations, as shown in Fig. 10.12. The original plots [99] have been somewhat simplified. The values obtained in the first and second heating are shown in the same

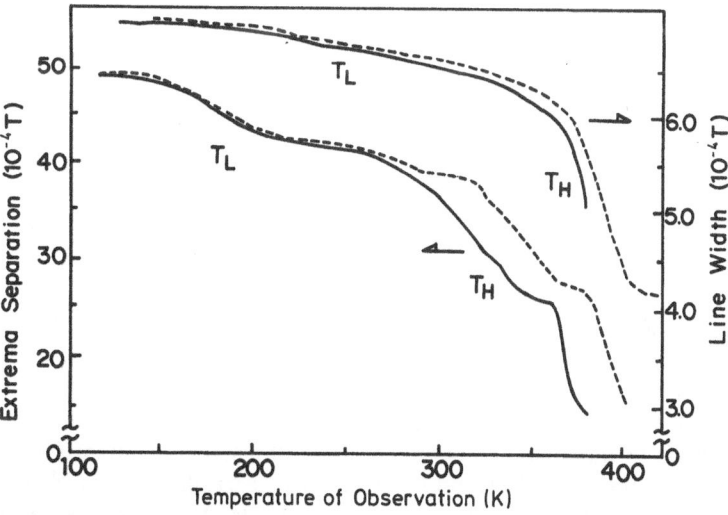

Fig. 10.12. Variations of line width and extrema separation with observation temperature of γ-irradiated materials labeled with nitroxide radical (Ref.[99])

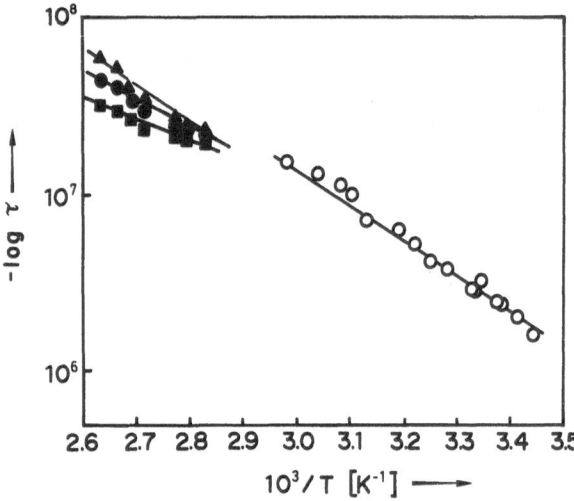

Fig. 10.13. Correlation time as a function of the inverse temperature: Estimated from line width (○); estimated from extrema separation based on Brownian diffusion (▲), moderate jump (●) and strong jump (■) (Ref. [99])

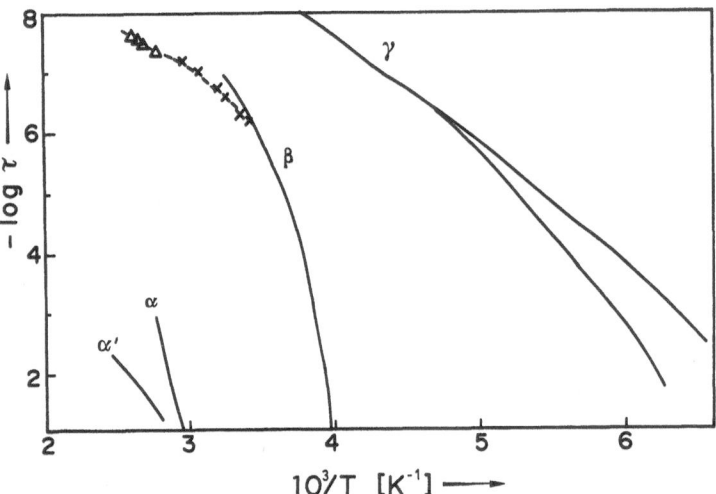

Fig. 10.14. Plots of correlation times estimated from line width (×) and extrema separation (moderate jump) (△) on the relaxation map arranged from mechanical and dielectric data (————) (Ref. [99])

figure. The influence of the first heating on the line width is similar to that on the extrema separation mentioned in the preceding section. It shows a similar temperature dependence both for line width and extrema separation; i.e., two narrowing regions, T_L and T_H, appear in both curves. It is seen that no change appears for T_L by the first heating, but T_H shifts to the lower temperature side in the second heating. This is a similar situation as seen in Fig. 10.11. In Fig. 10.12. there is a rather gradual change in the T_L region, while the change in T_H is quite remarkable. These two transition regions seem to correspond to the transitions in the regions of 200–350 K and 350 to 389 K, respectively, as observed by Bullock et al. [66].

In order to elucidate the character of the molecular motion associated with the narrowings of both extrema separation and line width in T_H, it is meaningful to discuss the correlation times calculated from extrema separation and line width. As mentioned in Sect. 9.1., correlation times can be estimated from the extrema separations by use of Eq. (9.5). The correlation time associated with the molecular motion reflected in the narrowing of the line width can be estimated by use of Eq. (8.1) if the distribution of the correlation time is not taken into consideration. Figure 10.13. shows the relation between correlation times calculated from the extrema separation and line width observed for the temperature region of T_H. As mentioned in Sect. 9.1, three models can be applied to calculate the correlation time with Eq. (9.5). However, the difference in the calculated values is relatively small and these values are in good accord with those obtained by extrapolating the correlation times estimated from the line width data shown in Fig. 10.13. Therefore, it can be considered that the decrease of extrema separation in the T_H region is caused by the same mechanism as that of the relaxation process reflected in the narrowing of the line width in T_H. The order of the correlation times plotted in Fig. 10.13. are seen to be located in the extrapolation of the curve for β-relaxation in the relaxation map arranged by Wada [19], as shown in Fig. 10.14. In the above discussions, no distribution of the correlation time has been considered. As referred to in Sect. 8, a distribution of the correlation time can be obtained from the data of the line width at various temperatures by taking the activation energy of the associated molecular motion as a parameter [see Eqs. (8.3) ∼ (8.6)]. Figure 10.15. shows a distribution obtained with various values of activation energies. It was reported that the distribution obtained from line width data discussed here is in good agreement with the distributions of relaxation time obtained from data of the dielectric dispersion and broad-line NMR studies when the activation energy was assumed to be 24 kcal/mol[99]. The value of 24 kcal/mol is quite reasonable when the molecular motion is considered to be associated with β-relaxation in polyethylene. According to the above results, it safely can be concluded that the transition reflected in temperature dependences of the line width and the extrema separation of the ESR spectrum originating from nitroxide radicals labelled onto the irradiated polyethylene is caused by a β-relaxation process.

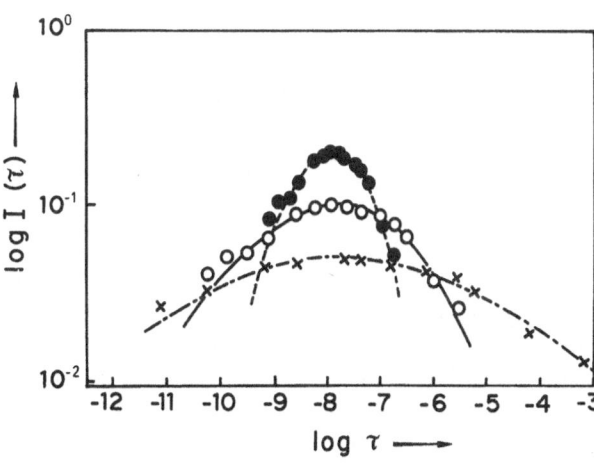

Fig. 10.15. Distribution of correlation time at T_r = 37.2 °C with the assumed values of activation energies of 42 kJ/mol (●), 84 kJ/mol (○), and 168 kJ/mol (×) (Ref.[99])

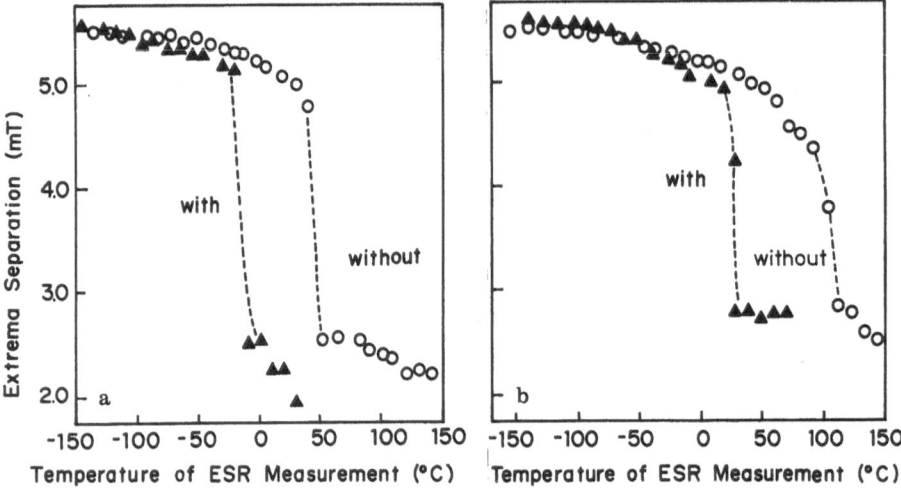

Fig. 10.16a and b. Effect of diluent on temperature dependence of extrema separation width; (a) for thermally degraded materials and (b) for γ-irradiated materials (Ref.[96])

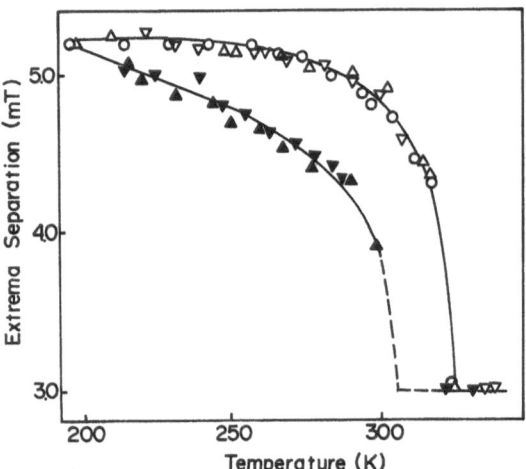

Fig. 10.17. Extrema separation vs. temperature plots for fractured low density polyethylene : ○, *in vacuo*; △, in toluene; ▽, in air; ▲, in CCl₄; ▼, in xylene (Ref. [91])

Several years ago, Kakizaki and Hideshima[100] reported the influence of diluent like tetrachloroethylene on the transition temperatures associated with relaxation processes in polyethylene. They showed that the temperature associated with β-relaxation shifts downward in the presence of the diluent and that the transition temperature for γ-relaxation shifts upward. These phenomena were interpreted in terms of an increase of the relative content of the amorphous part for the β-relaxation process and some freezing effect for γ-relaxation.

The effect of the diluent, tetrachloroethylene, to the narrowing curve of extrema separation for T_H was examined, leading to Fig. 10.16b.; i.e., the effect of the diluent is to shift the narrowing temperature downward. This is quite consistent with the results obtained by Kakizaki and Hideshima because the transition in T_H was interpreted in terms of β-relaxation as mentioned above. However, a similar diluent effect was also observed for thermally degraded materials as shown in Fig. 10.16a. The spectrum of nitroxide radicals in fractured polyethylene shows a very similar effect of the diluent (Fig. 10.17.), and a similar phenomenon was obtained also for fractured polypropylene. The motions observed for thermally degraded polyethylene or mechanically fractured polymers by the spin trapping technique are not attributed to β-relaxation. Therefore, it can be said that a plasticizing effect of the diluent may cause the influence of the diluent shown in Figs. 10.16. and 10.17., because narrowing temperatures in the present case are very high compared with those for the local relaxation in polyethylene.

11 Concluding Remarks and Acknowledgments

The present article is mainly a resume of rather recent studies of the authors' group though it contains results from another group and investigations completed jointly with others. The main intent is to show examples for using trapped or labelled free radicals having a small size, e.g., peroxy radicals for the study of molecular motion of solid polymers. Since ESR application to polymer physics has been motivating from a radiation chemical viewpoint, the authors have worked extensively in this field. Occasionally a close relation with the physical properties of solid polymers has been observed [9, 101]. This article focusses on the actual research including discussion of molecular motion. Estimation of the degree of orientation is also one of the important subjects in polymer physics and a few studies for applying ESR to this field have also been made. These shall be discussed elsewhere [37, 102].

The authors wish to acknowledge the excellent work done by Dr. S. Vivatpanachart described in Sect. 10.2. We are deeply indebted to Professors Y. Miyahara and H. Nomura for their sincere and stimulating contributions.

Several students and our technical staff also considerably contributed to our research. Especially, Mr. T. Tanigawa made an excellent contribution to the study mentioned in Sect. 8, the expertise of Miss T. Kitahara (now Mrs S. Araki) was essential in the research described in Sect. 10.3, and Mr. S. Aoyama was indispensable in completing the study shown in Sect. 7.4.

In the course of the research, the authors were awarded a research grant under the program "Grant-in-Aid for Scientific Research" of the Ministry of Education, Science, and Culture of the Japanese Government.

The authors would like to express their grateful thanks also to Miss T. Hiramatsu (now Mrs. T. Tanigawa), former secretary, for her help in preparing the manuscript of the authors' original papers (most of which are motivating the present article) and to Miss S. Kawai (now Mrs. M. Nakamura), secretary, for promptly completing the typescript.

12 References

1. Wilson, C. W., Pake, I. G. E.: J. Polym. Sci. *10*, 503 (1953)
2. Singer, J. R., Crooks, L. E.: Science *221*, 654 (1983)

3. Fraenkel, G. K., Hirshon, J. M.: J. Amer. Chem. Soc. *76*, 3606 (1954)
4. Abraham, R. J., Wiffen, D. H.: Trans. Faraday Soc. *54*, 1291 (1958)
5. a) Slichter, C. P.: Principles of Magnetic Resonance, 2nd Ed., Springer 1978
   b) Carrington, A., McLachlan, A. D.: Introduction to Magnetic Resonance, Harper and Row 1967
   c) Gordy, W.: Theory and Application to Electron Spin Resonance, J. Wiley and Son 1980
   d) Rånby, B., Rabek, J. F.: ESR Spectroscopy in Polymer Research, Springer 1977
6. Heller, C., McConnell, H. M.: J. Chem. Phys. *32*, 1535 (1960)
7. Shimada, S., Maeda, M., Hori, Y., Kashiwabara, H.: Polymer *18*, 19 (1977)
8. Kusumoto, N., Yamaoka, T., Takayanagi, M.: J. Polym. Sci. A-2 *9*, 1173 (1971)
9. Hori, Y., Kitahara, T., Kashiwabara, H.: Radiat. Phys. Chem. *19*, 23 (1982)
10. Kashiwabara, H.: Jpn. J. Appl. Phys. *3*, 302, 384 (1964)
11. Kashiwabara, H., Shimada, S., Sohma, J.: Nobel Symp. Series *No. 22*, 275 (1973)
12. Lawton, E. J., Balwit, J. S., Powell, R. S.: J. Chem. Phys. *33*, 395 (1960)
13. Cracco, F., Arvia, A. J., Dole, M.: J. Chem. Phys. *37*, 2449 (1962)
    Johnson, D. R., Wen, W. Y., Dole, M.: J. Phys. *77*, 2174 (1973)
14. Nara, S., Kashiwabara, H., Sohma, J.: J. Polym. Sci. A-2 *5*, 929 (1967)
15. Nara, S., Shimada, S., Kashiwabara, H., Sohma, J.: J. Polym. Sci. A-2 *6*, 1435 (1968)
16. a) Shimada, S., Kashiwabara, H.: Polym. J. *6*, 448 (1974)
    b) Shimada, S., Hori, Y., Kashiwabara, H.: Polymer *18*, 25 (1977)
17. Waite, T. R.: Phys. Rev. *107*, 463 (1957), J. Chem. Phys. *32*, 21 (1960)
18. Wada, Y.: J. Phys. Soc. Jpn. *16*, 1226 (1961)
19. Takayanagi, M., Matsuo, T.: J. Macromol. Sci. B *1*, 407 (1967)
20. Shimada, S., Hori, Y., Kashiwabara, H.: Polymer *22*, 1377 (1981)
21. Yoshida, H., Rånby, B.: J. Polym. Sci. A *3*, 1377 (1965)
22. Shimada, S., Hori, Y., Kashiwabara, H.: Polymer *19*, 763 (1978)
23. Monobe, K., Yokoyama, F.: J. Macromol. Sci. (B) *8*, 277 (1973)
24. Shimada, S., Fujiwara, A., Kashiwabara, H.: Polymer J. *13*, 769 (1981)
25. Bullock, A. T., Howard, C. H.: Mol. Phys. *27*, 949 (1974); J. Magn. Resonance *25*, 47 (1977)
26. Hori, Y., Shimada, S., Kashiwabara, H.: Polymer *18*, 567 (1977)
27. Hori, Y., Shimada, S., Kashiwabara, H.: Polymer *18*, 1143 (1977)
28. Matsuo, H., Dole, M.: J. Phys. Chem. *63*, 837 (1959)
29. Giberson, R. C.: J. Phys. Chem. *66*, 463 (1962)
30. Böhm, G. G. A.: J. Polym. Sci. A-2 *5*, 639 (1967)
31. Hori, Y., Shimada, S., Kashiwabara, H.: Polymer *18*, 151 (1977)
32. Ohnishi, S., Sugimoto, S., Nitta, I.: J. Polym. Sci. A, *1*, 605 (1963)
33. Michaels, A. S., Parker, R. B., Jr.: J. Polym. Sci. *41*, 53 (1959)
    Michaels, A. S., Bixler, H. J.: J. Polym. Sci. *50*, 393, 413 (1961)
    Bixler, H. J., Michaels, A. S., Salame, M.: J. Polym. Sci. A *1*, 895 (1963)
34. Hori, Y., Fukunaga, Z., Shimada, S., Kashiwabara, H.: Polymer *20*, 181 (1979)
35. Kashiwabara, H., Hori, Y.: Radiat. Phys. Chem. *18*, 1061 (1981)
36. Hori, Y., Makino, Y., Kashiwabara, H.: Polymer in press (1984) [preliminary report was published in: Rept. Prog. Polym. Phys. Jpn. *25*, 623 (1982)]
37. Shimada, S., Kotake, A., Hori, Y., Kashiwabata, H.: Macromolecules *17*, 1104 (1984)
38. Iwasaki, M., Sakai, Y.: J. Polym. Sci. Polym. Phys. Ed. *6*, 265 (1968)
39. Moriuchi, S., Nakamura, M., Shimada, S., Kashiwabara, H., Sohma, J.: Polymer *11*, 630 (1970)
40. Suryanarayana, D., Kevan, L., Schlick, S.: J. Am. Chem. Soc. *104*, 668 (1982)
41. Schlick, S., Kevan, L.: J. Am. Chem. Soc. *102*, 4622 (1980)
42. Suryanarayana, D., Kevan, L.: J. Phys. Chem. *86*, 2042 (1982)
43. Smith, A. E.: Acta Cryst. *5*, 224 (1952)
44. Hori, Y., Tanigawa, T., Shimada, S., Kashiwabara, H.: Polymer J. *13*, 293 (1981)
45. Van Vleck, J. H.: Phys. Rev. *74*, 1168 (1948)
46. Gutowsky, H. S., Pake, G. E.: J. Chem. Phys. *18*, 162 (1950)
47. Andrew, E. R.: J. Chem. Phys. *18*, 607 (1950)
48. Hori, Y., Aoyama, S., Kashiwabara, H.: J. Chem. Phys. *75*, 1582 (1981)
49. Chatani, Y., Anraku, H., Taki, Y.: Mol. Cryst. Liq. Cryst. *48*, 219 (1978)

50. Pemberton, R. C., Personage, N. G.: Trans. Faraday Soc. *61*, 2112 (1965), *62*, 553 (1966)
    Personage, N. G., Pemberton, R. C.: ibid. *63*, 311 (1967)
51. Gilson, D. F. R., McDowell, C. A.: Mol. Phys. *4*, 125 (1961)
52. Umemoto, K., Danyluk, S.: J. Phys. Chem. *71*, 3757 (1967)
53. Suryanarayana, D., Chamulltrat, W., Kevan, L.: J. Phys. Chem. *86*, 4822 (1982)
54. Hori, Y., Shimada, S., Kashiwabara, H.: Polymer *20*, 406 (1979)
55. Bloembergen, N., Purcel, E. M., Pound, P. V.: Phys. Rev. *73*, 679 (1948)
56. Miyake, A.: J. Polym. Sci. *28*, 476 (1958), Repts. Prog. Polym. Phys. Jpn. *3*, 119 (1960)
57. Gutowsky, H. S., Meyer, L. H.: J. Chem. Phys. *21*, 2122 (1953)
58. Shimada, S., Tanigawa, T., Kashiwabara, H.: Polymer *21*, 1116 (1980)
59. Gutowsky, H. S., Holm, C. H.: J. Chem. Phys. *25*, 1288 (1956)
60. Ohnishi, S., Sugimoto, S., Nitta, I.: J. Chem. Phys. *37*, 1283 (1962)
61. Shimada, S., Tanigawa, T., Kashiwabara, H.: Applied Spectroscopy *34*, 575 (1980)
62. Tamura, N.: J. Chem. Phys. *37*, 479 (1962)
63. Kivelson, D.: J. Chem. Phys. *33*, 1094 (1960)
64. Stone, T. J., Buckman, T., Nordio, P. L., McConnell, H. M.: Proc. Natl. Acad. Sci. U.S. *54*, 1010 (1965)
    Ohnishi, S., McConnell, H. M.: J. Am. Chem. Soc. *87*, 2293 (1965)
    Ogawa, S., McConnell, H. M.: Proc. Natl. Acad. Sci. U.S. *58*, 19 (1967)
65. Itzkowitz, M. S.: J. Chem. Phys. *46*, 3048 (1967)
66. Bullock, A. T., Butterworth, J. H., Cameron, G. G.: Europ. Polym. J. *7*, 445 (1971)
    Bullock, A. T., Cameron, G. G., Smith, P. M.: Polymer *13*, 89 (1972)
    Bullock, A. T., Cameron, G. G., Smith, P. M.: J. Polym. Phys. Ed. *11*, 1263 (1973)
67. Rabold, G. P.: J. Polym. Sci. A-1 *7*, 1187, 1203 (1969)
68. Törmälä, P., Savolainen, A.: J. Polym. Sci. Polym. Phys. Ed. *12*, 1251 (1974)
    Törmälä, P., Weber, G.: Polymer *19*, 1026 (1978)
69. Stryukov, V. B., Rozantsev, E. G.: Vysokomol. Soedin. A *10*, 626 (1968)
70. Janzen, E. G.: Account. Chem. Res. *4*, 31 (1971)
    Lagercrantz, C.: J. Phys. Chem. *75*, 3466 (1971)
71. Berliner, L. J. (ed.): Spin Labeling, Academic Press 1976
72. Törmälä, P.: J. Macromol. Sci.-Rev. Macromol. Chem. C *17*(2), 297 (1979)
73. Boyer, R. F. and Keinath, S. E. (eds.): Molecular Motion in Polymers by ESR, Harwood Academic Publishers 1980
74. Freed, J. H., Bruno, G. V., Polnaszek, C. F.: J. Phys. Chem. *75*, 3385 (1971)
    Goldman, S. A., Bruno, G. V., Freed, J. H.: J. Phys. Chem. *76*, 1858 (1972)
75. Kusumoto, N., Yonezawa, M., Motozato, Y.: Polymer *15*, 793 (1974)
76. Boyer, R. F.: Macromolecules *6*, 288 (1973)
77. Kumler, P. L., Boyer, R. F.: Macromolecules *9*, 903 (1976)
    Kumler, P. L., Boyer, R. F.: J. Macromol. Sci.-Phys. B *13*(4), 631 (1977)
78. Kusumoto, N., Sano, S., Zaitsu, N., Motozato, Y.: Polymer *17*, 448 (1976)
79. Sakaguchi, M., Kashiwabara, H.: J. Polym. Sci. Polym. Letter Ed. *18*, 563 (1980)
80. a) Sakaguchi, M., Yamakawa, H., Sohma, J.: J. Polym. Sci. Polym. Letter Ed. *12*, 193 (1974)
    b) Sakaguchi, M., Sohma, J.: J. Appl. Polym. Sci. *22*, 2915 (1978)
    c) Sohma, J., Sakaguchi, M.: Adv. Polym. Sci. *20*, 109 (1976)
81. Kurokawa, N., Sakaguchi, M., Sohma, J.: Polym. J. *10*, 93 (1978)
82. Vivatpanachart, S., Nomura, H., Miyahara, Y., Kashiwabara, H., Sakaguchi, M.: Polymer *22*, 132 (1981)
83. Kawashima, T., Shimada, S., Kashiwabara, H., Sohma, J.: Polymer J. *5*, 135 (1973)
84. Vivatpanachart, S., Nomura, H., Miyahara, Y., Kashiwabara, H., Sakaguchi, M.: Polymer *22*, 263 (1980)
85. Kurokawa, N., Sohma, J.: Polymer J. *11*, 559 (1979)
86. Yemni, T., McCullough, R. L.: J. Polym. Sci. *2*, 1385 (1973)
87. Kobayashi, M., Tadokoro, H.: Macromolecules *8*, 897 (1975)
88. Seto, T., Hara, T., Tanaka, K.: J. Appl. Phys. *7*, 31 (1968)
89. Sakaguchi, M., Kashiwabara, H.: J. Polym. Sci. Polym. Phys. Ed. *19*, 371 (1981)
90. Sakaguchi, M., Kashiwabara, H.: Polymer *23*, 1594 (1982)

91. Vivatpanachart, S., Nomura, H., Miyahara, Y., Kashiwabara, H., Sakaguchi, M.: Polymer *22*, 896 (1981)
92. Bullock, A. T., Cameron, G. G., Smith, P. M.: Eur. Polym. J. *11*, 617 (1975)
93. Hoffman, J. D., Williams, G., Passaglia, E.: J. Polym. Sci. C *14*, 173 (1966)
94. Keller, A., Priest, D. J.: J. Macromol. Sci. Phys. B *2*, 479 (1968)
95. Olf, H. G., Peterlin, A.: J. Polym. Sci. A-2 *8*, 771 (1970)
96. Kitahara, T., Shimada, S., Kashiwabara, H.: Polymer *21*, 1299 (1980)
97. Murabayashi, S., Shiotani, S., Sohma, J.: J. Phys. Chem. *83*, 844 (1979)
98. Terabe, S., Konaka, R.: J. Am. Chem. Soc. *93*, 4306 (1971)
99. Shimada, S., Kitahara, T., Kashiwabara, H.: Polymer *21*, 1304 (1980)
100. Kakizaki, M., Hideshima, T.: J. Macromol. Sci. B *8*, 367 (1973)
101. Shimada, S., Hori, Y., Kashiwabara, H.: Radiat. Phys. Chem. *19*, 33 (1982)
102. Shimada, S., Hori, Y., Kashiwabara, H.: Polymer J. *16*, 539 (1984)
103. Hori, Y., Kashiwabara, H.: J. Polym. Sci. Phys. Ed. *19*, 1141 (1981)
104. Nunome, K., Muto, H., Toriyama, K., Iwasaki, M.: Chem. Phys. Lett. *39*, 542 (1976)
105. Yoshida, H., Hayashi, K., Okamura, S.: Arkiv för Kemi *23*, 177 (1964)

Editor: W. Klöpffer
Received February 19, 1986

ESR Studies on Radical Polymerization

Mikiharu Kamachi

Department of Macromolecular Science, Faculty of Science, Osaka University, Toyonaka, Osaka 560, Japan

This article reviews typical applications of ESR spectroscopy to the radical polymerization of vinyl and diene compounds. ESR studies on the propagating radicals in both frozen and solid states are outlined, and the origin of complications attendent upon the analysis of ESR spectra are discussed. ESR observations of the intermediate species in the radical polymerization by the flow method and the spin-trapping method are surveyed, along with the chemical structure and reactivity of the propagating radicals. Special cavities designed for enhancing the sensitivity of ESR spectrometers to radical polymerization are described. These cavities allow determinations of the conformation of propagating radicals as well as the propagation rate constants for several monomers under conditions similar to usual radical polymerizations.

Advances in Polymer Science 82
© Springer-Verlag Berlin Heidelberg 1987

1 Introduction

Since electron spin resonance (ESR) phenomenon was first observed in 1945 by the Soviet physicist Zavoisky [1], the ESR spectroscopy has made a great contribution to the progress in physics, chemistry, and biology. Basically, it allows directly tracing the behavior of unpaired electrons, that is, the paramagnetism of free radicals, ions of variable valence, and triplet molecules. Especially, the following three features make ESR spectroscopy advantageous in acquiring physical and chemical information on organic free radicals. First, the total intensity of the absorption line reflects directly the number of paramagnetic species in the sample. Second, the hyperfine splitting constant (hfc) and the g-value for the spectrum make it possible to approach the detailed structure of the paramagnetic species. Third, the line shape of the spectrum is sensitive to the electron distribution, molecular orientation, and molecular motion of the paramagnetic species, and the nature of their environments.

Applications of ESR spectroscopy to polymer research began in the early 1950s, with the first ESR spectrum observed on X-ray irradiated poly(methyl methacrylate) [2]. Because the mobility of polymer chains is greatly reduced and hence radicals are effectively stabilized in the solid state, ESR studies on free radicals have since been mainly performed on γ-ray or light irradiated polymers in bulk.

In principle, it is possible by ESR spectroscopy to measure directly the structure, the electronic state, and the concentration of propagating radicals. Therefore, this spectroscopic technique has been considered the best method, at least potentially, to gain a clear understanding of the nature of the propagating radicals in the radical polymerization of vinyl compounds and to determine the propagation rate constants [3]. Unfortunately, the concentration of propagating radicals was too low to be detected by commercial ESR spectrometers. Thus, previous ESR studies on radical polymerization have been limited to such special conditions as the frozen state and crystaline state or have employed such a special technique as the flow method. However, these measuring conditions are far apart from the practical conditions for radical polymerization, and hence are of little use for quantitative estimation of the rate constants for usual radical polymerizations. Most of the current kinetic studies of radical polymerization still resort to the rotating sector method whose results depend on the theoretical model used for data analysis. It is therefore natural that there is a growing desire for expanding ESR spectroscopy to polymerization systems whose kinetic behavior cannot be accessed by the above-mentioned techniques.

The present paper summarizes and discusses the historical developments in ESR study on radical polymerization, putting emphasis on a new technique using a special cavity.

2 Scope

Chapter 3 outlines briefly the principle of ESR spectroscopy, the origin of hyperfine splitting, and its implications. This is followed by a discussion on the change in the spectrum due to interchange of conformations of free radicals.

Chapter 4 deals with ESR spectroscopy of the propagating radicals in both frozen and solid states. Complications attendant upon the analysis of ESR spectra in the solid state are discussed.

Chapter 5 is concerned with ESR observations of the intermediate species in radical polymerizations by the flow method. The chemical structure and conformation of the intermediates and the dependence of their spectra on their degree of polymerization are discussed. Further, kinetic studies of radical polymerizations are summarized.

Chapter 6 is concerned with the spin-trapping method. Its application to the identification of growing chain ends and the estimation of monomer reactivity are described.

Chapter 7 is devoted to direct ESR measurements of propagating radicals by ESR spectroscopy under conditions similar to the conventional radical polymerization. Special cavities used for this purpose are described. Directly measured rate constants are compared with the values obtained by other methods.

3 Electron Spin Resonance and Structure of Organic Free Radicals

3.1 Introduction

An ESR spectrum arises from the transition of an electron spin from one energy level to another. Interactions between spins or between spins and the external magnetic field give rise to various energy levels in a spin system. Sometimes, the transitions between different energy levels are affected by intramolecular motions of the paramagnetic species, yielding a spectrum which varies with temperature. For a correct interpretation of observed ESR spectra, we have to relate the transition of an electron spin to fundamental properties of the paramagnetic species.

This chapter reviews briefly the principle of ESR spectroiscopy and the influence of the dynamic behavior of paramagnetic species on their ESR spectra. For details of this spectroscopy, the reader is referred to the books cited in Ref. [4-7].

3.2 Magnetic Resonance Phenomenon

A free radical has an unpaired electron characterized by a spin vector $\mathbf{S}$. The magnetic moment μ_e of the electron is proportional to the spin angular momentum $h\mathbf{S}$, i.e.,

$$\mu_e = -\gamma h \mathbf{S} = -g\beta \mathbf{S} \tag{3-1}$$

where the proportionality constant γ is referred to as the magnetogyric ratio of the electron and equals $1.76 \times 10^7 \text{ sec}^{-1} \text{ G}^{-1}$, g the electron g factor which equals 2.0023 for a free electron and is about 2.0 for most organic free radicals, and $\beta = eh/2mc = 0.927 \times 10^{-20} \text{ erg G}^{-1}$ referred to as the electronic Bohr magneton.

When a free radical is in an applied magnetic field $\mathbf{H}$, its magnetic moment vector has two possible orientations, in which the spin vector $\mathbf{S}$ is parallel and antiparallel to the magnetic field $\mathbf{H}$. If the field $\mathbf{H}$ is in the z direction and the z component of $\mathbf{S}$ is denoted by m_s, the energy E of the spin is represented by

$$E = g\beta H m_s \tag{3-2}$$

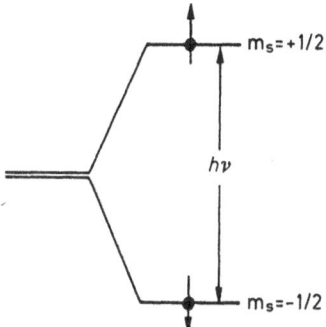

$m_s = +1/2$

$h\nu$

$m_s = -1/2$

Fig. 1. Energy levels for a singlet system showing ESR absorption

Since $S = 1/2$ for the electron, m_s has two values $\pm 1/2$, giving the two energy levels shown in Fig. 1; the lower state has spin down ($m_s = -1/2$), that is, the spin is anti-parallel to the applied magnetic field. The transition between the upper and lower energy levels is induced by a sinusoidal microwave irradiation of the frequency ν which satisfies the relation

$$h\nu = g\beta H \tag{3-3}$$

This transition gives rise to an ESR spectrum.

3.3 Line Shape and Relaxation

A spin system undergoes a radiationless transition by a fluctuating magnetic field of the surroundings which are conventionally called the lattice. The inverse rate of this type of transition is referred to as the spin-lattice relaxation time and is denoted by the symbol T_1. Since only the paramagnetic species in the fluctuating field which satisfy Eq. (3-3) undergo transitions between different energy levels, the ESR line shape depends on T_1.

In addition to the spin-lattice relaxation, there exists a spin-spin relaxation by which energy is redistributed within the spin system. In both liquid and solid phases, the net local magnetic field H rapidly varies owing to molecular motion. At high spin concentrations, direct spin-spin exchanges also can occur. The relaxation time associated with these exchanges is denoted by T_2.

In the liquid state, ESR lines are generally Lorenzian in shape, as theoretically predicted by Bloch's theory; i.e.,

$$I = I_0 \times \frac{(\Delta H_{1/2})^2}{(\Delta H_{1/2})^2 + (H - H_0)^2} \qquad \text{(Lorenzian)}$$

where I_0 is the absorption intensity at the centroid in the magnetic field H_0 and $\Delta H_{1/2}$ the line half-width.

In the solid state where molecules are either essentially in rest or tumbling slowly, each electron spin experiences a local static field due to dipolar interaction with the surrounding nuclear spins. In a non-viscous solution, this field is averaged to zero by rapid molecular motion, while, in a viscous solution, this does not occur and the resonance frequency spreads. This phenomenon, referred to as inhomogeneous broadening, should be distinguished from pure relaxation broadening. In general, ESR lines in the solid state have a Gaussian or a near Gaussian shape; i.e.,

$$I = I_0 \exp\left[-(H - H_0)^2/\Delta H^2\right] \qquad \text{(Gaussian)}$$

3.4 Hyperfine Structure

The interaction of an electron spin with nuclear spins gives rise to an ESR spectrum with many lines referred to as the hyperfine structure. The energy difference between adjacent line peaks is defined as the hyperfine splitting.

To evaluate the magnetic energy levels of the electron spin, the following spin-Hamiltonian H is used:

$$H = g\beta H \cdot S - g_N\beta_N H \cdot I + \sum_n a_n I_n \cdot S \qquad (3\text{-}4)$$

where a_n is the hyperfine splitting constant, with n denoting the number of nuclear spins in the system. The first term is the electron-Zeeman interaction, and if the external field is in the z direction, this equals $g\beta HS_z$. The second term, which represents the nuclear-Zeeman interaction, generally does not affect the position or the number of ESR lines. Hence, it can be neglected in analyzing ESR spectra. The third term is related to the hyperfine structure, and each term of the sum can be written $a(I_xS_x + I_yS_y + I_zS_z)$. The I_xS_x and I_yS_y operators affect the energy level only to a second order. Thus, to a first order, Eq. (3-4) may be written

$$H = g\beta HS_z + \sum_n a_n I_z S_z \qquad (3\text{-}5)$$

The energy levels of a spin system with $m_s = \pm 1/2$ and $m_I = -I, -I + 1, \ldots, I - 1$, I are given by

$$E = g\beta H m_s + m_s \sum_n a_n m_I \qquad (3\text{-}6)$$

Figure 2 shows the energy-level diagram for CH_3 calculated by use of Eq. (3-6). Transition does not always occur between every pair of these energy levels. By calculating transition probabilities, the following selection rules can be derived:

$$\Delta m_s = \pm 1 \qquad (3\text{-}7)$$

$$\Delta m_I = 0 \qquad (3\text{-}8)$$

where Δm_s and Δm_I denote the differences in the quantum number of electron spins and nuclear spins, respectively. The allowed transitions which obey the rules are shown in Fig. 2. The hyperfine structure for a free radical with n equivalent protons appears as a series of lines whose intensities are proportional to the expansion coefficients of the binomial $(1 + X)^n$. In a free radical with two sets of equivalent protons A and B

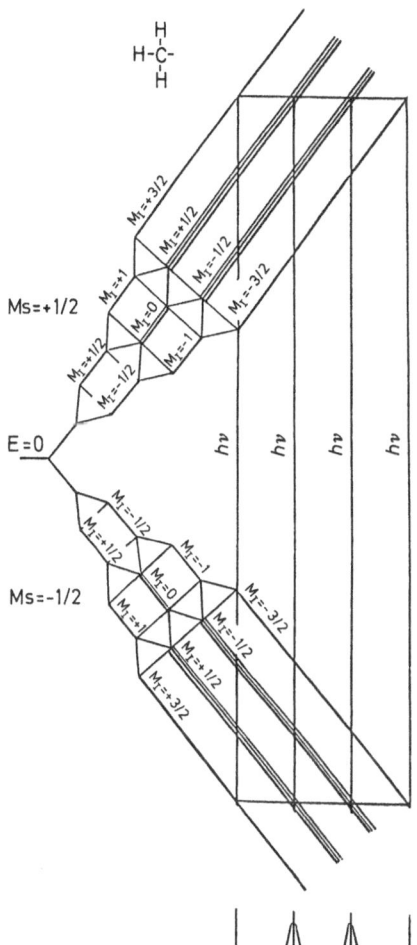

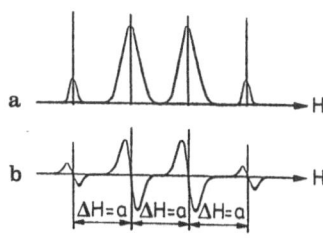

Fig. 2a and b. Energy levels for methyl radical in variable external magnetic field and at constant frequency [8]. (**a**) Absorption spectrum, (**b**) First derivative spectrum

consisting of n_1 and n_2 protons respectively, the hyperfine structure due to A protons consists of $(n_1 + 1)$ lines, each of which is further split into $(n_2 + 1)$ lines owing to B protons. Thus, the resulting ESR spectrum consists of $(n_1 + 1)(n_2 + 1)$ lines, though some of them sometimes overlap [Fig. 3(a–c)].

There are two types of interaction leading to a hyperfine structure: (1) the contact interaction, which is isotropic and results from the delocalization of the unpaired electron on the nucleus, and (2) the dipolar interaction between electron and nucleus spins, which is directional and thus anisotropic. For free radicals tumbling rapidly in a non-viscous solution, the latter is averaged to zero by thermal motion.

Many organic radicals have unpaired electrons localized in their 2p-orbitals.

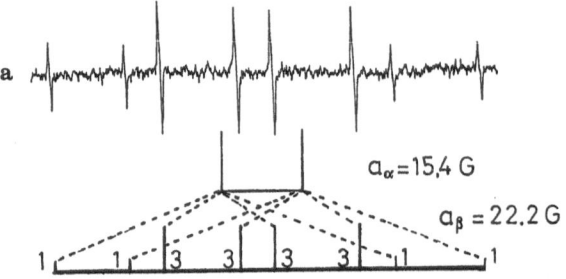

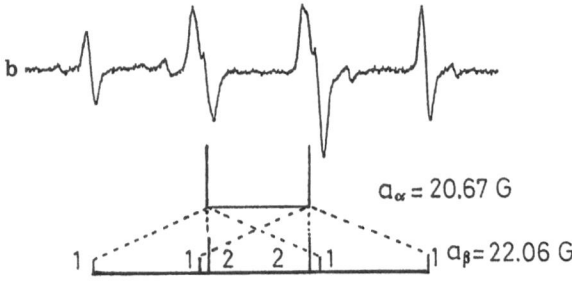

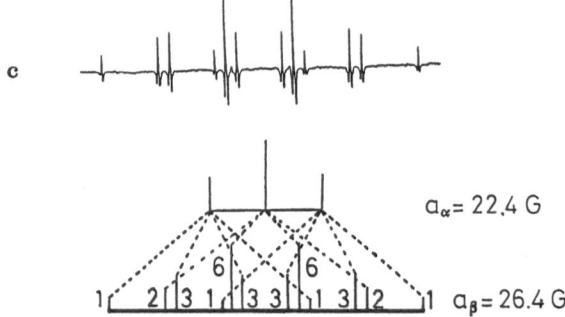

Fig. 3a–c. Organic radicals with two equivalent sets of hydrogen. (a) $CH_3\dot{C}HOH$ [9]; (b), HO—
—$[CH_2CH(COOH)]_nCH_2\dot{C}H(COOH)$ [10]; (c), $CH_3\dot{C}H_2$ [11]. (a) and (b), first derivative spectra;
(c), second derivative spectrum

Consider the ethyl radical as an example (Fig. 4). Since its 2p-orbital has a node at the
nucleus, no contact interaction is likely to take place in this organic free radical.
However, the spin exchange interaction of the unpaired electron of the 2p-orbital
with the electrons of the σ-orbitals with some partial s character can produce a net
spin density at the nucleus. Furthermore, this interaction causes ·the electron spins
to polarize at each α-proton and leads to a hyperfine structure. Accordingly, the hyperfine splitting constant a_H for the α-proton is proportional to the unpaired electron
density ϱ_c at the radical center. Thus we obtain the relation [12]

$$a_H = Q_{CH} \cdot \varrho_c \qquad\qquad (3\text{-}9)$$

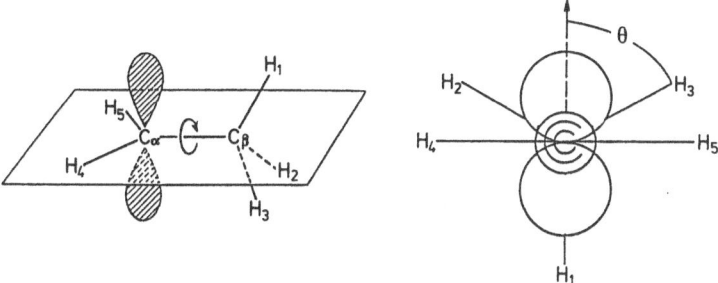

Fig. 4. Structure of ethyl radical and definition of the dihedral angle θ for hyperfine splitting due to α-hydrogen

where the parameter Q_{CH} corresponds to the splitting for which the unpaired electron density at the carbon atom bearing the electron becomes unity.

β-protons also give a hyperfine structure by their interactions with the unpaired electron, which are of the same order as those of α-protons with the unpaired electron (Fig. 3). The spin-polarization has to pass through two bonds to reach protons, so that the hyperfine splitting constant for the β-proton should be smaller than that for the α-proton. Therefore, the hyperfine interaction of the electron spin with β-protons may be explained by hyperconjugation, which allows some spin density to produce in the β-proton 1s orbitals, and its magnitude depends on the angle between the C_β—H bond and the $2p_z$ orbital, both projected on the plane perpendicular to the C_α—C_β bond bearing the unpaired electron (Fig. 4). The hyperfine splitting constant a_β associated with the β-proton is given by

$$a_\beta = (B_0 + B \cos^2 \theta) \varrho_C \qquad (3\text{-}10)$$

where B_0 and B are constants, and ϱ_C the spin density at the radical center [13]. Sometimes, B_0 is neglected in the evaluation of θ, since it is much smaller than B.

3.5 Dynamical Effect on ESR Spectra

The ESR spectrum of a free radical is affected by intramolecular motions such as internal rotation and ring inversion. If a magnetic nucleus changes its conformation from A to B by rotational motion, it experiences different effective magnetic fields and has time-dependent hyperfine splitting constants $a_A(t)$ and $a_B(t)$, where t denotes time. The ESR spectral pattern depends on how a(t) is averaged out. For a radical which undergoes a slow rotational motion, the lifetimes of the nucleus at conformations A and B, τ_A and τ_B, are longer than the reciprocal of the frequency difference between the corresponding resonance lines at v_A and v_B, and the ESR signals associated with A and B are observed. The line-width Γ_i of signal i in the limit of slow rotational exchange is given by [14]

$$\Gamma_i = \Gamma_0 + (2\tau_i \gamma)^{-1} \qquad (3\text{-}11)$$

where Γ_0 is the line-width in the absence of the rotational motion, and τ_i denotes τ_A or τ_B.

If the rate of rotation is much faster than the reciprocal of $v_A - v_B$, the nucleus experiences an effective field which is a weighted average of the magnetic fields. Hence, there is observed a single signal whose position is between those of the signals at v_A and v_B. In this case, the line-width Γ is given by [14]

$$\Gamma = \Gamma_0 + \gamma\tau(\delta B_0)^2/4 \qquad (3\text{-}12)$$

where δB_0 is the separation between the two resonance lines in the absence of the exchange, and $\tau = \tau_A\tau_B/(\tau_A + \tau_B)$. For very fast rotational motion, the spectrum narrows to the limiting line width Γ_0.

At intermediate rates of rotation, the spectrum is not quite simple. The widths of the respective resonance lines depend not only on the hyperfine splitting constants a_A and a_B but also on the nuclear spin quantum numbers m_A and m_B, each given by [14]

$$\Gamma = \Gamma_0 + \gamma\tau(a_A - a_B)^2 (m_A - m_B)^2 \qquad (3\text{-}13)$$

Figure 5 shows the hypothetical spectra of a free radical having two sets of two protons which exchange at slow, intermediate, and fast rates. At a slow rate, the observed ESR spectrum has a triplet of triplets due to the two sets of two protons. At an intermediate rate, no broadening occurs in the lines whose positions are either determined from the hyperfine splitting caused by the two sets with the same nuclear spin quantum number ($m_I = 2$ or -2 and then $m_A - m_B = 0$) or correspond to the special condition $m_A = m_B = 0$. Other lines are broadened to varying extents, depending on the nuclear spin quantum number. At a fast rate, the observed spectrum is

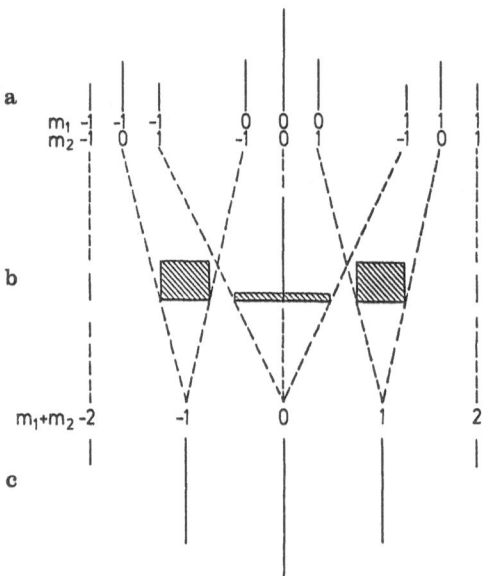

Fig. 5a–c. Effect of exchange rate on spectrum for two equivalent sets of two hydrogens with hyperfine splitting constants a_1 and a_2 [14]. (**a**) slow, (**b**) intermediate, and (**c**) fast exchange rates

narrowed to a quintet because all the four protons are completely indistinguishable in the time average.

For free radicals with more than two conformations, information about their dynamical processes such as internal rotation and ring inversion can be obtained by measuring the spectrum over a range of temperature. However, analysis of the spectrum becomes more difficult and uncertain.

4 Solid State Polymerization

4.1 Introduction

Radical polymerization is initiated by a free radical, which subsequently adds to a vinyl or diene monomer to produce a propagating radical. To obtain information about the structure and concentrations of initiating and propagating radicals in radical polymerizations, use of ESR spectroscopy has called the interest of physical or polymer chemists. However, ESR measurements on these radicals in solution polymerization were found to be difficult, except for the case where polymers precipitated, because otherwise the concentrations of the radicals were too low. Thus, these measurements had to be limited to polymerization systems in highly viscous solutions or in the solid state, where the disappearance of free radicals by bimolecular reactions is suppressed. Bresler et al. [15–17] succeeded for the first time in obtaining ESR spectra of free radicals which were produced in homogeneous bulk polymerization of methyl methacrylate (MMA), methyl acrylate (MA) and vinyl acetate (VAc) at conversions of 50–60% (in the gel state).

Generally, ESR spectra of propagating radicals in a gel or in the solid state are not well resolved, and sometimes their intensity distributions depend on measurement conditions such as temperature and the nature of a matrix used to obtain the frozen state. Thus, the spectra have led to the proposal of various interpretations.

4.2 Radical Polymerization of Methacrylic Acid and its Derivatives in the Amorphous State

A 9-line spectrum shown in Fig. 6 was obtained in a photo-initiated radical polymerization of MMA in its polymer at room temperature [15]. The 9 lines consist of strong 5 lines, which are characterized by a hyperfine splitting constant of 23.4 G and whose intensity distribution is 1:4:6:4:1, and weak 4 lines, each appearing between the 5 lines. The intensity distribution of these 9 lines seemed abnormal for a single species. Similar ESR spectra were independently obtained for the x-ray or γ-ray decomposition of polyMMA [18–22] (see Fig. 7a), and various interpretations were proposed for them.

Piette [23] found that the intensity of the 5-line component decayed more rapidly than that of the 4-line component when the temperature was raised from room temperature to about 100 °C. This finding led him to consider that the 9-line spectrum may be attributed to two kinds of radicals, one being a species of high concentration giving the 5-line component and the other a species giving the 4-line component.

However, Symons [22] showed that the abnormal intensity distribution in the 9-line

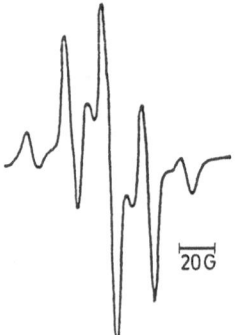

Fig. 6. ESR spectrum of radical trapped in polyMMA in polymerization of MMA at room temperature [15]

spectrum and its temperature dependence can be explained without invoking two radicals. His argument is as follows. If the C—H bonds of the β-methylene of a propagating radical is symmetrical about the p-orbital in the projection on the radical plane, two β-hydrogens H_a and H_b differ in that one is close to the methyl group and the other to one of the oxygen atoms of the carbomethoxy group as illustrated in Fig. 7c. Perhaps, such different environments somewhat distort the symmetrical conformation. If H_a and H_b are not equivalent and their dihedral angles between the C—H bonds and the p-orbital of the unpaired electron are $\theta_a = 65°$ and $\theta_b = 55°$ (Fig. 7c), a 13-line spectrum should be obtained, because the hyperfine splitting constant a_H calculated from

$$a_H = B \cos^2 \theta \qquad (4\text{-}1)$$

with $B = 46$ G is 8.1 G for H_a and 15.1 G for H_b; this value of B was estimated by using a value of 23 G for the hyperfine splitting constant of the three protons of the

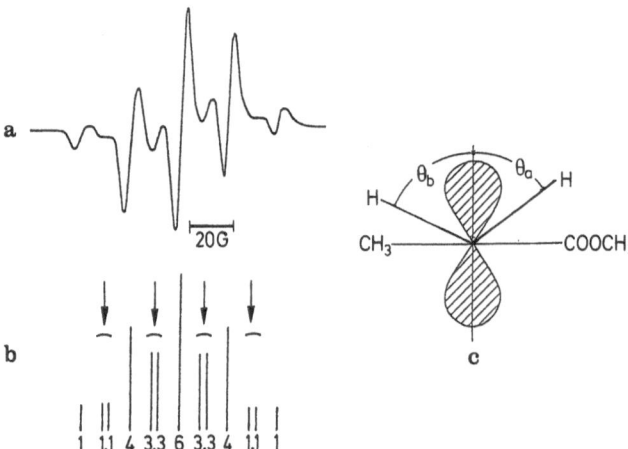

Fig. 7a–c. Possible reconstruction of 9-line spectrum [22]. (**a**) observed ESR spectrum of γ-irradiated polyMMA; (**b**) suggested reconstruction; (**c**) projections (θ_a, θ_b) of the C—H bonds of β-methylene on the plane perpendicular to the C_α—C_β bond

freely rotating methyl group. When the neighboring two lines of the inner 8 lines, indicated with arrows in Fig. 7b, are close enough to appear as a single broad line, the 13-line spectrum reduces to a 9-line spectrum consisting of main 5 lines and broad 4-lines, as observed experimentally (Fig. 7a).

For a photo-initiated bulk polymerization of MMA in the frozen state, Sohma et al. [24] did not find the 9-line spectrum during the initial stage, in which most molecules either remain monomeric or have low degrees of polymerization. As the polymerization proceeded, the 9-line spectrum gradually appeared. These findings suggest that the 9-line spectrum is not due to the monomer radical but to highly polymerized radicals. Moreover, Sohma et al. [24] showed that the abnormal distribution of the 9-line spectrum can be reproduced by a weight superposition of the spectra for two stable conformations of a single propagating radical, which are characterized by $\theta_a = \theta_b = 60°$, and $\theta_a = 45°$ and $\theta_b = 75°$.

Harris et al. [25] were concerned with $FeCl_3$-photo-initiated polymerizations of MAA, MMA. and other methacrylates in rigid glass of methanol or acetone at the liquid nitrogen temperature. The propagating radical of methacrylate initially gave a 5-line spectrum at -160 °C, which gradually changed to a 9-line spectrum as the temperature was raised to -140 °C. Harris et al. explained the 9-line spectrum as due to the existence of two polymer radicals, one having a conformation that gives 5 lines and the other having a conformation that generates 4 lines.

4.3 Radical Polymerization of Methacrylic Acid and its Barium Salt in the Crystalline State

O'Donnell et al. [26] investigated γ-ray initiated polymerization of a single crystal of barium methacrylate dihydrate at room temperature. When the crystalline monomer was irradiated at -196 °C, no polymer was produced. When the irradiated crystal was allowed to stand at 25 °C, the conversion after 670 hr was only about 3%, but at 50 °C, the conversion reached 58% after 16 hr, yielding an amorphous polymer. These findings suggest that the molecules should have a certain mobility, since for the polymerization in the crystal to occur the monomer must be transported to the growing chain by transposition or diffusion.

A single crystal of barium methacrylate dihydrate γ-irradiated at -196 °C was examined by ESR [26]. The recorded spectrum was isotropic and consisted of 7 lines spaced at 23 G (Fig. 8a). It was consistent with the hyperfine splitting constant for equivalent 6 hydrogens, and can be assigned to the species $(CH_3)_2\dot{C}COO^-$ formed by the addition of a hydrogen to the double bond of the methacrylate ion. When the irradiated crystal was warmed to -80 °C, the spectrum changed to one due to another species. The change completed in about 2 hr at 20 °C. The final spectra at 20 °C are shown in Fig. 8b, 8c, and 8d. Anisotropic 9-line spectra are seen for the magnetic field pararell to the a- or b-axis of the pseudo-orthombic monomer crystal, and a 13-line spectrum for the magnetic field pararell to the c-axis. These spectra were attributed to the following propagating radical:

$$RCH_2\dot{C}H(CH_3)COO^- \qquad\qquad 4\text{-}I$$

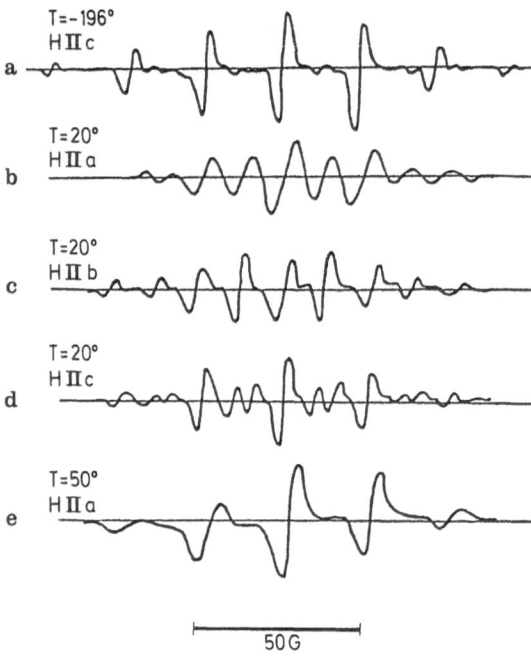

Fig. 8a–e. ESR spectra of single crystal of barium methacrylate [26]. **(a)**, taken at −196 °C, **(b)**, **(c)**, and **(d)** taken after warming the sample for 2 hr at room temperature, and **(e)** taken after 200 hr at 50 °C [26]

When polymerization was carried out to a high conversion at 50 °C, the anisotropic 9-line spectra all gradually changed to an isotropic 5-line spectrum (Fig. 8e). The anisotropy of the ESR spectra at 20 °C was ascribed to a dimeric radical produced by addition of the primary radical to one monomer unit. Preferred orientation of the dimeric radical in the crystallographic direction of the parent monomer was not considered advantageous for the addition of the dimer radical to the monomer. The dimer radical transformed at a significant rate to longer chain radicals with random orientation only at 50 °C. The 5-line spectrum was explained in terms of the conformation in which one of the methylene hydrogens is magnetically equivalent to the three hydrogens of the α-methyl group and the unpaired electron has a negligible interaction with the other methylene hydrogen.

Later, O'Donnell et al. [27] pointed out from a computer simulation that the temperature dependence of the above-mentioned spectra arises from a variation in the conformational angles of the two methylene carbon-hydrogen bonds with the p-orbital of the radical. These angles are affected by (1) the degree of polymerization, (2) the temperature, and (3) the environment of the radical end.

4.4 The Influence of the Environment of a Radical on its ESR Spectrum

Bamford et al. [28] showed that the solid state polymerization of crystalline methacrylic acid (MAA) can be initiated by UV irradiation below 250 nm, instead of γ-ray. Polymerization hardly occurred below −20 °C, but radical formation below −20 °C

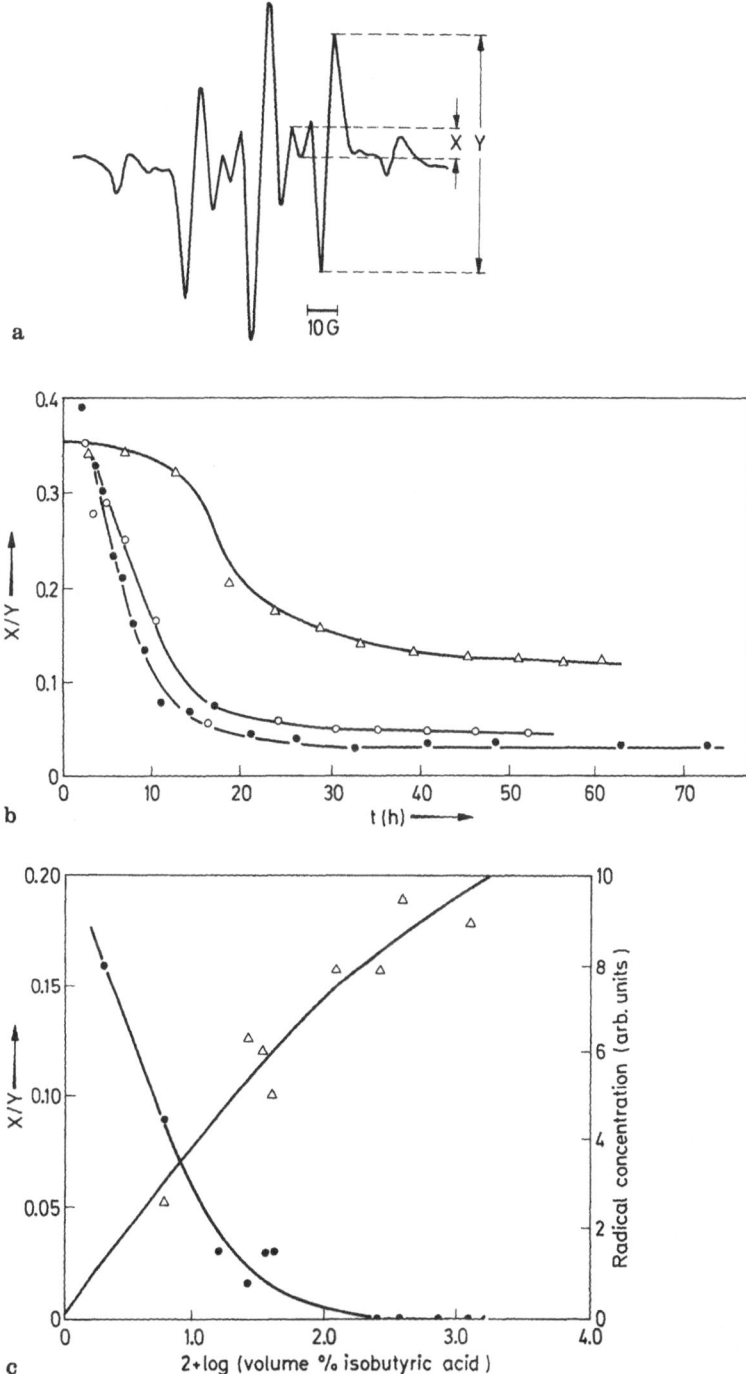

Fig. 9a–c. (a) ESR spectrum of irradiated MAA crystal, with definition of X and Y [28]. The spectrum is a mixture of 9- and 13-line components. (b) Variation of X/Y with irradiation time at; ●, −20 °C; ○, −30 °C; and △, −50 °C. (c) Dependence of radical concentration [R'] (△) and X/Y (●) on isobutyric acid concentration; 11 hr irradiation at −14 °C

was detected as a 13-line ESR spectrum (Fig. 9a), which differs from the well-known 9-line spectrum of polyMMA radical. The 13-line spectrum eventually changed to a mixture of 13-line and 9-line spectra on prolonged irradiation even below −20 °C, and the mixed spectrum depended on temperature. The intensity ratio X/Y of these two spectra varied with irradiation time, as shown in Fig. 9b.

When the specimen irradiated below −20 °C was maintained at 0 °C where poly-merization occurred at a significant rate, the 13-line spectrum changed in a much shorter time to the conventional 9-line spectrum. It was concluded that the radicals formed by irradiation initially have a specific conformation giving the 13-line spectrum and eventually relax to the usual conformation giving the 9-line spectrum. Both the rate of polymerization and the molecular weight of resulting polyMAA greatly increased by the addition of a low concentration of chemically inactive isobutyric acid to the monomer. The influence of the concentration of isobutyric acid in the MAA crystal on the radical center was studied by ESR spectroscopy. Figure 9c shows that the total concentration of the radicals formed by 11 hr UV irradiation at −14 °C increases rapidly with an increase in the initial concentration of isobutyric acid. This figure also illustrates the ratio X/Y to decrease markedly and vanish at an iso-

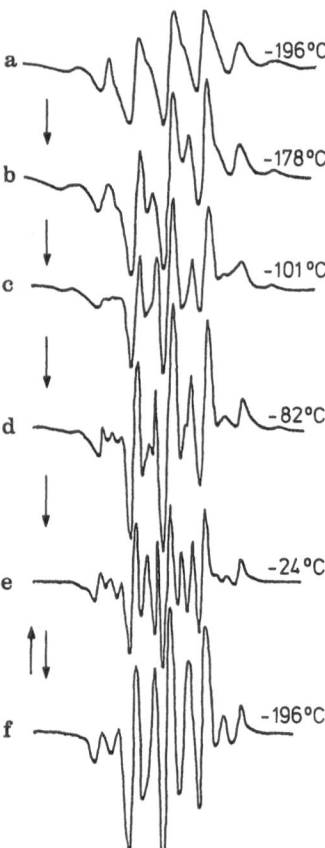

Fig. 10a–f. Temperatures variation of ESR spectrum for MAA γ-irradiated at −196 °C [29]. Change between curves (e) and (f) is reversible

butyric acid concentration of one percent by volume, suggesting that the initiating
and propagating radicals be formed in defect regions of the crystals in which the
monomer can move rather freely.

Iwasaki et al. [29], studying more systematically the temperature dependence of the
ESR spectra of free radicals trapped in solid MAA γ-irradiated at −196 °C, obtained
the data illustrated in Fig. 10. The 7-line spectrum at −196 °C corresponds to the
monomer radical which was formed by addition of a hydrogen atom to the double
bond. This spectrum gradually changed to a 9-line spectrum with increasing temperature from −196 to −178 °C (Fig. 10b), indicating that the monomer radical began
to react with neighbouring monomers to form a propagating radical. As the temperature was increased further to −101 °C, the monomer radical almost disappeared,
leaving the 9-line spectrum shown in Fig. 10c. At −82 °C, each line of the 4-line
component of the 9-line spectrum began to split into a doublet (Fig. 10d), and the
9-line spectrum changed to a 13-line spectrum at −24 °C. The 13-line spectrum is
essentially the same as the one obtained for UV irradiated MAA below −5 °C [28].
On cooling the sample very slowly (20–30 °C/hr) from −24 to −196 °C, the 13-line
spectrum again changed to the 9-line spectrum shown in Fig. 10f. This change was
thermally reversible.

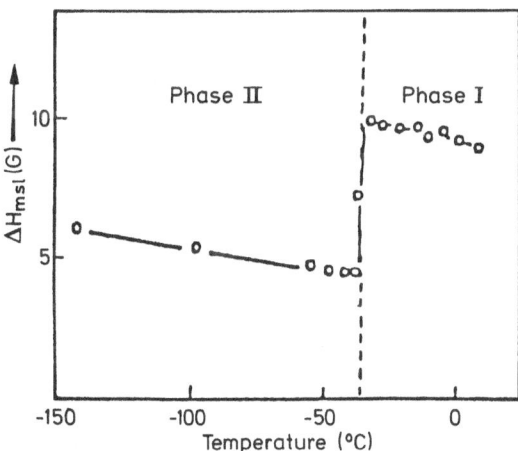

Fig. 11. Temperature variation of the
maximum slope distance (ΔH_{msl}) in
broad-line NMR spectra of MAA [29]

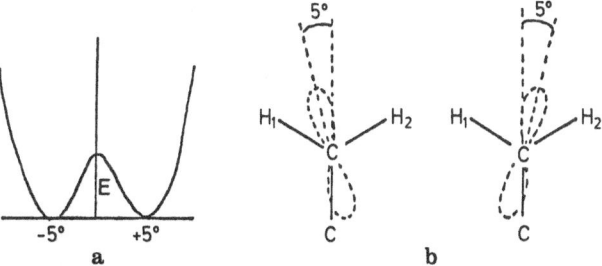

Fig. 12a and b. (a) Schematic potential energy curve of MAA radical for hindered oscillation about
the $C_\alpha - C_\beta$ bond; (b) two stable conformations which are mirror images [29]

Parallel with this ESR study, broad-line NMR measurements were carried out at various temperatures. Figure 11 shows the temperature change in the maximum slope distance ΔH_{msl} in the NMR spectrum of MAA. When the monomer was rapidly cooled from 20 to -196 °C, ΔH_{msl} gradually decreased with increasing measurement temperature, and was suddenly narrowed by about 5 G when the temperature reached about -40 °C, as indicated in Fig. 11. This sharp change in ΔH_{msl} was reversible, suggesting that there occurs a crystalline transition at about -40 °C. By making the assumption that the potential barrier in phase I of Fig. 11 is higher than that in phase II, Iwasaki et al. [29] interpreted this temperature dependence of ESR spectrum as arising from the fact that the two β-methylene carbon-hydrogen bonds are interchanged by hindered oscillation about the C_α—C_β bond of the single radical $\sim$CH$_2$Ċ(CH$_3$)—
—COOH (see Fig. 12).

4.5 Computer Simulation of the 9-line and the 13-line Spectra

Iwasaki et al. [29] simulated an ESR spectrum by solving a modified Bloch equation, with the interchange of two conformations of a free radical assumed to be induced by hindered oscillation about the C_α—C_β bond of the radical end (see Fig. 12). They also assumed that the potential curve has a mirror image and that the potential barrier E between the two conformations depends on temperature. Figure 13 shows the calculated ESR spectra corresponding to several average lifetimes τ of the two conformations. They change from 9 to 13 lines with increasing τ, and can be compared favorably with the observed spectra in Fig. 10(a–e). However, the dose dependence of the spectrum shown in Fig. 14 cannot be explained by the hindered oscillation model of Iwasaki et al.

Computer simulation was also performed on another model. Thus, O'Donnell et al. [26] and Iwasaki et al. [30] independently illustrated that the 13-line, 9-line, and 5-line spectra can be simulated by assuming a statistical distribution of conformational angles about the most probable conformation of the propagating radical. When the conformation with $\theta_a = 65°$ and $\theta_b = 55°$ is assumed as the most stable, the 13-line spectrum shown in Fig. 15a is obtained. When a Gaussian distribution of conformational angles is assumed about this conformation, the 13-line spectrum changes to a 9-line one, with an intensity distribution very sensitive to the half-width of the Gaussian

Table 1. Variation of the conformation of the propagating radical with environment and temperature [27]

Temperature °C	A Dimer radical in monomer matrix	B Polymer radical in monomer matrix	C Polymer radical in polymer matrix
−154	58–62°	56–64°	53–67°ᵃ
−80	58–62°	57–63°	
0		57.5–62.5°	
50		58–62°	53–67°

ᵃ Asymmetry of conformation greater than 53–67° will produce similar spectra.

distribution, $\Delta_{1/2}$, as illustrated in Fig. 15 b–d. The 4-line component of the spectrum decreases with an increase in $\Delta_{1/2}$.

Propagating radicals are surrounded by a regular matrix of molecules at low extents of polymerization, but, as the polymerization proceeds further, they tend to be embedded in a matrix of randomly oriented molecules owing to the presence of polymers.

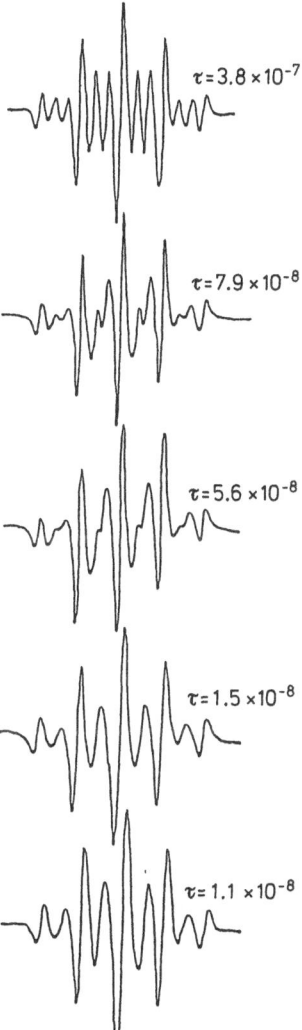

$\tau = 3.8 \times 10^{-7}$

$\tau = 7.9 \times 10^{-8}$

$\tau = 5.6 \times 10^{-8}$

$\tau = 1.5 \times 10^{-8}$

$\tau = 1.1 \times 10^{-8}$

Fig. 13. Simulated spectra for various average lifetimes of two conformations [29]

This concept seems to make a reasonable explanation possible for the dose dependence of the ESR spectrum of the propagating radical.

O'Donnell et al. [26] attempted to interpret the temperature and time dependence of the ESR spectrum of the propagating radical of barium methacrylate in terms of changes in its stable conformation as well as in the distribution of conformations about it. The results are shown in Table 1.

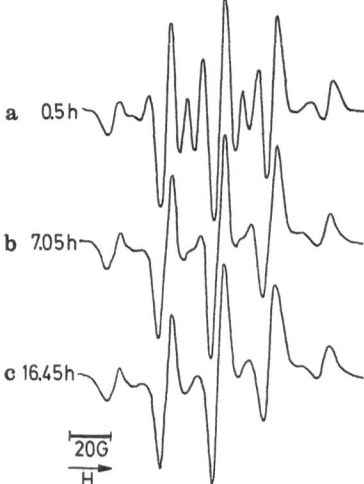

Fig. 14a–c. Dose dependence of ESR spectrum for propagating radical in MAA irradiated at 0 °C [30]. (a) 1.5×10^4 R; (b) 21.2×10^4 R; and (c) 49.5×10^4 R

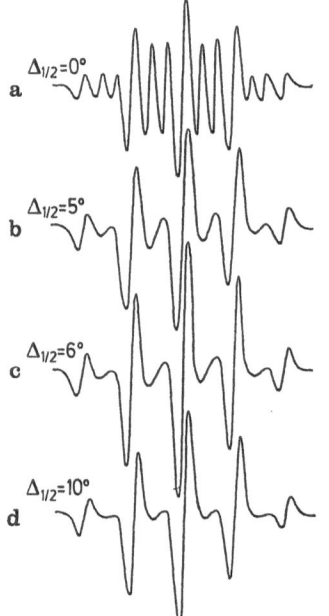

Fig. 15a–d. Simulated ESR spectra for radical $-CH_2-$ $-\dot{C}(CH_3)COOR$ with Gaussian conformation distributions about the most probable positions $\theta_1 = 55°$ and $\theta_2 = 65°$, with Half-height widths $\Delta_{1/2}$ indicated [30]

4.6 9-Line Spectrum and two Conformations

Kamachi et al. [31], making an ESR study of the BPO- initiated polymerization of methacrylates in frozen aromatic solvents with irradiation of light, obtained well-resolved 9-line spectra spaced at about 11.5 G such as shown in Fig. 16. The intensity distributions of these spectra were found to depend on the temperature and the nature of the solvent. The radical polymerization of MMA-β-d_2 was investigated by ESR

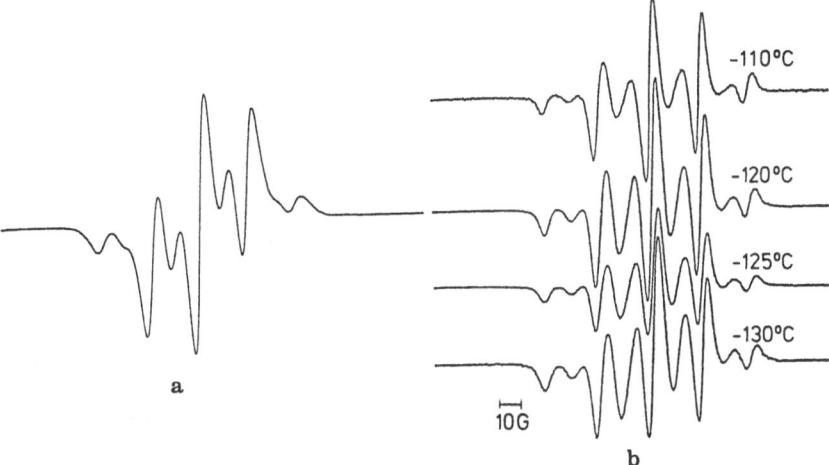

Fig. 16a and b. Temperature dependence of intensity distribution of polyMMA radical [31]. (a) in anisole at −120 °C; (b) in benzonitrile at indicated temperatures

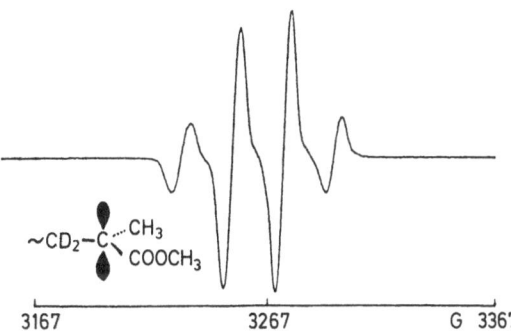

Fig. 17. ESR spectrum of polyMMA-β-d$_2$ in benzonitrile at −120 °C [31]

to understand the origin of the 9-line spectrum, on the expectation that the hyperfine splitting constant of deuterium is so small that the lines due to deuterium are concealed in the 4 lines split by three-methyl protons. A well-resolved 4-line with the intensity distribution 1:3:3:1, shown in Fig. 17, was obtained, and it was assigned to the species $\sim CD_2\dot{C}(CH_3)COOCH_3$. The intensities themselves decreased with raising temperature, but their distribution remained unchanged until getting undetectable. This 4-line spectrum can be ascribed to a single radical, because no other singnal with different lifetime was recorded.

Kamachi et al. [32] calculated the conformation energy of the propagating radical of MMA by modeling it as a short sequence of the radical end unit and a penultimate monomer unit, as shown in Fig. 18a. They found that two stable conformations which have different dihedral angles between the C—H bond and the p-orbital are possible (Fig. 18b).

There are two possible ways of monomer addition to a propagating end, i.e.,

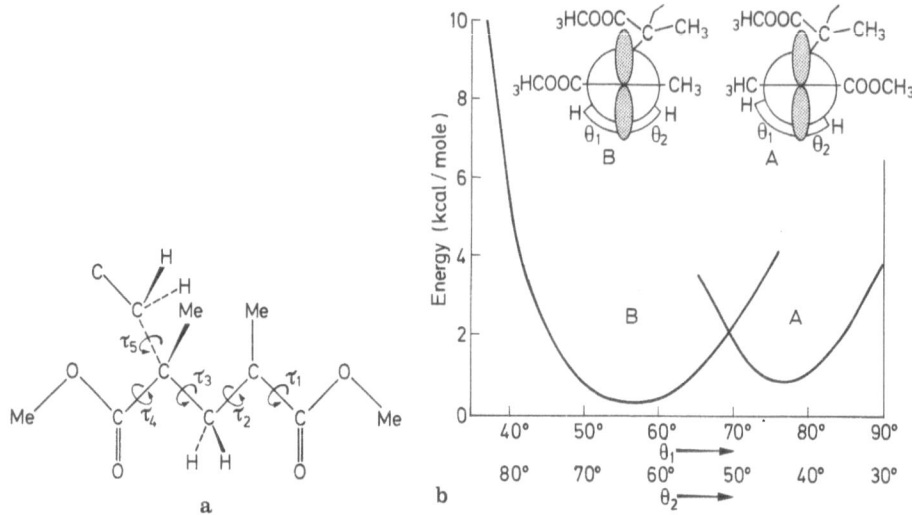

Fig. 18a and b. (a) Schematic diagram of poly(MMA) radical. τ_1, τ_2, τ_3, τ_4, and $\tau_5 = 180°$ define the planar zigzag conformation. (b) Conformation energy of poly(MMA) radical as function of the dihedral angle of β-protons. A and B interchange by free rotation of radical end [32]

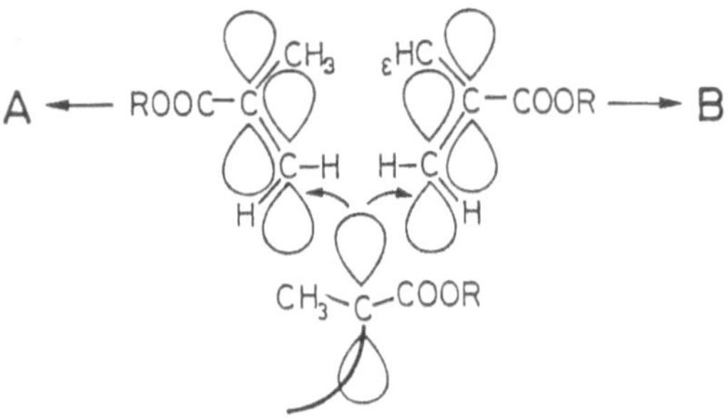

Fig. 19. Two ways of monomer addition to a propagating radical in the frozen state

isotactic and syndiotactic configurations, as shown in Fig. 19. Kamachi explained the 9-line spectrum of the propagating radical in the frozen state, where no internal rotation about C_α—C_β bond takes place, as due to an overlap of the spectra corresponding to these two configurations. The difference in intensity distribution between the 9-line spectra in anisole and benzonitrile may be due to a solvent effect on the poly-MMA radical, because the propagation rate constants for MMA were found to differ for these two solvents [34, 35].

4.7 Radical Polymerization of Triphenylmethyl Methacrylate

In a study on the radical polymerization of various methacrylates in 2-methyltetrahydrofuran (MTHF) rigid glass, Kamachi et al. [32] observed even at room temperature the ESR spectrum of the poly(triphenylmethyl methacrylate) (polyTPMA) radical, which has a bulky side group. They investigated the temperature dependence of the

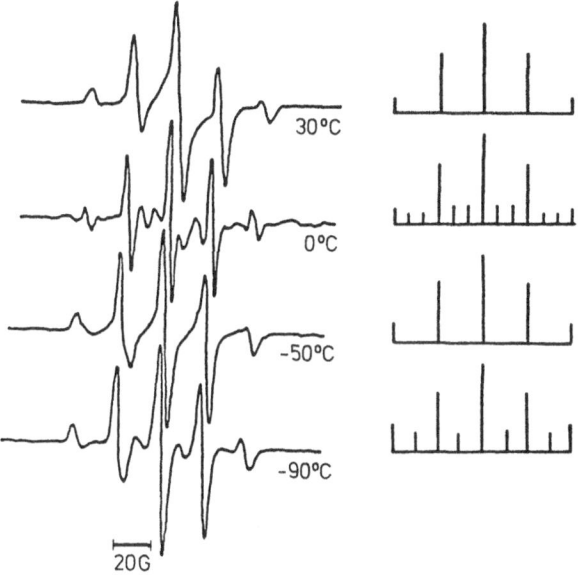

Fig. 20. Temperature variation of ESR spectrum for poly(TPMA) radical in 2-methyltetrahydrofuran (MTHF). [TPMA] = 2.0 M and [DTBP] = 0.1 M

ESR spectrum of this radical over a range from —90 to 30 °C, with the result shown in Fig. 20.

The 9-line spectrum recorded at —90 °C agreed with that for MMA in anisole or benzonitrile, except for the intensity distribution (see Fig. 16). The 4 lines in the 9-line spectrum became weaker with raising temperature, and a 5-line spectrum spaced at about 23 G was obtained at —50 °C. The latter changed to a 13-line one at about 0 °C. As the temperature was raised further to 30 °C, the 8 lines of the 13-line spectrum

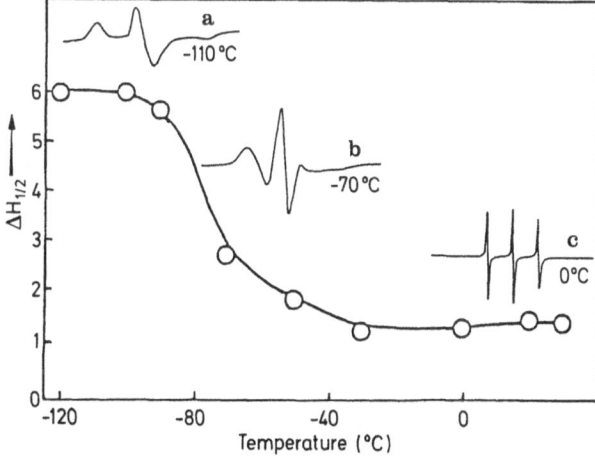

Fig. 21. Temperature variation of line width $\Delta H_{1/2}$ of centroid for Tempol in mixture of TPMA (2 M) and MTHF. [Tempol] = 10^{-4} M. Spectra (**a**), (**b**), and (**c**) at 0°, —70°, and —110°, respectively [32]

became weaker and a 5-line spectrum with an equal spacing of 23 G appeared again at 30 °C. The signal intensity decreased with raising temperature and became indistinguishable from noises at about 40 °C.

The line width $\Delta H_{1/2}$ of the central line in the ESR spectrum of 2,2,6,6-tetramethyl piperizin-1-oxyl (TEMPOL) added to the polymerization system at a concentration of 10^{-4} M was investigated as a function of temperature, taking advantage of the known fact that the ESR spectrum of TEMPOL changes markedly with the state of its surrounding medium. The results are shown in Fig. 21. The ESR spectrum below −85 °C in Fig. 21a exhibits a broad signal, which is referred to as the rigid glass spectrum. This signal and the value of $\Delta H_{1/2}$ of the central line undergo appreciable changes in the temperature region from −85 to −40 °C, leading to the broad 3-line spectrum shown in Fig. 21b at −50 °C. Although the line widths of these spectra remained invariant above −30 °C, the intensity distribution of the 3 lines changed with an increase in temperature and approached the 1:1:1 ratios, as shown in Fig. 21c. This temperature dependence of the intensity distribution of the 3 lines suggests that TEMPOL becomes mobile at higher temperature.

The following three concepts have been independently proposed for interpreting the 9-line and 13-line spectra of the polyMMA radical or the polyMAA radical: 1) a Gaussian angle distribution about the most stable conformation of a single species [25, 29], 2) exchange broadening due to the interchange of two β-methylene protons [30], and 3) overlap of the two conformations [24, 25].

Kamachi et al. [32] pointed out that the temperature dependence of the ESR spectrum for the poly(TPMA) radical can be systematically explained in terms of an appropriate combination of these concepts. The observed spectra from 0 to 30 °C clearly show that the 8 lines of the 13-line spectrum disappears owing to the exchange broadening caused by the rotation of two β-methylene protons about the C_α—C_β bond of the radical end. Accordingly, two conformations should be taken into account to explain the 9-line spectrum at −90 °C. However, the free rotation of two β-methylene protons at the radical end is not likely to take place in a rigid glass. Hence, simulation was attempted by superposing the spectra for two stable conformations A and B′ of Fig. 22 in the ratio 9:1. The result obtained is shown in Fig. 23. Thus, it appears that the change in the spectrum from 9 to 5 lines accompanying the temperature change from −90 to −50 °C can be explained by the exchange broadening. However, this explanation assumes the exchange broadening to occur twice in the same system. Hence, another explanation should be sought.

The 5-line spectrum may be considered due to slow oscillation about each stable

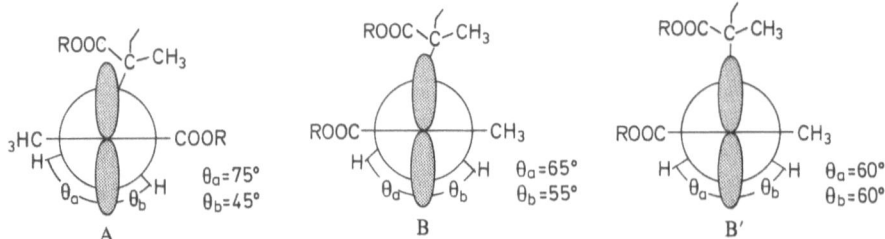

Fig. 22. Conformations used for simulation [32]

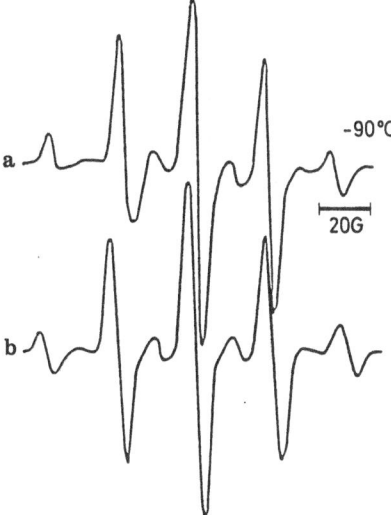

Fig. 23a and b. ESR spectra of poly(TPMA) radical. (**a**) Observed; (**b**) simulated with conformation A:conformation B = 9:1 [32]

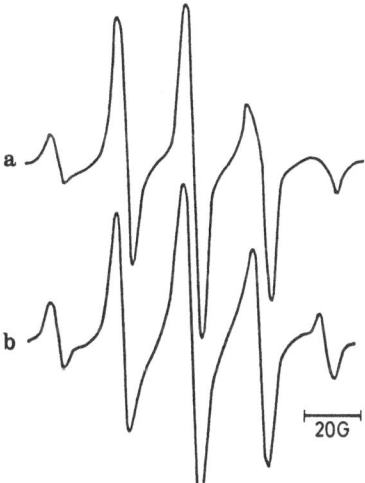

Fig. 24a and b. ESR spectra of poly(TPMA) radical. (**a**) Observed; (**b**) Simulated assuming a Gaussian distribution about two conformations, with half-height width = 10°, and line width = 4 G

conformation, since this consideration is consistent with the high viscosity of the medium and the bulkiness of the side group. Thus, simulation was attempted on the assumption that the angle distribution about the two stable conformations A and B is Gaussian with a half-height width of 10°, and the 5-line spectrum in Fig. 24b was obtained at —50 °C.

The oscillation rate increases with raising temperature. Therefore, only two stable conformations could be measured at 0 °C as averages of the multiconformations. Simulation was made at 0 °C by overlaping the conformations A and B in the ratio 8:3, and the 13-line spectrum shown in Fig. 25 was obtained. Since the propagating radicals more strongly interact with solvent molecules in the glassy state, the stable

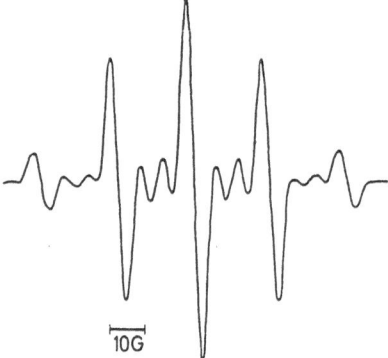

10G

Fig. 25. Simulated ESR spectrum of poly(TPMA) radical, assuming an overlap of two conformations [32]. Conformation A:conformation B = 8:3, and line width = 4 G

conformation in the liquid state may be somewhat different from that in the glassy state.

In this way, Kamachi et al. succeeded in explaining the temperature dependence of the ESR spectrum of the poly(TPMA) radical on the basis of two conformations; their calculated results are summarized in Table 2. Later, they substantiated the simulated spectra by ESR measurements on the radical polymerization of several methacrylates at room temperature. The details will be described in Section 5.4.

Table 2. Assignments of poly(TPMA) radical spectra [32]

Temperature °C	Spectrum	Assignment
30	5 lines	Exchange broadening between A and B
0	13 lines	Overlap of A and B (8:3)
−50	5 lines	Distribution (Slow interchange)
−90	9 lines	Overlap of A and B′ (9:1)

4.8 Radical Polymerization of Other Monomers in the Solid State

Radiation-induced solid-state polymerizations of vinyl and diene compounds were studied by ESR spectroscopy. Important examples are discussed in this section.

Ohnishi et al. [35] irradiated single crystals of acrylic acid(AA) with γ-ray at −196 °C and investigated the dependence of ESR spectrum on reaction time and temperature. The spectrum due to $CH_3\dot{C}HCOOH$ appeared as a doublet of quartets at −196 °C. Its angular dependence showed that the radical orients in the crystalline lattice in the same way as does the undamaged molecule. A broad 3-line spectrum due to the polymer radical $-CH_2\dot{C}HCOOH$, which takes no preferred orientation in the crystal, appeared at temperatures above −100 °C. This indicates that polymerization takes place at the phase boundaries to form an amorphous polymer. Similar results were obtained in ESR studies [36, 37, 38] on the polymerization of irradiated single crystals of acrylamide.

Harris et al. [39] concerned themselves with ferric chloride photosensitized polymerization of AA, its esters, and acrylamide in rigid glasses of methanol, ethanol, isopropanol, and acetone near the liquid nitrogen temperature. 5-line and 3-line ESR spectra whose intensity distributions depended on temperature as well as the nature of the rigid glass were recorded, and they were explained in terms of two stable conformations of propagating radicals.

Free radicals [40] produced by γ-irradiation of VAc in the solid state were examined by ESR. A broad triplet with a coupling constant 22 G became observable on elevating temperature from −196 to −160 °C, and it remained unchanged up to −100 °C. This triplet was assigned to $CH_2=CHOCOCH_2 \cdot$. Unfortunately, the ESR study on the γ-irradiated glassy sample of vinyl acetate gave no information on the growing chain radical, in contrast with that on the BPO-initiated polymerization of VAc in frozen benzonitrile which is described below [41].

An ESR spectrum was observed by Shida et al. [42] for 1,3-butadiene γ-irradiated in butyl chloride glass at −196 °C. It was attributed to the butadiene cation-radical which is responsible for the cationic polymerization mechanism. The successive addition of butadiene monomers to the cationic end produced a polymer of butadiene, with an allylic radical at its other end. The allylic radical is considered too stable to induce radical polymerization of butadiene at low temperature. Thus, γ-ray polymerization of butadiene should take place through ionic propagating species.

In an ESR study on the light-irradiated BPO-initiated polymerization of VAc as well as the diene compounds shown below in frozen benzonitrile at −120 °C, Kamachi et al. [41] obtained well-resolved ESR spectra of the propagating radicals depicted in Fig. 26.

$$\begin{array}{cc}
\overset{\displaystyle OC_2H_5}{\underset{\displaystyle COOC_2H_5}{CH_2=C-CH=CH}} & \overset{\displaystyle CH_3}{\underset{\displaystyle COOC_2H_5}{CH_2=C-CH=CH}} \\[2em]
\text{EEP} & \text{EMP}
\end{array}$$

$$\begin{array}{cc}
\underset{\displaystyle COOC_2H_5}{CH_2=CH-CH=CH} & \underset{\displaystyle OCOCH_3}{CH_2=CH-CH=CH} \\[2em]
\text{EP} & \text{AB}
\end{array}$$

No spectrum was obtained in the absence of BPO or without irradiation. The observed spectra were assigned to the propagating radicals produced by addition of the initiator radical to monomers.

The 3-line spectrum for VAc was simulated, assigning hyperfine splitting constants of 22.0, 21.0, and 9.0 G to the α-proton, one β-proton, and the other β-proton, respectively, and assuming a Gaussian intensity distribution with a linewidth of 14.0 G (Fig. 26a).

The 6-line spectrum for EMP was simulated with $a_H1 = 10.90$ G, $a_H2 = 4.50$ G, $a_{CH_3} = 10.70$ G, $a_H4 = 10.60$ G, and $a_H5 = 2.80$ G, assuming a Gaussian intensity

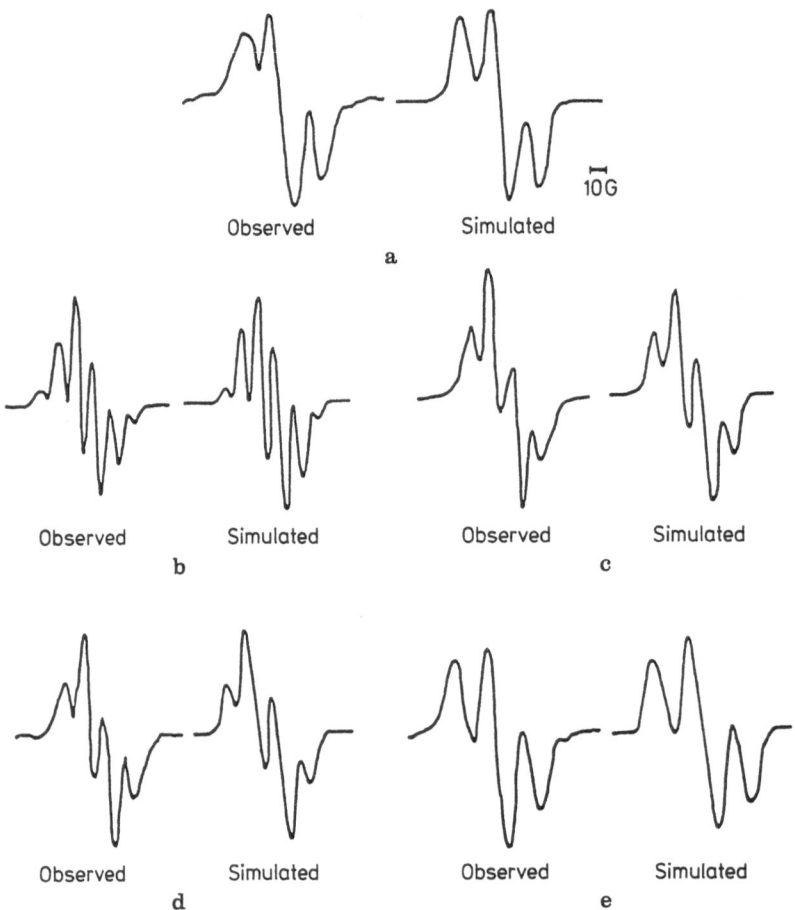

Fig. 26 a–e. Observed and simulated ESR spectra of propagating radicals in frozen benzonitrile at −120 °C [41]. (**a**) VAc, (**b**) EMP, (**c**) EP, (**d**) AB, and (**e**) EEP

distribution with a linewidth of 8.4 G (Fig. 26b) for the following radical with X = OCH_3. In this spectrum, the hyperfine splittings due to the β-proton and one of the δ-methylene protons were smeared out because of the assumed broad linewidth. The assumed values of $a_H 1$ and a_{CH} are almost one-half the hyperfine splitting constants due to the α-proton (20.4 G) of the polyEA radical and the methyl protons (23 G) of the polyMMA radical, respectively, implying the above radical to be allylic; in other words, the unpaired electron densities in the α and γ carbons of the poly(EMP) radical

Table 3. Hyperfine splitting constants and linewidths (Gauss) [41]

Monomer	a_H1	a_H2	a_x	a_H4	a_H5	Linewidth
EEP[a]	11.00	4.50	—	10.50	3.00	8.4
EMP[a]	10.90	4.50	10.70	10.60	2.80	8.4
EP[a]	10.80	4.50	10.85	16.00	0.20	8.4
AB[a]	13.80	5.50	13.30	16.00	0.20	9.0

[a] X, $-OC_2H_5$ (EEP), $-CH_3$ (EMP), and $-H_3$ (EP, AB).

Table 4. Monomer reactivity ratios[a] and Q—e values at 60 °C [41]

Monomer	r_1	r_2	Q	e
EEP[b]	0.08 ± 0.01 (0.07)	10.48 ± 0.00 (10.10)	8.94	−0.38
EMP[b]	0.06 ± 0.00 (0.08)	10.59 ± 0.00 (11.26)	9.57	−0.16
EP[b]	0.09 ± 0.01 (0.09)	5.75 − 0.00 (5.65)	5.86	0.02
AB[b]	0.45 ± 0.06 (0.43)	1.78 ± 0.03 (1.47)	3.19	−1.26
Butadiene (Bu)[25]	0.78	1.39	2.39	−1.05

[a] M_1, styrene; M_2, diene.
[b] The values in parentheses were obtained by the Kelen-Tüdös method

is about one-half that of the α carbon of growing polyEA and polyMMA chain ends.

The 4-line spectra shown in Fig. 26b and 26c were observed for EP and AB. Simulation gave the hyperfine splitting constants given in Table 3. In both cases, the value of a_H1 was the same as that of a_H3, indicating that the unpaired electron of the propagating radical is completely delocalized over the allylic three carbons.

The 3-line spectrum shown in Fig. 26d was observed for EEP and simulated by using the hyperfine splitting constants listed in Table 3. Since the values of a_H1 and

Table 5. Rate constants for radical polymerizations of diene compounds at 25 °C [43]

Monomers	$\dfrac{k_p}{M^{-1} s^{-1}}$	$\dfrac{k_t}{M^{-1} s^{-1}}$
EEP	10	0.9×10^7
EMP	30	2.3×10^7
EP	31	1.9×10^7
AB	18	28.0×10^7
St	44*	4.7×10^7*

* Ref. [71]

a_H4 of the EEP radical are similar to those of EP and EMP, the propagating radical was concluded to be allylic.

Diene compounds were copolymerized with styrene in benzene at 60 °C, using AIBN as an initiator. The resulting reactivity ratios are listed in Table 4. The radical reactivities of diene compounds relative to that of styrene were estimated in terms of $1/r_1$ and r_2. From the finding that $1/r_1 > 1$ and $r_2 > 1$, it was concluded that the radical reactivities of diene compounds are higher than that of styrene. The radical polymerization rate constants of these diene compounds were estimated by the rotating sector method (Table 5) and compared with that of styrene. The k_p values obtained were smaller than that of styrene in spite of the higher reactivities of these compounds toward a free radical [43]. Thus, it was concluded that the propagating radicals of diene compounds have lower reactivities than the polystyryl radical. This conclusion may be explained by assuming delocalization of the unpaired electrons of the propagating radicals over the allylic three carbons.

Radical polymerizations of N-substituted dimethacrylamides are known to proceed with complete cyclization to form a five-membered ring as a main repeating unit and a six membered ring as a minor repeating component [44-46]. Various mechanisms have been proposed for the cyclization polymerization of unconjugated dienes [44, 45, 46]. Kodaira et al. [45, 46] examined by ESR the cyclization polymerization of N-substituted

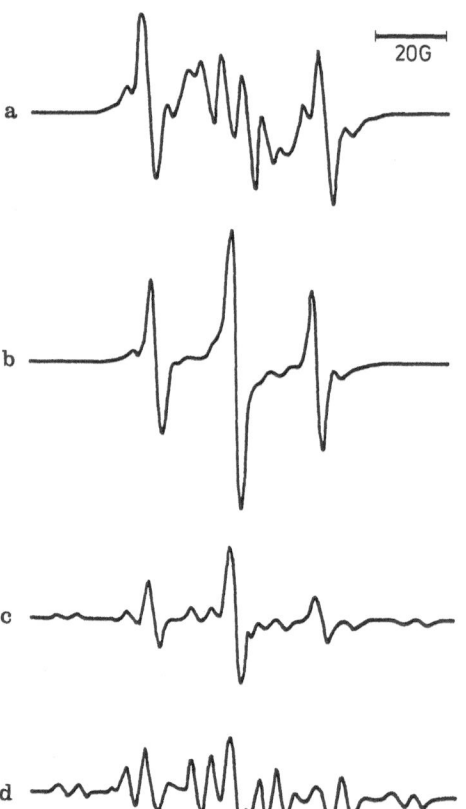

20G

a

b

c

d

Fig. 27 a–d. ESR spectra of MDMA single crystal irradiated at −196 °C to 5.34 Mrad and heat-treated for 5 min at −78 °C [46]. Magnetic field is parallel to the c′ axis in the b′c′ plane, and recorded at (**a**), −196 °C; (**b**), −90 °C; (**c**), −60 °C; and (**d**), −30 °C. Spectra (**b**), (**c**), and (**d**), taken after heating to the respective temperatures for 20 min

dimethacrylamides initiated by γ-ray irradiation. The results on an N-methyl dimethacrylamide (MDMA) single crystal irradiated at −196 °C are presented in Fig. 27. The spectrum in Fig. 27a is from the measurement at −196 °C. With increasing temperature from −196 to −78 °C, it changes to a triplet with an intensity of 1:2:1, in which the hyperfine splitting constant is 26.5 G. Various spectral patterns were recorded at different orientations of the single crystal. This finding indicates that the hyperfine splitting is due to the two α-protons. The triplet spectrum was assigned to radical III:

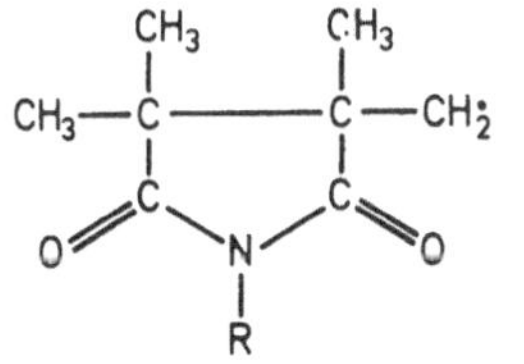

4-III

The triplet changes to 12-line spectra shown in Fig. 27c and d with raising temperature, and the latter is ascribed to uncyclic radical IV:

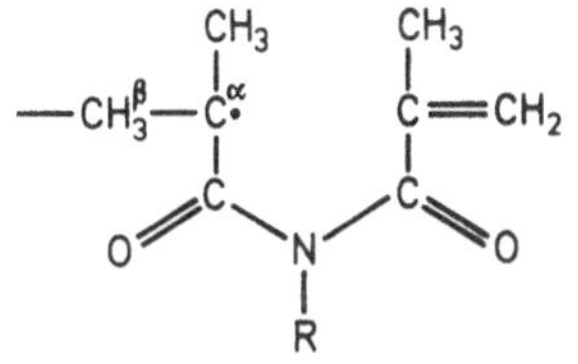

4-IV

The above assignements were confirmed by ESR measurement of an MDMA-d$_4$ single crystal [46]. Kodaira et al. [46] concluded that the rate-determining step of cyclization of RDMA in the solid state is intramolecular. The nonclassical radical suggested by Butler et al. [39] was not observed by ESR.

Kodaira et al. [47] compared the cyclopolymerizability of acryloylmethyl allylamine (AMAL), which gives polymers containing a considerable amount of pendant allyl groups, with that of methacryloylmethyl allylamine (MAMAL), which is known to yield a highly cyclized polymer containing a small amount of pendant methallyl groups.

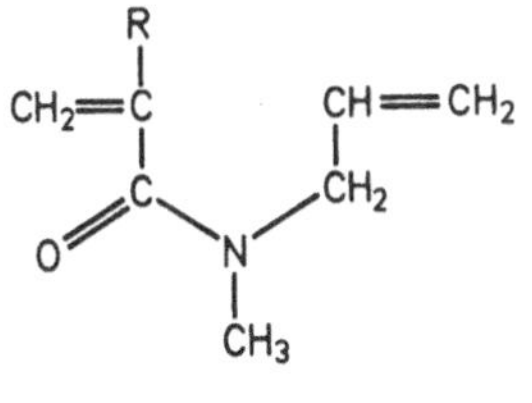

4-V

R = H: AMAL, and R = CH$_3$: MAMAL

The ESR spectrum of the former revealed only the acryloyl radical as the propagating radical, while that of the latter consisted of cyclized and uncyclized radicals. This finding may be accounted for by considering that the lower the polymerizability of the monofunctional counterparts of unconjugated dienes, the higher their cyclopolymerizability becomes.

5 ESR Studies of Radical Polymerization by the Flow Method

5.1 Introduction

One of the fascinating features of the ESR technique is that it allows direct observation of growing chain radicals in radical polymerization. But, it is usually difficult to do this under usual polymerization conditions, because the growing radicals have a very short lifetime and their concentrations are very low. Therefore, the use of ESR for the study of radical polymerization has had to be limited to solid-state polymerization.

Dixon and Norman [50] developed a rapid-mixing continuous flow method to observe ESR spectra of short-lived radicals in the liquid state. Applying it to the radical polymerization of AA, MAA, and itaconic acid (ITA), Fischer et al. [10, 51-56] observed ESR spectra of monomer, dimer, and polymer radicals and discussed the conformations of these radicals in terms of the hyperfine splitting constants for their β-methylene protons. Ranby et al. [57-61] extended its application to the radical polymerization of several monomers such as vinyl esters and butadiene and also to the copolymerization of binary monomer systems. However, the use of the thermal-redox radical-generating method has been chiefly restricted to reactions in aqueous solution.

Smith et al. [62-65] and Ranby et al. [66] independently employed a photo-flow method to observe radicals resulting from the addition of a photolytically generated primary radical to vinyl monomers, and also to investigate the polymerization kinetics. However, the conditions for their kinetic studies considerably differed from those encountered in usual radical polymerizations. Hence, the results obtained are of little practical interest.

5.2 Redox Initiators

Fischer et al. [10, 51-57] were the first to use ESR for the study of radical polymerization of AA, MAA, and itaconic acid (IA) with Ti^{+3}/H_2O, Ti^{3+}/t-butylhydroperoxide, Ti^{+3}/NH_2OH or $Fe^{+2}/EDTA/H_2O_2$. Figure 28 shows the ESR spectrum obtained in the redox system $Fe^{+2}/EDTA/H_2O_2$ at pH 7. At low monomer concentrations, the spectrum can be regarded as an overlap of those due to the following two isomeric monomer radicals mixed at a molar ratio of 80/20:

$$M_a^{\cdot}:\quad HO^{\cdot} + \underset{\overset{|}{COOH}}{CH_2}{=}CH \rightarrow HO{-}CH_2{-}\underset{\overset{|}{COOH}}{\dot{C}H}$$

and

$$M_b^{\cdot}:\quad HO^{\cdot} + \underset{\overset{|}{COOH}}{CH_2}{=}CH \rightarrow HO{-}\underset{\overset{|}{COOH}}{CH}{-}\dot{C}H_2$$

At medium monomer concentrations, there appear signals due to the dimer radical

$$D^{\cdot}:\quad M_a^{\cdot} + \underset{\overset{|}{COOH}}{CH_2}{=}CH \rightarrow HO{-}CH_2{-}\underset{\overset{|}{COOH}}{CH}{-}CH_2{-}\underset{\overset{|}{COOH}}{\dot{C}H}$$

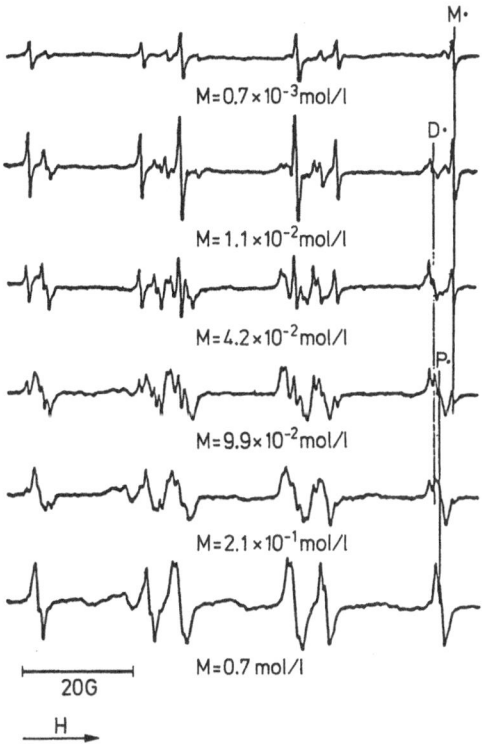

M = 0.7 × 10⁻³ mol/l

M = 1.1 × 10⁻² mol/l

M = 4.2 × 10⁻² mol/l

M = 9.9 × 10⁻² mol/l

M = 2.1 × 10⁻¹ mol/l

M = 0.7 mol/l

20 G

H

Fig. 28. ESR spectra of polymerizing aqueous solution of AA as function of monomer concentration [52, 56]. Redox system: Fe^{+2}-EDTA/H_2O_2, pH = 7

At very high monomer concentrations, the spectrum is ascribed to the polymer radicals

$$P^{\cdot}: \quad D^{\cdot} + n(CH_2{=}\underset{\displaystyle COOH}{CH}) \rightarrow HO{-}(CH_2{-}\underset{\displaystyle COOH}{CH})_{n-1}{-}CH_2{-}\underset{\displaystyle COOH}{\overset{\displaystyle \cdot}{C}H} \quad (n \geqq 1)$$

in which the hyperfine splitting constants for β-methylene protons are somewhat different from those for the monomer and dimer radicals. Such differences among the spectra for monomer, dimer, and polymer radicals were interpreted by Fischer et al. [10, 52, 55, 56] on the basis of the conformations of the growing radicals.

Figure 29 shows the concentrations of $M_a^{\cdot}$, $M_b^{\cdot}$, $D^{\cdot}$, and $P^{\cdot}$ as functions of the monomer concentration. When the equations describing these functions were fitted to the experimental data, different ratios of the rate constants were obtained. The propagation rate constants k_p thus obtained were greater by one or two orders in magnitude than those by the conventional methods.

The well-resolved 16-line spectrum shown in Fig. 30a was obtained in the polymerization of MA in aqueous solution at pH 1. It consists of a doublet of doublets of quartets for unequal β-methylene protons and methyl protons, in which the hyperfine splitting constant is 13.75 and 11.04 G for the former and 22.46 G for the latter. Therefore, it may be concluded that the spectrum is due to the propagating radical of MAA. Fischer et al. [52, 56] found this spectrum to be similar to the 9-line spectrum for MAA polymerized in the solid state, shown in Fig. 30b, and indicated that the 9-line spectrum probably resulted from the broadening of each line of the 16-line spectrum.

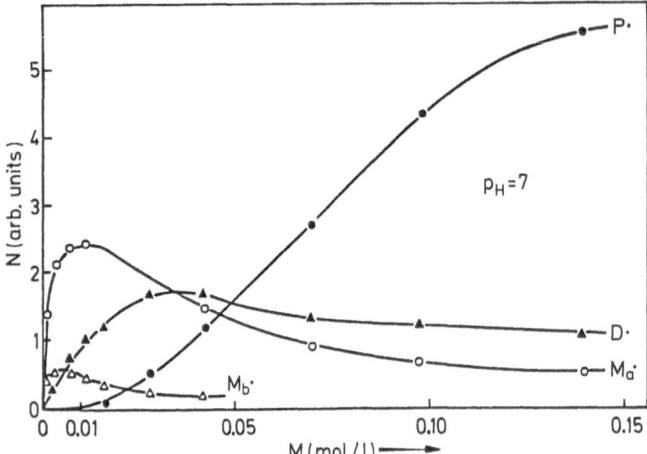

Fig. 29. Polymerization of AA with Fe^{2+}-EDTA/H_2O_2; concentrations of detectable radicals as functions of monomer concentration [52, 56]

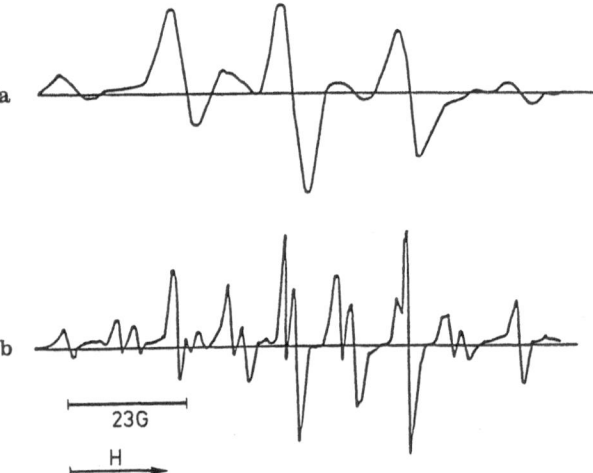

Fig. 30 a and b. ESR spectra of polyMMA radical [51], (**a**) in the solid phase, (**b**) in solution

Ranby et al. [58] observed a well-resolved spectrum shown in Fig. 31 in the polymerization of VAc with Ti^{+3}/H_2O_2, and described it as a doublet of triplets of narrow quartets, in which the hyperfine splitting constant was 21.0, 12.5, and 1.4 G for the α-proton, two β-methylene protons, and methyl protons of the ester group, respectively. The spectrum was assigned to

$$OHCH_2\dot{C}H(OCOCH_3)$$

For VAc, this was the only spectrum observed even at high concentration, in marked contrast to the finding by Fischer et al. [53, 55] for AA and MAA. Ranby et al. [60]

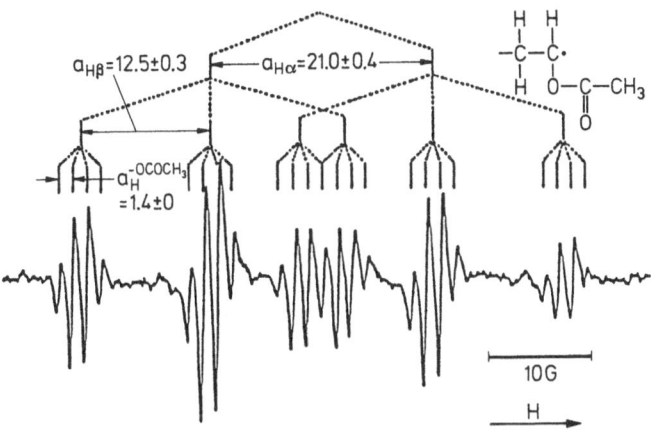

Fig. 31. ESR spectra of VAc in aqueous redox reaction system; VAc concentration 0.12 M [58]

discussed this difference in terms of the reactivity of the growing radical has a lower resonance stabilization power than AA and MAA ones do, the former should be more reactive than the latter. In consideration of this difference, Ranby et al. [60, 61] explained the monomer concentration effect on the ESR spectra of these monomers on the basis of the idea that the highly reactive VAc radical is preferentially deactivated by chain

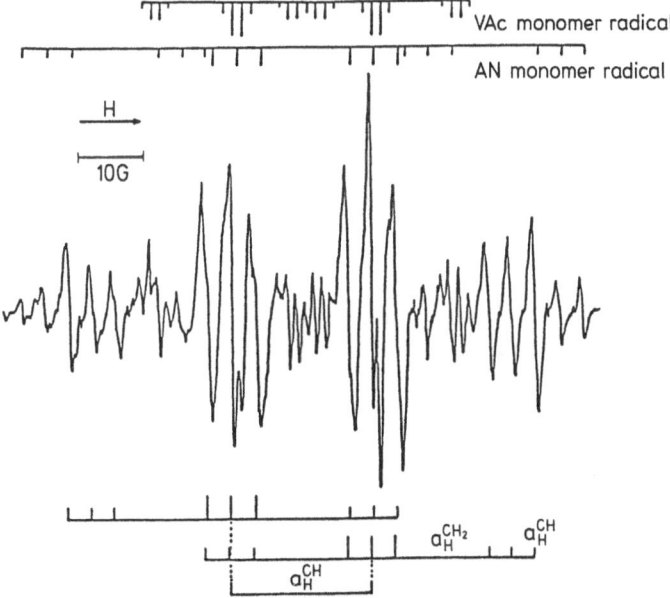

Fig. 32. ESR spectrum for VAc copolymerized with AN in aqueous solution [60, 61]. Dominant spectral component is assigned to radical $HOCH_2CH(OCOCH_3)CH_2CH(CN)$ [60, 61]. Concentration, 5.5 × 10⁻² M for VAc and 1.9 × 10⁻² M for AN; molar ratio of VAc/AN, 75/25. Stick spectra show expected relative intensities of the hyperfine structures for VAc and AN monomer radicals

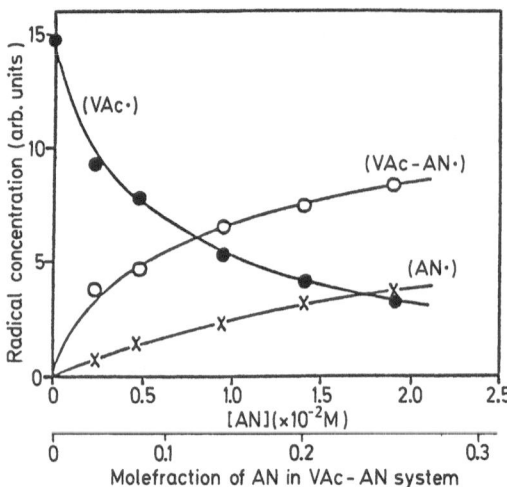

Fig. 33. Concentrations of different radicals in copolymerization of VAc with AN at different molar concentrations of AN [60, 61]. VAc$^{\cdot}$ and AN$^{\cdot}$, monomer radicals; VAc-AN$^{\cdot}$, copolymer radical. [VAc] = 5.5 × 10^{-2} M, and [AN] = variable

termination with $^{\cdot}$OH before the monomer radical propagates to the dimer radical.

When a small amount of a second monomer was added to VAc, the ESR spectrum for the copolymer radical HO-VAc-M$_2$$^{\cdot}$ and that for the monomer radical of the comonomer, i.e., HO—M$_2$$^{\cdot}$ overlapped on that for the monomer radical. As an example, the ESR spectrum obtained for the polymerization system VAc-AN is shown in Fig. 32. The relative concentrations of the three radical species as functions of [AN], shown in Fig. 33, indicate that with the addition of small amounts of the comonomer M$_2$(M$_2$/M$_1$ < 0.5), the concentration of VAc$^{\cdot}$ decreases sharply and that of VAc-M$_2$$^{\cdot}$ correspondingly increases. The concentration ratio [VAc-M$^{\cdot}$]/(VAc$^{\cdot}$) plotted against comonomer concentration [M$_2$] gives the characteristic slopes as shown in Fig. 34.

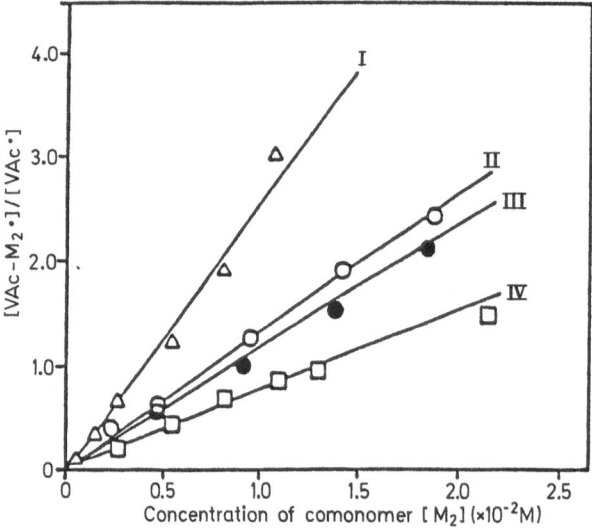

Fig. 34. Concentration ratio [VAc-M$_2$$^{\cdot}$]/[VAc$^{\cdot}$] as a function of concentration of comonomer [M$_2$] for (I), VAc-FA$^{\cdot}$; (II), VAc-AN$^{\cdot}$; (III), VAc-AA$^{\cdot}$; (IV), VAc-MAA$^{\cdot}$; [VAc] = 5.5 × 10^{-2} M [60, 61]

The different slopes indicate the relative rates of conversion of VAc$^\cdot$ to VAc-M$_2$$^\cdot$. Such rates determined for several monomers were found to increase in the order MA (0.8×10^2) < AA (1.2×10^2) < AN (1.3×10^2) < FA (2.5×10^2), where the values in parentheses are the slope values. These findings are in good agreement with the monomer reactivity ratios r_1 for the copolymerization of VAc with these comonomers.

Generally speaking, however, the complexities inherent to the flow-mixing method and due to the presence of more than two types of radicals make an exact analysis difficult. Furthermore, the use of the thermal-redox system for the generation of initiating radicals has been limited chiefly to reactions in aqueous solutions.

Beckwith et al. [67, 68] applied the flow method to cyclization polymerizations of 1,6-diene compounds

in order to clarify the reaction mechanism and the structure of polymer radicals.

When hydroxyl radicals are generated in the flow cell in the presence of diallyl-malonic acid (R = H in 5-I), the ESR spectrum shown in Fig. 35 was recorded, which consists of a doublet of triplets; the hyperfine splitting constants for the doublet and the triplets are 22.0 and 24.5 G, respectively. This multiplicity requires that the radical formed has a structure in which the unpaired electron interacts with three protons, two of which are equivalent. Of the following possible species (5-IV, -V, -VI, -VII, and -VIII), which can be formed initially, only two (3-V and -VII) fulfil this requirement.

The formation of 5-V compatible with the spectral evidence involves addition of the initiator radical to the double bond of 5-I in the sense opposite to the way normally found with monoolefins. Therefore, it may be concluded that the species present is a five-membered cyclic radical. Another evidence for the formation of a five-membered cyclic radical from diallylmalonic acid was obtained from the studies of its reactions with amino and phenyl radicals in the flow systems.

ESR spectra similar to Fig. 35 were observed for the polymerization systems of

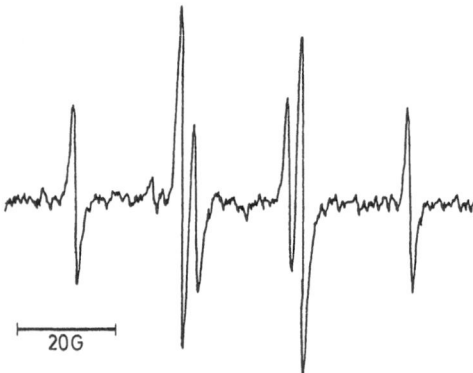

Fig. 35. ESR spectrum for radical formed from diallylmalonic acid and ˙OH [67]

diallylamines and diallylethers [67]. They were assigned to five-membered cyclic radicals formed by the following reaction sequence:

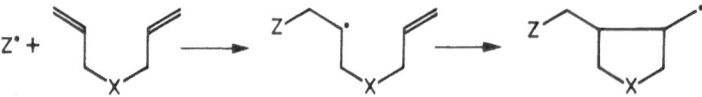

where X denotes $C(COOH)_2$, NH, or O.

Solomon et al. [69, 70] investigated the AIBN-initiated polymerizations of allylamines and its derivatives by ESR, and confirmed the observed spectra by the isolation and identification of the products.

5.3 Photolytic Radical-Generating Method

Smith et al. [62, 63, 64] used a photolytically radical-generating continuous-flow-method for ESR measurement of transient radicals in the radical polymerization of vinyl compounds. This method is free from the problems associated with the thermal-redox radical-generating method. Figure 36 shows the ESR spectra at 25 and 40 °C obtained by its application to photolysis of azocompounds RN:NR in neat methyl methacrylate. Some of the observed lines are due to the initiator radical R˙ but the contribution of R˙ to the entire spectrum is quite small; the stick plot in the Fig. 36a shows the spectrum of R˙. The main spectrum at either 25 or 40 °C (Fig. 36b or c) consists of 13 lines each of which is split into 4 lines owing to a weak, long-range interaction of the unpaired electron with methyl protons. It may be interpreted as a quartet of doublets of doublets of quartets in which the hyperfine splitting constant is 22.19 G for the α-methyl group, 14.18 and 9.27 G for respective β-methylene protons, and 1.13 G for methyl protons of the ester group. The intensities and line shapes of the spectra b and c, and also the positions and widths of the lines were essentially independent of the flow rate at either temperature. The inner 8 lines, each situated between the other 5 lines, are narrower at 40 °C than at 25 °C. According to Smith et al. [62], this narrowing arises because preferential conformations of the growing chain ends

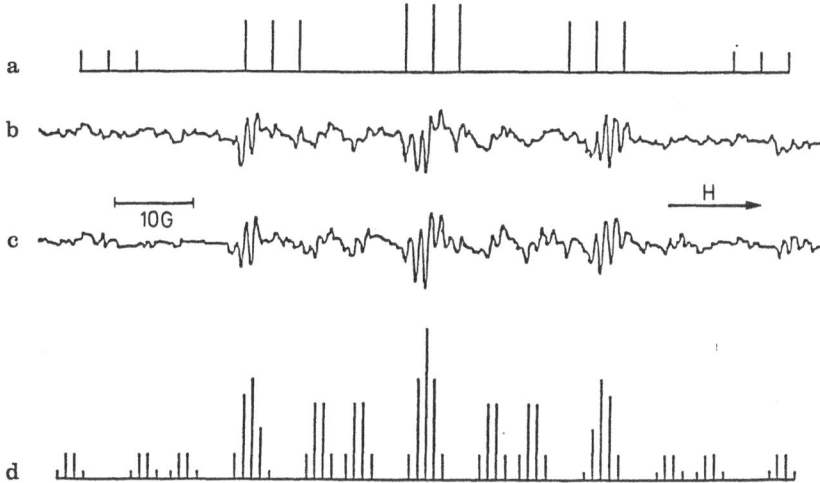

Fig. 36a–d. (a), Stick plot for the central 15 lines of the spectrum for R radical. (b), ESR spectrum from photolysis of RN:NR in neat MMA at 25 °C, [RN:NR] = 0.2 M, flow rate 0.10 ml s⁻¹. (c), ESR spectrum under the same reaction conditions with the same spectrometer settings as in (b) except for the temperature which was 40 °C. (d), Stick plot for the quartet of doublets of doublets of quartets from terminal structure —CH$_2$C(COOCH$_3$)CH$_3$ [62]

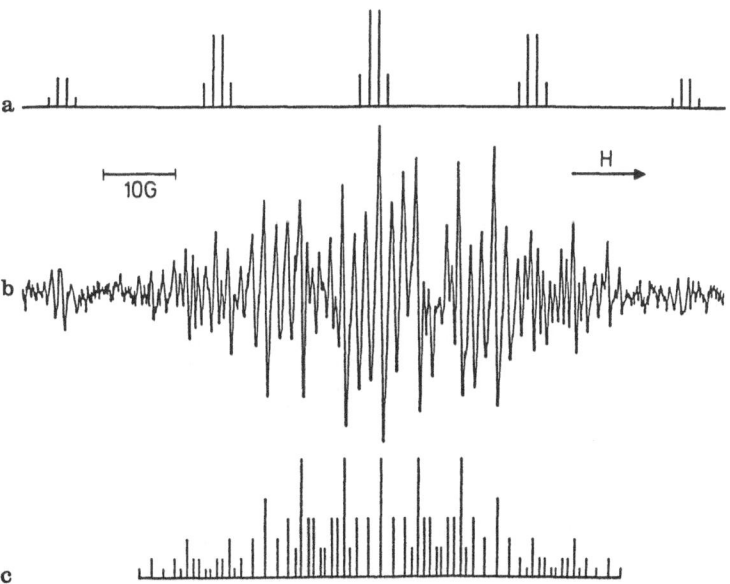

Fig. 37a–c. (a), Stick plot for central 20 lines of the ESR spectrum for Rc˙ ((CH$_3$)$_2$CCOOCH$_3$), a septet of quartets. (b), ESR spectrum from photolysis of (CH$_3$)$_2$C(COOCH$_3$)N:NC(COOC$_3$)(CH$_3$)$_2$, 0.20 M, in neat styrene; temperature, 25 °C; flow rate, 0.10 ml s⁻¹. (c), Stick plot for a quartet of doublets of triplets of small triplets consistent with addition radicals of terminal structure —CH$_2$CHPh [63]

at 25 °C become longer-lived than expected from the difference in the two β-methylene proton hyperfine splitting constants.

Smith et al. [63] made a similar study on the polymerization of styrene photoinitiated by the following azo-compounds:

$$(CH_3)_2C(CN)-N:N-C(CN)(CH_3)_2 \,,$$

$$\underline{CH_2(CH_2)_4C(CN)}-N:N-\underline{C(CN)(CH_2)_4CH_2}$$

and

$$(CH_3)_2C(COOCH_3)-N:N-C(COOCH_3)(CH_3)_2$$

which are denoted by $R_aN:NR_a$, $R_bN:NR_b$, and $R_cN:NR_c$, respectively. Figure 37 b shows a typical spectrum at 25 °C. Some of these lines are due to the initiator radical and are shown by the stick plot in Fig. 37a. The main lines shown with the stick plot in Fig. 37a. The main lines shown with the stick plot in Fig. 37c were interpreted as a quartet of doublets of triplets of small triplets, which may be associated with the interactions of the unpaired electron with an α-proton and two equivalent β-methylene protons, a para-proton, two ortho protons, and two metha protons. Table 6 lists the hyperfine splitting constants and the g values for the respective initiator systems. The concentrations of initiator radical $R^{\cdot}$ and propagating radical $A^{\cdot}$ in each system are presented in Table 7.

Table 6. ESR data for $R-[CH_2-CHPh]_n-CH_2-\dot{C}HPh$ [63]

R—	Proton couplings (G)					
	α	β	ortho[b]	meta[b]	para[b]	g Value
R_a—[a]	15.62	15.62	4.92	1.68	5.85	2.0027
R_b—	15.64	15.64	4.94	1.70	5.83	2.0027
R_c—	15.60	15.60	4.89	1.67	5.82	2.0027

[a] At 25 ± 1 °C. All protons of a given designation are magnetically equivalent. The g and a values have maximum uncertainties of 0.0001 and ca. 0.05 G, respectively. The peak-to-peak linewidth was 0.6 G regardless of R—, with no evidence of linewidth alternation.

Smith et al. [63] analyzed the polymerization kinetics of styrene, assuming the usual simple mechanism for radical polymerization, i.e., photolysis of an azoinitiator to give $R^{\cdot}$, followed by initiation, propagation, and termination with rate constants k_i, k_p, and k_t, respectively. If the steady state of radical concentration $[A^{\cdot}]$ is assumed, it follows that the kinetic chain length ν is given by

$$\nu = k_p[M]/(2k_t[A^{\cdot}]) \tag{5-1}$$

where [M] refers to the concentration of styrene. The radical size distribution F_x is given by

$$F_x = [R(M)_x^{\cdot}]/[A^{\cdot}] = v^{-1}(1 + v^{-1})^{-x} \tag{5-2}$$

where $R(M)_x^{\cdot}$ indicates the propagating radical with a number-average polymerization degree x.

When the primary radical termination cannot be neglected, we have insted of Eqs. (5-1) and (5-2)

$$v = \frac{k_p[M]}{2k_t([A^{\cdot}] + [R^{\cdot}])} \tag{5-3}$$

and

$$F_x = [R(M_1)^{\cdot}]/([A^{\cdot}] + (R^{\cdot}] = v^{-1}(1 + v^{-1})^{-x} \tag{5-4}$$

In this case, if $RM_x^{\cdot}$ and $R^{\cdot}$ are assumed to be kinetically equivalent, it follows from Eq. (5-4) that

$$[A^{\cdot}]/(R^{\cdot}] = v \tag{5-5}$$

From Eqs. 5-3 and 5-5 it follows that

$$k_p/2k_t = ([A^{\cdot}] + [R^{\cdot}]) [A^{\cdot}]/([R^{\cdot}] [M]) \tag{5-6}$$

$k_p/2k_t$ was estimated by substituting the experimental values of $[A^{\cdot}]$ and $[R^{\cdot}]$ (Table 7), and [M] = 8.66M into Eq. (5-6), and found to range from 5×10^{-7} to 2.3×10^{-6}. These values are smaller than the literature data estimated by the rotating sector method [71]. The discrepancy may be accounted for by the fact that k_t depends on the chain length of the reacting radical, while k_p does not. The signal intensity due to $[R^{\cdot}]$ seen in Fig. 37b is comparable to noises, suggesting that it is difficult to obtain kinetic data which are sufficiently reliable to make a meaningful discussion on the elementary rate constants possible.

Table 7. Steady state spin concentrations at $25 \pm 1\ ^{\circ}C$ [63]

R	$[A^{\cdot}] \times 10^7$ (M)	$[R^{\cdot}] \times 10^7$ (M)
$R_a^{\cdot}$	8.7	0.92
$R_b^{\cdot}$	9.0	<2.5
$R_c^{\cdot}$	13.4	0.97

5.4 The Rate of Addition of Vinyl Monomers to α-Acetoxyethyl Radicals

Ranby et al. [66] used the photolytically-generating flow method to measure the addition rates of vinyl monomers. They found that the α-acetoxyethyl radical, a model of

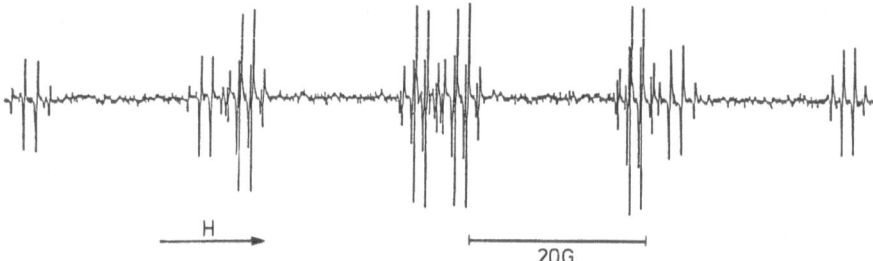

Fig. 38. ESR spectrum for 0.3 M DTB photolyzed in ethyl acetate at 27 °C. Spectrum consists of a quartet of doublets of quartets, and can be assigned to α-acetoxyethyl radical. Second order splittings can be seen for the lines corresponding to the inner two components of the larger quartet [60)]

the VAc radical, was preferentially formed when di-tertbutylperoxide (TBPO) was photolyzed in ethyl acetate; the spectrum obtained is shown in Fig. 38. The addition of the α-acetoxyethyl radical to three typical vinyl monomers AN, MA, and VAc was studied by ESR, with the kinetic parameter estimated from the signal intensity of the α-acetoxyethyl radical as a function of the monomer concentration. The reaction scheme in the presence of a monomer may be written as follows:

$$\text{DTBP} \xrightarrow{\text{hv}} 2\,\text{tBuO}^{\cdot}$$

$$\text{tBuO}^{\cdot} + \text{EtAc} \longrightarrow \text{tBuOH} + \text{R}^{\cdot}$$

$$2\,\text{R}^{\cdot} \xrightarrow{2k_r} \text{P}_1$$

$$\text{R}^{\cdot} + \text{M} \xrightarrow{k_a} \text{RM.} \; (= \text{A}^{\cdot})$$

$$\text{RM}^{\cdot} + \text{R}^{\cdot} \xrightarrow{k_{r'}} \text{P}_2$$

$$2\,\text{RM}^{\cdot} \xrightarrow{2k_{r''}} \text{P}_3$$

where $R^{\cdot}$ denotes the α-acetoxyethyl radical, M the monomer, and $RM^{\cdot}$ the addition radical, and P_1, P_2, and P_3 the respective reaction products. Here, the direct addition of the t-butoxy radical to each monomer is ignored.

New radicals appearing in the spectrum upon addition of a monomer are assignable to addition radicals formed by the attack of the α-acetoxyethyl radical. They are only the monomer addition radicals with the acetoxyethyl group, though the monomer concentration used is considerably higher than those in the kinetic experiments described below. Thus, the propagation of the monomer radical will be neglected in the subsequent kinetic considerations.

The rate equations are as follows:

$$d[R^{\cdot}]/dt = kI - 2k_r[R^{\cdot}]^2 - k_a[R^{\cdot}][M] - k_r'[R^{\cdot}][RM^{\cdot}] \tag{5-7}$$

$$d[A^{\cdot}]/dt = k_a[R^{\cdot}][M] - k_r'[R^{\cdot}][A^{\cdot}] - 2k_r''[R^{\cdot}]^2 \tag{5-8}$$

where kI is the production rate of the α-acetoxyethyl radical.

Kinetic measurements were made at low monomer concentrations, where $[RM^{\cdot}]$

is so small compared to [R'] that the addition rate constant can be evaluated without the measurement of absolute radical concentration. In the steady state, $d[R']/dt = d[RM']/dt = 0$, and [R'] is given by

$$[R'] = (2k_t)^{-1}\{-k_a[M] + (k_a^2[M]^2 + 2k_r kI)^{1/2}\} \qquad (5\text{-}9)$$

Representing $(kI/2k_r)^{1/2}$ by $[R']_0$ and $[R']/[R']_0$ by R we may rewrite Eq. (5-9) as

$$(1 - R^2)/2R = k_a(2k_r[R']_0)^{-1}[M] \qquad (5\text{-}10)$$

where the subscript zero indicates the absence of a monomer.

Figures 39a and b show changes in α-acetoxyethyl radical concentration with the concentration of the added monomer. From the slopes of the dashed lines together with the estimated values of $(2k_r \cdot [R']_0)$, the following values were obtained for k_a at 25 °C: $k_a = 1.1 \times 10^5$ M^{-1} s^{-1} for acrylonitrile, 9.4×10^4 M^{-1} s^{-1} for methyl acrylate, and 3.8×10^3 M^{-1} s^{-1} for vinyl acetate. Since the α-acetoxyethyl radical

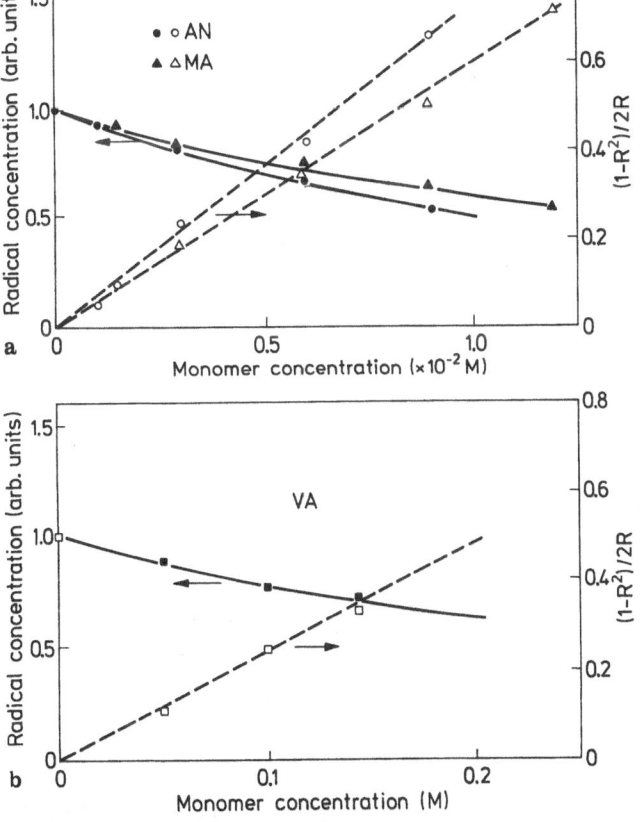

Fig. 39 a and b. Change in α-acetoxyethyl radical concentration with addition of monomer at 25 °C [66]. (a) ●, ○, AN; ▲, △, MA; (b) ■, □, VAc. [DTB] = 0.3 M

is an analog of the propagating radical of VAc and k_p is not as sensitive to the chain-length as k_t. k_a for VAc was compared to the reported k_p for viny acetate, and found to be consistent in order-of-magnitude with the latter.

6 Applications of the Spin Trapping Method to Radical Polymerization

6.1 Introduction

Short-lived free radicals formed during radical polymerization have been detected either by the flow method or by the trapping method in the frozen state. However, these methods are not applicable under usually employed polymerization conditions. According to Janzen et al. [72], short-lived radicals change to stable radicals, spin adducts, by reaction with appropriate radical scavengers, spin trapping agents. This means that the structure of a precursor radical may be inferred from the ESR spectrum of the spin adduct. This method, which is referred to as the spin trapping method, has been used to identify transient radicals formed in radical reactions with the aid of such spin trapping agents as:

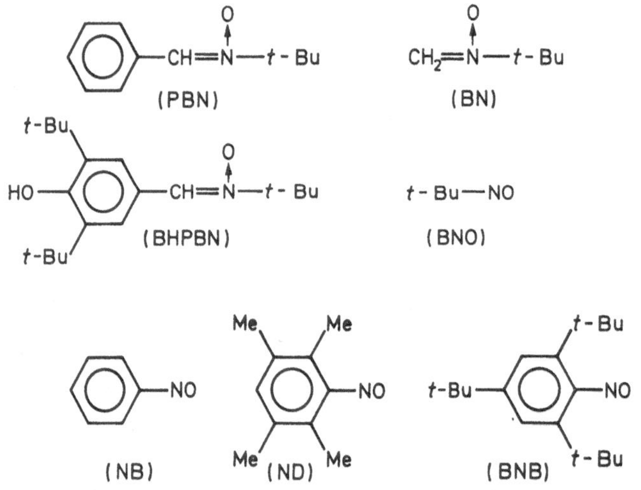

where PBN stand for phenyl-N-t-butylnitrone, BN t-butylnitrone, BHPBN 3,5-di-t-butyl-4-hydroxyphenyl-N-t-butylnitrone, BNO 2-methyl-2-nitrosopropane, NB nitrosobenzene, ND nitrosodurene, and BNB 2,4,6-tri-t-butylnitrosobenzene.

6.2 ESR Spectra for the Spin Adducts of Polymer Radicals

Chalfont et al. [73] were the first to apply the spin trapping method to radical poly-merization. They allowed styrene to radical-polymerize in the presence of 2-methyl-2-nitrosopropane (BNO) and observed the trapping of the styryl radical. Figure 40

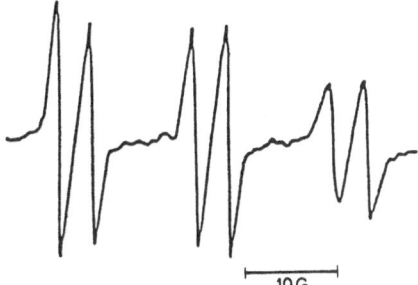

Fig. 40. ESR spectrum of styrene polymerized in the presence of BNO [73]. [Styrene] = 8.3 M and [BNO] = 0.001 M

shows a typical ESR spectrum obtained. The 6-line spectrum, in which a_N = 14.5 G for the nitrogen nucleus and a_H = 3.3 G for the β-proton attached to the N(O) group, was ascribed to the adduct of the polystyryl radical formed by the reaction

Kunitake et al. [74, 75] applied the spin trapping method to radical polymerization and copolymerization of some vinyl and diene monomers, and obtained 6-line spectra similar to Fig. 40 for MA, VAc, and divinyl ether. They concluded that the propagating radicals of vinyl monomers are trapped according to the reaction:

where X = $-COOCH_3$, $-OCOCH_3$, and $-OCH=CH_2$.

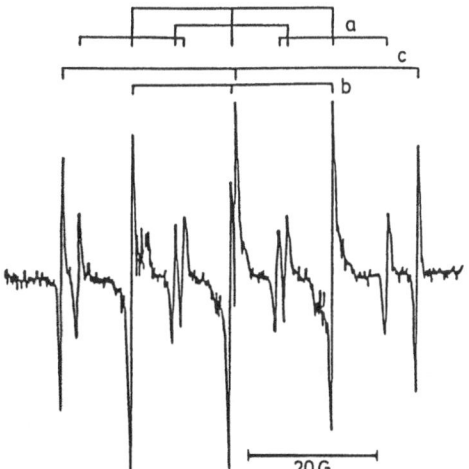

Fig. 41. ESR spectrum for polymerization mixture of butadiene [75]. Polymerization conditions: butadiene, ca. 4 M; TBPO, 0.05–0.1 M; BNO, ca. 0.02 M; room temperature; 10 min; solvent, toluene

Figure 41 shows the ESR spectrum obtained in tert-butylperoxide initiated poly-merization of butadiene. It consists of overlapped 9-line, 3-line, and 3-line spectra, whose splitting patterns are illustrated by stick plots. The 9-line spectrum is a triplet of triplets in which the hyperfine splitting constant is 15.3 G for nitrogen nucleus and 8.40 G for two methylene protons, and it can be assigned to the trapped 1,4 structure 6-II below.

$$t\text{-BuON}\overset{\displaystyle t\text{-Bu}}{\underset{\displaystyle O\cdot}{\diagdown}} \qquad \sim\!\!\sim\!\!\sim CH_2CH\!\!=\!\!CHCH_2N\overset{\displaystyle t\text{-Bu}}{\underset{\displaystyle O\cdot}{\diagdown}} \qquad \sim\!\!\sim\!\!\sim CH_2\underset{\displaystyle \underset{\displaystyle CH=CH_2}{|}}{CH}N\overset{\displaystyle t\text{-Bu}}{\underset{\displaystyle O\cdot}{\diagdown}}$$

$$\text{6-I} \qquad\qquad\qquad \text{6-II} \qquad\qquad\qquad \text{6-III}$$

One of the 3-line spectra is a triplet with the hyperfine splitting of 15.5 G for the nitrogen nucleus, and can be assigned to the spin adduct of the initiating radical 6-I. The other 3-line signal still remains unassignable, but it is probably due to the spin adduct of the t-butyl radical formed by the reaction of 6-I with DBN. The absence of a 6-line spectrum suggests that no spin adduct 6-III of the 1,2-addition radical is formed in this polymerization system. However, as is well known, polybutadiene obtained by radical polymerization contains a considerable amount of 1,2 structure. This dis-crepancy may be accounted for by a steric effect which retards addition of the 1,2-addition radical to BNO.

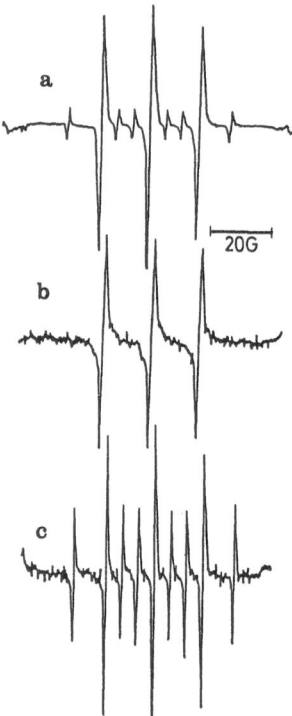

Fig. 42a–c. ESR spectra for (a), polymerization mixture; (b), poly(MMA); (c), reaction mixture, with no AIBN add-ed [75]. Polymerization conditions: MMA, 4 M; AIBN, 4×10^{-2} M; BNO, 2×10^{-2} M; 60 °C; 30 min; solvent, benzene

Figure 42a is the ESR spectrum observed in AIBN-initiated polymerization of MMA in the presence of BNO at 60 °C. It consists of a stronger-line spectrum, which is probably a complete overlapping of the spectra of the trapped initiating radical and polymer radicals, and a weaker 9-line spectrum. When the polymer was purified by repeated reprecipitation from benzene and methanol, the spectrum exhibited only the triplet characterized by a hyperfine splitting constant of 15.1 G for the nitrogen nucleus, and was considered as due to the spin adduct 6-IV of the propagating radical formed according to

$$\underset{\overset{|}{COOCH_3}}{\overset{\overset{CH_3}{|}}{\sim\sim CH_2-C\cdot}} + t\text{-BuNO} \longrightarrow \underset{\overset{|}{COOCH_3}}{\overset{\overset{CH_3\ t\text{-Bu}}{|\ \ \ |}}{\sim\sim CH_2C-N-O\cdot}} \qquad \text{6-IV}$$

The intensity of the 9-line spectrum relative to that of the triplet decreased with decreasing t-BuNO˙ concentration, and vanished at 4×10^{-3} M of t-BuNO. When a mixture of MMA and t-BuNO in benzene was heated to 60 °C, there was observed a clear 9-line spectrum, which was a triplet (intensity ratio 1:2:1) of nitrogen triplets (intensity ratio 1:1:1) for which the hyperfine splitting constants are 10.1 and 15.7 G, respectively (see Fig. 42c). Similar 9-line spectra, in addition to the expected 3-line spectrum, were obtained in the polymerization of other α-methyl-substituted monomers: MAA, α-methylstyrene, and methacrylonitrile. Hence, we may concluded that the 9-line spectrum should be due to the α-methyl group. This spectrum was assigned to the spin adduct 6-V of the allylic radical formed by the following hydrogen abstraction process:

$$t\text{-BuO}\cdot + \underset{\overset{|}{COOCH_3}}{\overset{\overset{CH_3}{|}}{CH_2=C}} \rightarrow t\text{-BuOH} + \underset{\overset{|}{COOCH_3}}{\overset{\overset{CH_2\cdot}{|}}{CH_2=C}}$$

$$\underset{\overset{|}{X}}{\overset{\overset{CH_2\cdot}{|}}{CH_2=C}} + t\text{-BuNO} \rightarrow \underset{\overset{|}{X}}{\overset{}{CH_2=CCH_2N}}\diagdown_{O\cdot}^{\diagup t\text{-Bu}} \qquad \text{6-V}$$

In an AIBN or BPO intiated polymerization of methyl [β-d₂]methacrylate in the presence of BNO, Kamachi et al. [76] observed a 15-line ESR spectrum along with another nitrogen spectrum which is displayed in Fig. 43. The intensity distribution indicated that the spectrum is an overlap of the 15-line spectrum being a triplet of quintets with the 3-line one due to the spin adduct 6-IV formed by "teil addition". No 9-line spectrum due to the spin adduct of the allylic radical was found. The 15-line spectrum is due to the adduct which has two-deuterium atoms attached to the NO˙ group. Kamachi et al. [76] considered "head addition" of the initiating or propagating radical to the monomer to be responsible for the 15-line spectrum. Since the radical formed by the "head addition" is not delocalized on the substituents at the α-position, it may be more reactive than that formed by "tail addition", which has carboethoxy group at the α-position. Therefore, the addition of the former to BNO

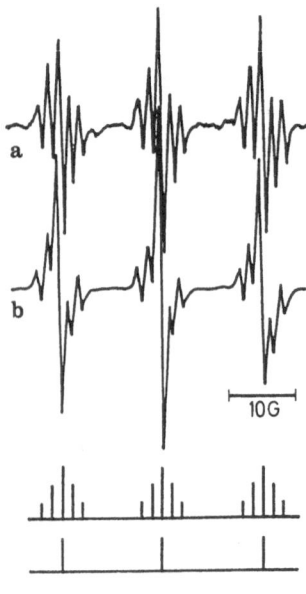

Fig. 43 a and b. ESR spectra for reaction mixtures of MMA-β-d$_2$ [76]. (a), BNO, 5×10^{-3} M, DISOP, 4×10^{-3} M; (b), BNO, 2×10^{-2} M, AIBN, 4×10^{-2} M

is probably much faster than that of the latter, and the radical obtained by "head addition" may be more effectively trapped by BNO than that by "tail addition".

Recently, a 15-line spectrum was found by Tabner et al. [77] during the reaction of methyl [β-d$_2$]methacrylate with BNO in the absence of radical initiator. These authors proposed to explain the finding as well as the 9-line spectrum obtained for MMA in terms of the products from the following "ene" reaction between the monomer and the trapping agent, i.e.,

6.3 Radical Reactivity of Vinyl Monomers

When AIBN is used as the initiator for radical polymerization of vinyl compounds, the cyanopropyl radical either reacts with BNO or adds to the monomer to form a propagating radical. Therefore, assuming that all the propagating radicals are eventually trapped by BNO, we may estimate approximately the relative ease of formation of trapped propagating radicals from ESR measurements in which the trapped cyanopropyl radical is used as an internal reference. This idea was used by Kunitake et al. [75], who found the relative reactivity of vinyl monomers with an initiating radical to vary as follows: AN > MA > VAc > styrene > divinyl ether.

A more quantitative study on the radical reactivity of vinyl monomers was performed by Sato et al. [78, 79)] using the spin trapping method. They showed that the initiation step in the radical polymerization of a monomer consists of two successive reactions: i) radical decomposition of the initiator R—R and ii) initiation reaction of the resulting initiator radical $R^\cdot$ with the monomer M. Schematically,

$$R\text{--}R \; \rightharpoonup \; 2R^\cdot$$

$$R^\cdot + M \rightarrow R\text{--}M^\cdot \,(= M^\cdot)$$

When di-tert-butyl peroxalate(DBPOX)-initiated polymerization of a vinyl compound was carried out in the presence of xylene (X) and BNO (the former is an internal standard), the following reactions took place competitively:

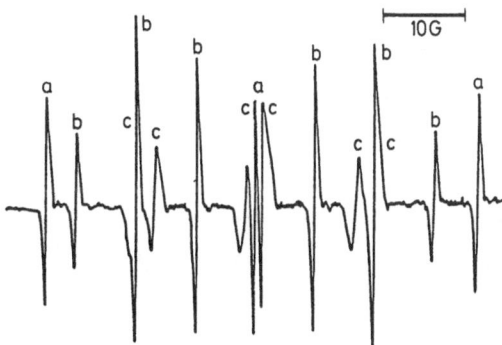

$$((H_3C)_3COOCO)_2 \longrightarrow 2\,CO_2 + 2\,(H_3C)_3CO\cdot$$
$$(DBPOX)$$

$$(H_3C)_3CO\cdot \;+ CH_2\!\!=\!\!CH \xrightarrow{\;k_1\;} (H_3C)_3CO\text{--}CH_2\text{--}\overset{\cdot}{C}H$$
$$\qquad\qquad\qquad\quad\; X \qquad\qquad\qquad\qquad\qquad X$$

(reaction with p-xylene, rate k_2)   6-VII

6-IV + (CH₃)₃CNO $\xrightarrow{\text{fast}}$ ...   6-VIII
(BNO)

6-VII + BNO $\xrightarrow{\text{fast}}$ ...   6-IX

Figure 44 shows an ESR spectrum obtained when MA was polymerized. It consists of an overlap of spectra a, b, and c, which may be assigned to the spin adducts of the monomer radical, the p-methylbenzyl one formed by hydrogen abstraction of the initiator radical from xylene, and the t-butoxyl one, respectively. The MA to xylene relative reactivity with respect to the addition of the t-butoxyl radical was evaluated from the relation

$$\frac{[6\text{-VIII}]}{[6\text{-IX}]} = \frac{k_1[M]}{k_2[X]}$$

Fig. 44. ESR spectrum for DBPOX/ MA/ p-xylene(X)/ BNO at 25 °C [79)]; [MA]/[X] = 1.38, [DBPOX] = 4.3 $\times 10^{-3}$ M, [BNO] = 4.5×10^{-2} M. (a), (b), and (c) indicate spectra of spin adducts $(CH_3)_3CO^\cdot$, 6-IX, and 6-VIII, respectively

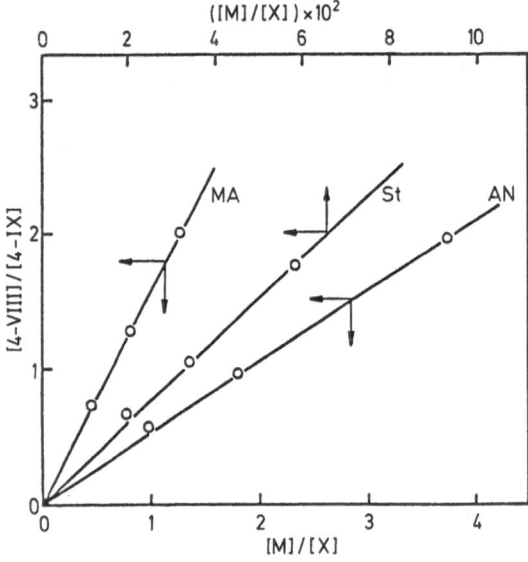

Fig. 45. Relation between [6-VIII]/[6-IX] and [M]/[X] [79)]

Figure 45 shows plots of [6-VIII]/[6-IX] against [M]/[X] for the DBOX-initiated polymerizations of styrene, MA, and AN in p-xylene in the presence of BNO at 25 °C. The relative reactivities of other monomers were also evaluated in a similar way. The resulting values of the relative rate constant k_1/k_2 are given in Table 8 together with those reported for other initiators.

Sato et al. [79)] found that relative reactivities of monomers toward the t-butoxyl radical are related to the value of e rather than that of Q. This finding indicates that the polar effect plays a more important role than the resonance effect in these reac-

Table 8. Relative reactivities (k_1/k_3) of various monomers toward addition of tert-butoxyl, benzoyloxyl, methyl, and phenyl radicals [79)]

Monomer	$(H_3C)_3CO^\cdot$ (25 °C)	$C_6H_5COO^\cdot$ (60 °C)	$CH_3^\cdot$ (65 °C)	$C_6H_5^\cdot$ (60 °C)
Isobutyl vinyl ether	9.60 (0.32)[a]	—	—	—
α-Methylstyrene	6.90 (0.23)	—	1.17	1.24
tert-butyl vinyl sulfide	7.43 (0.24)	—	—	—
Styrene	29.8 (1.00)	1.0	1.0	1.0 (1.0)
Vinyl acetate	0.28 (0.009)	0.36	0.04	0.23 ($\geq$0.08)
Methyl methacrylate	1.73 (0.05)	0.12	1.82	1.78 (1.7)
Methyl isopropenyl ketone	2.47 (0.08)	—	—	—
Methyl acrylate	1.54 (0.05)	—	1.30	0.78
Methyl vinyl ketone	2.68 (0.08)	—	2.40	—
Methacrylonitrile	0.71 (0.023)	—	2.68	2.46
Acrylonitrile	0.52 (0.017)	$\leq$0.05	2.18	— (0.8)

[a] Reactivity relative to styrene.

tions. Comparing the reactivities of styrene, MMA, MA, and VAc with the radicals listed in Table I, we observe the following order of reactivity:

AN > MMA > St > VAc   To methyl and phenyl radicals
St > VAc > MMA > AN   To benzoyloxyl radical
St > MMA > AN > VAc   To t-butoxyl radical

It appears possible to interpret the observations with the carbon radicals, which are electron-donating, in terms of the domination of the resonance effect over the polar effect. On the other hand, those with the electron-accepting benzoyloxyl radical may be attributed only to the polar effect, as has been pointed out by Bevington et al. [80]. The order of reactivity with the t-butoxyl radical differs from that with either carbon or benzoyloxyl radical. Probably, the resonance effect plays some role in the monomer reaction with this radical.

The spin trapping method has also been applied to the detection and identification of intermediate primary radicals produced from initiator systems such as triethylborane/oxygen [81], benzoyl peroxide/N,N-dimethylaniline [82], and N-chlorosuccinimide/p-toluenesulfonic acid [83], and provided important information on the generation of free radicals in initiator systems.

7 Direct Measurement of Active Species in Radical Polymerization

7.1 Introduction

As mentioned in Sect. 2.1, the stationary concentration of propagating radicals in radical polymerization is too low to be detected by commercially available ESR spectrometers. Hence, it is necessary for ESR measurement of growing radicals in radical polymerization to have a spectrometer which is sensitive to very low concentrations, say, down to at least 10^{-7} M. The ESR spectrometer improved by Bresler et al. [84, 85, 86] using a balance resonator permitted evaluation of the rate constants of radical polymerization for styrene, methyl methacrylate, and vinyl acetate. However, it should be noted that their measurements were on bulk polymerization, not on solution polymerization. Kamachi et al. [87-90] enhanced the sensitivity of the ESR spectrometer by making a TM_{110} cavity, designed specially for photoreaction, and obtained ESR spectra of the propagating ends not only in bulk polymerization but also in solution polymerization. These spectra allowed the conformation of propagating radicals as well as the propagation rate constants for several monomers to be estimated.

7.2 Special Cavity for ESR Measurement of Radical Polymerization

The sensitivity of an ESR spectrometer may be made higher in two ways: (i) to use the microwave of larger amplitude, because the ESR signal is proportional to the square root of the microwave power in the absence of power saturation, and (ii) to increase the sample volume up to an optimum value depending on the dielectric loss of the sample.

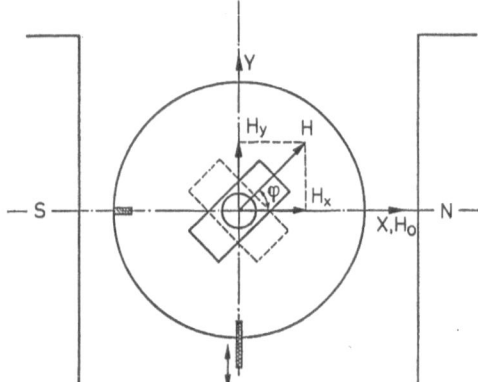

Fig. 46. Balance transmission cavity [84]

Bresler et al. [84] placed a cylindrical transmission cavity with an oscillation mode H_{111} in a stationary magnetic field H_0, with the cylinder axis normal to the direction of the stationary field, (see Fig. 46). The wave guide was mounted in such a way that the magnetic field H_0 of the microwave was 45° inclined to the stationary field H_0. The microwave oscillation in the cavity can be resolved into two components — one with its plane parallel with and the other with its plane normal to the external field H_0. The receiving wave guide was connected to the cavity through a coupling hole. The design of the transmission cavity is illustrated in Fig. 47. It consists of two parts: part 1 accommodating the sample and part 2 coupled with the receiving wave guide by a disphragm.

The commercially available TE_{011} mode cavity is cylindrical as shown in Fig. 48a. The magnetic field H_0 caused by a microwave introduced into it has a maximum on the cavity axis. Therefore, a given sample is placed on the cavity axis, so that its effective volume is limited by the magnetic field.

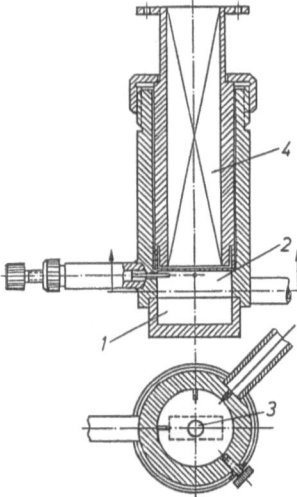

Fig. 47. Design of the balance transmission cavity [84]
1: cell for sample, 2: resonator, 3: diaphragm, 4: receiving guide

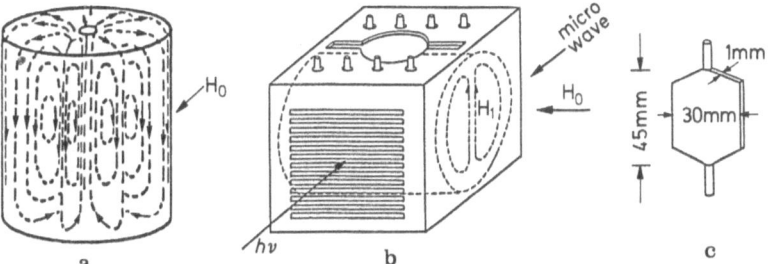

Fig. 48 a–c. Cavities and sample tubes [87]. (a) TE_{011} cavity, (b) TM_{110} cavity improved for photoreaction, and (c) Sample tube

Kamachi et al. [88] turned the TE_{011} cavity sideways, as shown in Fig. 48 b. In this TM_{110} cavity, the magnetic field of the irradiating microwave is normal to a horizontal plane including the cavity axis. Although the magnetic flux density in the TM_{110} cavity is weaker than that in the TE_{011} cavity, the sample volume effective for ESR measurement is greatly increased. Accordingly, a wider flat cell of Fig. 48c can be used for the measurement. The increase in sample volume compensates the lowering in sensitivity due to the decrease in magnetic flux density. The cavity sensitivity for detection of free radicals became 10 times as high as that of the TE_{011} cavity because of a great increase of the irradiated surface, much wider for the former than for the latter.

7.3 Thermally-Initiated Polymerization

In the above-cited study, Bresler et al. [84, 85, 86] obtained the ESR spectra shown in Fig. 49. The spectrum (a) for styrene consists of 12 lines. Theoretically, the hyperfine structure for the growing chain radical should reflect the interaction of the unpaired

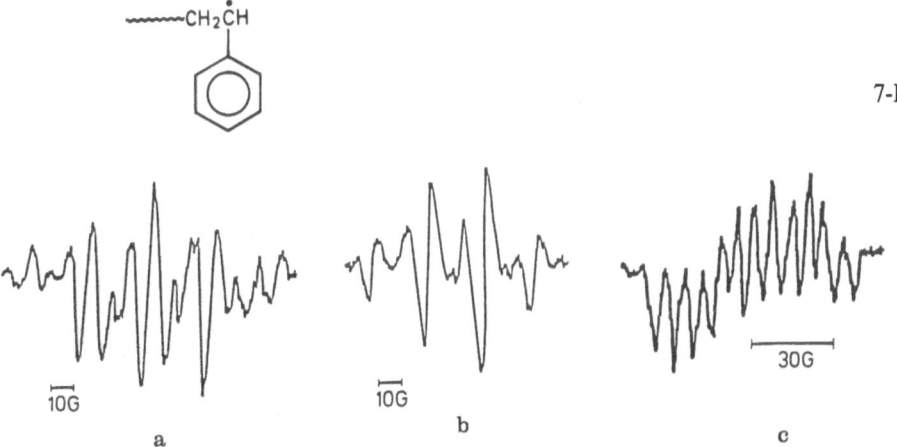

Fig. 49 a–c. ESR spectra of growing radicals [85, 86]. (a) MMA, (b) VAc, and (c) styrene

electron with the α- and β-protons of the terminal monomer unit. The protons of the phenyl ring induce no additional splitting because the spin density of the unpaired electron in this ring is very low. Therefore, the hyperfine interaction of the unpaired electron with one α- and two β-protons would generate 6-lines in the spectrum. Thus, the observed 12-line spectrum may be explained as due to overlapping of two 6-line spectra, each associated with two radicals which lead to either isotactic or syndiotactic configuration by reacting with the monomer.

The spectrum (b) in Fig. 49 is for the growing radical of polyMMA. Bresler attributed it to the overlap of 5 lines spaced at 23 G with the intensity distribution 1:4:6:4:1 and 4 lines with the intensity distribution 1:3:3:1. Their simultaneous appearance was ascribed to two possible monomer additions.

The ESR spectrum (c) in Fig. 49 consists of 4 lines spaced at 21 G with the intensity ratio 1:3:3:1. It was assigned to the growing radical of polyVAc whose unpaired electron interacts equivalently with the three protons, one α-proton and two β-protons of the methylene group.

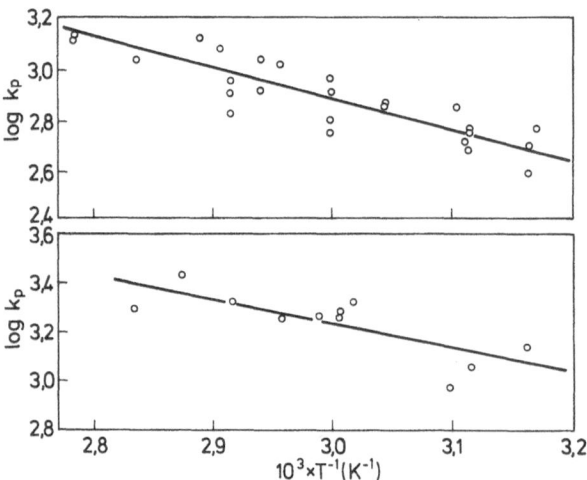

Fig. 50. Temperature dependence of propagation rate constant k_p [86] (a) MMA, and (b) VAc

Bresler et al. determined the propagation rate constant k_p using the equation for stationary polymerization kinetics, i.e.,

$$-d[M]/dt = k_p[R\,^{\cdot}]\,[M] \tag{7-1}$$

where [M] and [R$^{\cdot}$] are the monomer and macroradical concentrations, respectively, and d[M]/dt the overall rate of polymerization.

The k_p values for MMA, VAc, and styrene determined at different temperatures

gave the Arrhenius plots shown in Fig. 50. The indicated straight lines are represented
by

$$k_p = 2.5 \cdot 10^6 \exp\left(-\frac{5400 \pm 600}{RT}\right) \quad \text{(methyl methacrylate)}$$

$$k_p = 2.0 \cdot 10^6 \exp\left(-\frac{4700 \pm 700}{RT}\right) \quad \text{(vinyl acetate)}$$

$$k_p = 2.4 \cdot 10^8 \exp\left(-\frac{9000 \pm 400}{RT}\right) \quad \text{(styrene)}$$

The termination rate constant k_t was calculated at different temperatures from the
equation

$$k_t = \frac{-d[M]/dt}{P \cdot [R^\cdot]^2} \tag{7-2}$$

where P is the number-average degree of polymerization. The k_t value was 7.2×10^8
for MMA and 2.75×10^7 M^{-1} s^{-1} for VAc, independent of temperature.

7.4 Detection of Propagating Radicals by TM_{110} Cavity

Radical polymerization of vinyl compounds was performed under UV irradiation
through slotted openings of the cavity shown in Fig. 48b, and ESR spectrum measurements were made immediately after irradiation. The ESR spectra in Fig. 51 are for
benzoyl-peroxide initiated polymerizations of MMA, isobutyl methacrylate (IBMA),
benzyl methacrylate (BzMA), and TPMA at 30 °C. No ESR spectrum was obtained
without an initiator, while an ESR spectrum of the initiating radical was found in the
absence of methacrylates.

The spectrum for MMA [87] in Fig. 51 consists of 13 lines with the intensity distribution 1:2:1:4:3:3:6:3:3:4:1:2:1 for a single conformation, in which the dihedral
angles of β-methylene protons with the p-orbital of the unpaired electron are 55° and
65°. Similar 13-line ESR spectra were obtained in the radical polymerization of
IBMA and BzMA. However, the intensities of the 8 lines, each between the other
5 lines, relative to those of the 5 lines were weaker in these methacrylates than in MMA.
The 8 lines became weaker with an increase in the bulkiness of the ester group. For
TPMA, which has a bulky ester group, the 8 lines scarecely appeared, and hence the
spectrum consisted of only 5 lines spaced at 23 G.

The ESR spectra for methacrylate ester groups higher than the carbetoxy methyl
group reveal the exchange broadening of the 8 lines due to the interchange of β-
methylene protons. Therefore, it is reasonable to attribute the 13 line spectrum for
MMA to an average of two stable conformations which convert to each other by free
rotation about the C_α—C_β bond of the growing radical end.

A well-resolved 4-line spectrum shown in Fig. 52 was observed in the polymerization
of VAc in benzene [88]; similar spectra were obtained in other solvents. The 4 lines are

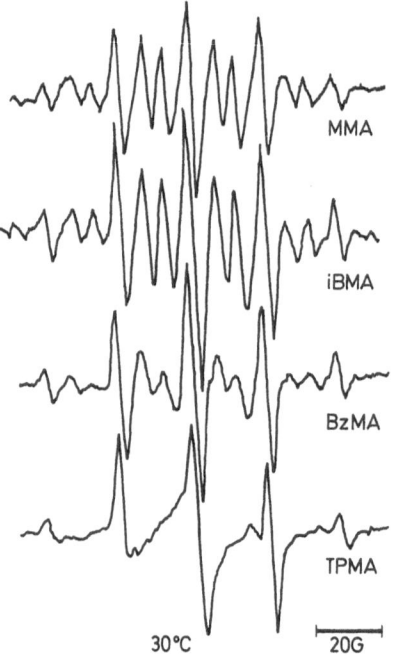

MMA

iBMA

BzMA

TPMA

30 °C |———— 20G ————|

Fig. 51. ESR spectra of propagating radicals of methacrylates in benzene at 30 °C [87]. MMA, 4.26 M; IBMA, 3.11 M; BzMA, 2.92 M, and TPMA, 1.01 M. [BPO] $= 1.0 \times 10^{-2}$ M

spaced at 21 G, with the intensity distribution being $1:3:3:1$. This finding shows that hyperfine splitting constants for two β-methylene protons happen to be the same as that for α-proton. From the fact that hyperfine splitting constant is equal for the two β-methylene protons, free rotation of the propagating radical about the C_α–C_β bond may be concluded.

The ESR spectrum Fig. 53a was obtained in the radical polymerization of iso-propenyl acetate (IpAc), and can be compared favorably with the spectrum of Fig. 53b, which was simulated with a doublet of doublets of quartets due to unequal β-methylene protons and β-methyl protons, in which the hyperfine splitting constant is 15.0 and 13.6 G for the former and 23 G for the latter [89]. Thus, the observed spectrum can be attributed to the propagating radical of IpAc, i.e.,

$$
\begin{array}{c}
CH_3 \\
| \\
\text{------}CH_2\overset{\displaystyle}{C}\bullet \\
| \\
OCOCH_3
\end{array}
\qquad\qquad 7\text{-II}
$$

However, the observed inner 4 lines indicated by arrows in Fig. 53a are somewhat weaker and broader than the corresponding simulated ones. This difference may be ascribed to: (1) an incomplete overlap of the two 4 lines and (2) an exchange broadening of the 4 lines due to the interchange of C—H bonds between the two conformations. As the temperature was lowered from 32.5 to —50 °C, the 4 lines were broadened and their intensities were decreased. This finding implies the exchange broadening to occur in the ESR spectra for methacrylic esters, and hence two stable conformations

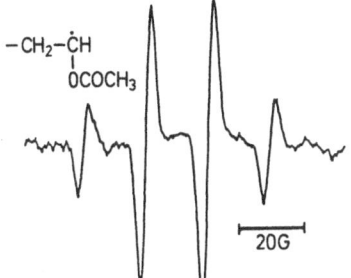

$-CH_2-\overset{\cdot}{C}H$
$\quad\quad OCOCH_3$

Fig. 52. ESR spectrum of poly(VAc) radical in bulk at 32.5 °C [BPO] = 2.35×10^{-2} M

of the radical end convertible by free rotation about the C_α—C_β bond to exist. Conformation analysis evidenced such stable conformations.

Radical bulk polymerization of tert-butyl vinyl ethers performed at 32.5 °C [90] yielded a well-resolved 6-line spectrum shown in Fig. 54a. This spectrum can be assigned to the propagating radical —$CH_2\overset{\cdot}{C}HOC(CH_3)_3$, in which the hyperfine splitting constant is 16.1 G for one proton and 10.6 G for two methylene protons. The well-resolved spectrum shown in Fig. 54b was obtained in a radical polymerization of isobutyl vinyl ether. This is more complicated than that for tert-butyl vinyl ether, reflecting a long-ranged interaction of the unpaired electron with the protons of the substituent of the compound. It is assignable to the propagating radical —$CH_2\overset{\cdot}{C}HO$— —$CH_2CH(CH_3)_2$, in which the hyperfine splitting constant is 16.8, 7.4, and 2.2 G for α-proton, two equivalent β-methylene protons, and two equivalent protons of the side group, respectively. The equal hyperfine splitting constants of the two β-methylene protons indicate the propagating radicals of vinyl ethers to rotate freely about the C_α—C_β bond, as in the case for vinyl acetate and methyl methacrylate.

It has long been considered that a monovinyl ether undergoes no radical polymerization [91], but ESR observation has evidenced the occurrance of a radical reaction of

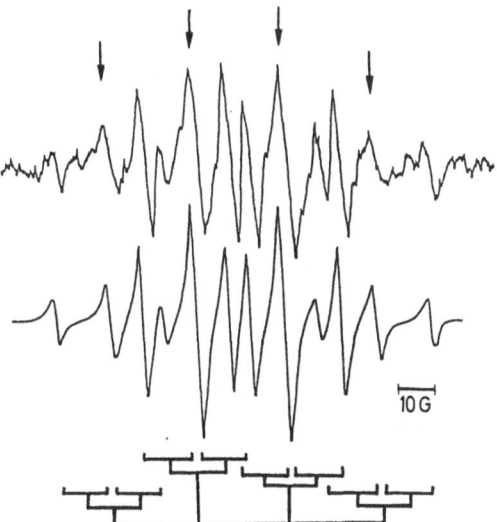

Fig. 53. ESR spectra for a propagating radical of IpAc at 32.5 °C [89]. (a) observed, (b) simulated. [BPO] = 1 $\times 10^{-1}$ M

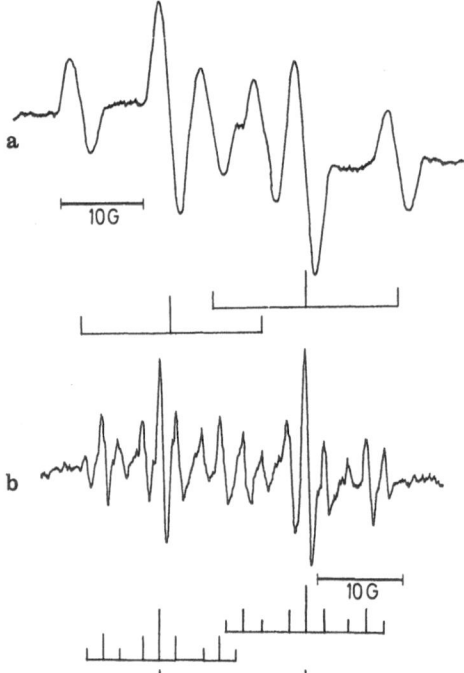

Fig. 54a and b. ESR spectra with TM_{110} cavity of propagating radicals of tert-butyl vinyl ether and isobutyl vinyl ether at room temperature [90]. [TBPO] = 1.3 M. (**a**) tert-butyl vinyl ether, (**b**) isobutyl vinyl ether

the initiating radical with vinyl ether. Recently, Matsumoto et al. [92] found radical polymerization of butyl vinyl ether to an oligomer when a large amount of radical initiator was used.

7.5 Kinetic Studies

Kamachi et al. [87, 88, 89] used the TM_{110} cavity to study the kinetics of peroxide-initiated, UV irradiated radical polymerization of several vinyl monomers. The change in radical concentration with reaction time was determined from that of the height of a resonance line in the ESR spectrum. Generally, the radical concentration leveled off after 20 second irradiation and then remained stationary for a certain period of time which depended on monomer and initiator concentrations. The radical rapidly decayed after the irradiation had been stopped. A typical example of this behavior is shown in Fig. 55 for the bulk polymerization of VAc.

The radical concentration [R˙] was determined from the area under the integrated ESR spectrum by use of the relation calibrated with the spectrum for 4-hydroxy-2,2,6,6-tetramethylpiperidin-1-oxyl of known concentration dissolved in either the monomer or mixtures of monomer and solvent. The polymerization rate —d[M]/dt was determined from the extent of polymerization after a given time. The values obtained for [R˙] and —d[M]/dt were substituted into Eq. (7-1) to evaluate k_p. When [R˙] changed with reaction time, k_p was estimated from the relation

$$\ln \frac{[M]_0}{[M]} = k_p \int [M^\bullet]\, dt \qquad (7\text{-}3)$$

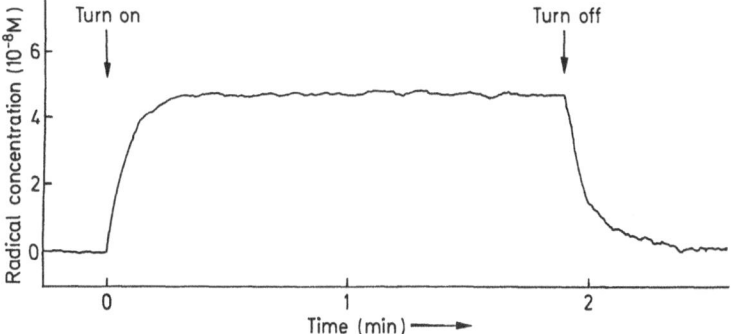

Fig. 55. Time-dependence of propagating radical in radical polymerization of VAc in benzene [88]. [VAc] = 5.4 M, [BPO] = 5.2×10^{-2} M. 32.5 °C

Table 9. Rate constants for VAc at 32.5 °C

[VAc] M	[BPO] mM	Solvent	$R_p \times 10^6$ Ms^{-1}	$[M] \times 10^8$ M	$k_p \times 10^{-2}$ $M^{-1} s^{-1}$	$k_t \times 10^{-8}$ $M^{-1} s^{-1}$
5.41	52.2	Benzene	134.7	4.7	5.3	6.3
5.43	32.3	Benzene	114.0	3.5	6.0	—
5.43	24.9	Ethyl acetate	257.9	2.5	19	2.8
10.9	23.5	—	549.3	3.6	14	—

Rotating sector method. Solvent: benzene, $k_p = 1.2 \times 10^2$ M^{-1} s^{-1} and $k_t = 3.1 \times 10^8$ M^{-1} s^{-1}; Solvent: ethyl acetate, $k_p = 6.37 \times 10^2$ M^{-1} s^{-1} and $k_t = 0.9 \times 10^8$ M^{-1} s^{-1} [34].

The k_p values so obtained for vinyl acetate in several solvents are presented in Table 9 along with those determined by the rotating sector method. The former agree in order-of-magnitude with the latter. The ESR k_p values in bulk and in ethyl acetate solution are about four times as large as that in benzene, which is consistent with the solvent effect on k_p previously found by the rotating sector method [34, 93]. Kamachi et al. [34, 93] found that the solvent effect on k_p for vinyl esters, which have negative e values, is opposite to that for methacrylates, which have positive e values; the k_p values for these monomers were determined in benzene or monosubstituted benzene by the rotating sector method. (see Fig. 56). They explained the solvent effect by assuming that the effective concentration of propagating radicals decreased with formation of charge transfer complexes of some of the radicals with aromatic solvents. To obtain direct information on the complex formation, ESR measurements were made on the radical polymerization of VAc in aromatic solvents. The ESR spectrum obtained in chlorobenzene, shown in Fig. 57, is composed of an overlap of sharp and broad 4 lines with the intensity distribution 1:3:3:1. The former agrees with that obtained in bulk polymerization (see Fig. 52), while the latter shows additional hyperfine splitting due to weak interaction of unpaired electron with several protons. The additional hyperfine splitting suggests the possibility of complex formation of some propagating radicals with aromatic solvents.

The ESR k_p values for methacrylates are compared in Table 10 with those by the

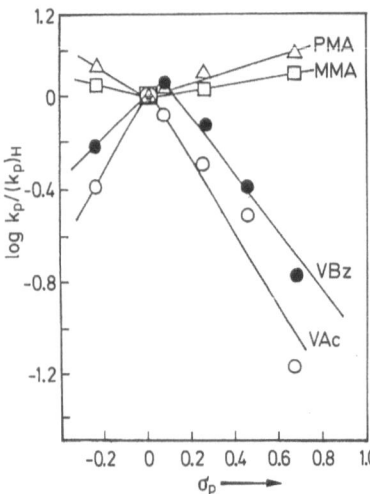

Fig. 56. Hammett's plot of solvent effect on k_p for VAc vinyl benzoate (VBz), MMA, and phenyl methacrylate (PMA) [34]

rotating sector method [93, 94]. For MMA the two methods give consistent results in order-of-magnitude. The k_p for TPMA is smaller than that for MMA, which is probably due to the steric effect of the bulky triphenylmethyl group. The k_p for benzyl methacrylate is about 1000 M^{-1} s^{-1}. Using the rotating sector method, Yokota et al. [95] obtained a k_p value for BzMA which was larger than those for other methacrylates, while Schulz et al. [96] found no significant differences. Kamachi et al.'s k_p value for BzMA is a little larger than those for MMA and TPMA.

Work on the radical polymerization of IpAc did not observe the production of a high-molecular-weight polymer, and this finding was explained in terms of the degra-

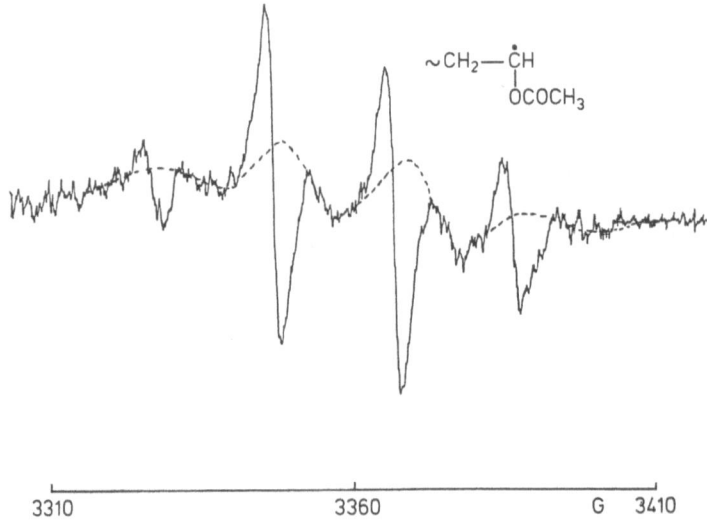

Fig. 57. ESR spectrum of polyVAc radical in chlorobenzene at 32.5 °C; [VAc] = 5.4 M, [BPO] = 6.5×10^{-2} M

Table 10. Kinetic data for methacrylates at 32.5 °C

Monomer	Radical Concentrations $\times 10^7$ M	$R_p \times 10^5$ Ms^{-1}	k_p M^{-1} s^{-1}	$k_t \times 10^{-5}$ M^{-1} s^{-1}
MMA	2.7	21.5	187 (450[a])	721 (556[a])
TPMA	10.5	2.73	26	3.01
BzMA no. 1	3.34	221	110 (1250[b]) (510[c])	800 (419[b]) (289[c])
no. 2	3.01	70.2	895	400

[a] Data from Ref. [34]; [b] Data from Ref. [95]; [c] Data from Ref. [96]

dative chain transfer due to the hydrogen abstraction of the propagating radical from the α-methyl group of the monomer. Bywater et al. [97], however, reported the formation of a high polymer in the polymerization of IpAc under high pressure. Thus, it is likely that under atmospheric pressure the formation of a high-molecular-weight polymer is greatly retarded by a slow propagation coupled with an effective depropagation. The determination of k_p for IpAc by the conventional method still remains untouched for the reason that the polymerization rate does not follow the usual kinetic equation for radical polymerization, i.e.,

$$R_p = k_p(fk_d/k_t)^{0.5} [M] [I]^{0.5} \tag{7-4}$$

in which I is the initiator concentration, k_d the rate constant for initiator dissociation, and f the initiator efficiency. However, the k_p for IpAc was succesfully determined to be 280 M^{-1} s^{-1} by ESR, which is somewhat larger than that for MMA [87] and about one fifth that for VAc [88]. This k_p value would be large enough to yield a high-molecular-weight polymer if no chain transfer reaction took place. Since this was not the case in actuality, it may be concluded that the formation of a high polymer from IpAc is hampered by degradative chain transfer.

The termination rate constant k_t can in principle be determined by analyzing the decay in radical concentration after irradiation has been stopped. However, this decay

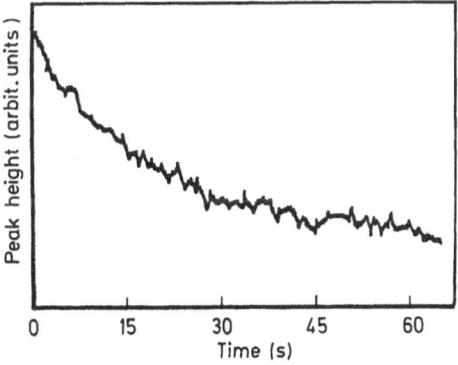

Fig. 58. Decay curve of poly(TPMA) radical after irradiation is off [87]. [BPO] = 4.1 $\times 10^{-2}$ M, [TPMA] = 1.01 M, solvent: benzene

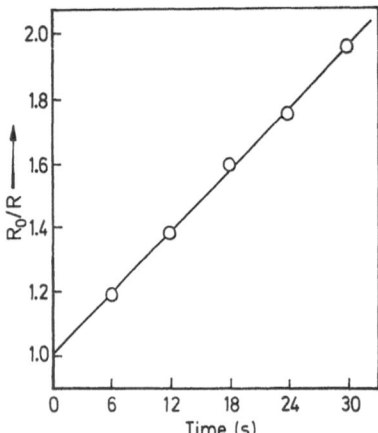

Fig. 59. Second order plot of peak height after irradiation is off [87]

is usually too fast to be measured by the conventional 100 KHz magnetic-field modulation method. Therefore, Eq. (7-2) was used. The resulting k_t values for VAc, BzMA and MMA are given in Tables 9 and 10.

For TPMA the radical decay was so slow that it was followed by a change in the peak height of the central line of an ESR spectrum. Figure 58 shows the result obtained, and its second-order plot (Fig. 59) gives 3.01×10^5 M^{-1} s^{-1} for k_t. It is reasonable to ascribe this k_t value, significantly small compared with k_t for MMA and BzMA, to the bulkiness of the side group, which interferes the termination reaction.

7.6 ESR Observation of Active Species by Other Cavity

When the stationary concentration in the radical polymerization of vinyl compounds is larger than 10^{-6} M, ESR measurement of growing radicals does not require use of the special cavities described in Sect. 7.2. In fact, Kamachi et al. [32] were able to observe, even at room temperature, a well-resolved 5-line spectrum in BPO-initiated, UV-irradiated polymerization of TPMA by use of the commercially available TE_{011}

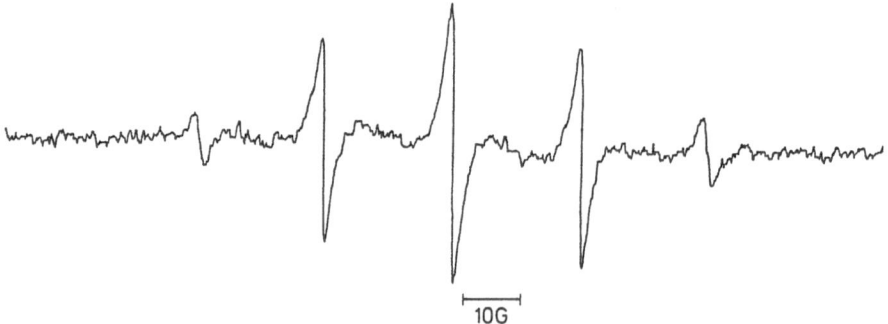

10G

Fig. 60. ESR spectrum of poly(diphenylmethyl methacrylate) radical at 20 °C [101]. [Monomer] = 4.0 M, [BPO] = 1.0×10^{-2} M, solvent: benzene

cavity (see Fig. 48 in Chapt. 7). Figure 60 shows the ESR spectrum observed in a similar way for the radical polymerization of diphenylpyridylmethyl methacrylate. Yamada et al. [100] observed ESR spectra for propagating radicals produced by TBPO-initiated polymerizations of ortho-substituted phenyl methacrylates in MTHF over the temperature range from -15 to -53 °C. All these observations may be accounted for by an increase in the stationary concentration due to a lowering in k_t which results from a steric repulsion of the bulky ester group.

Recently, Otsu et al. [98,99] found that some radical polymerization of alkyl fumarates yielded high polymers and that the polymerization rate increased as the ester group became bulkier. Although this finding is in contrast to the general concept that high polymers are difficult to be obtained by radical polymerization of 1,2-di-substituted olefins, Otsu et al. [99] explained it as due to the fact that the termination reaction is more suppressed than the propagation reaction by steric hindrance of the bulky ester group. Their explanation was confirmed by an ESR observation as well as by the determination of the lifetimes of the propagating radicals of the fumarates.

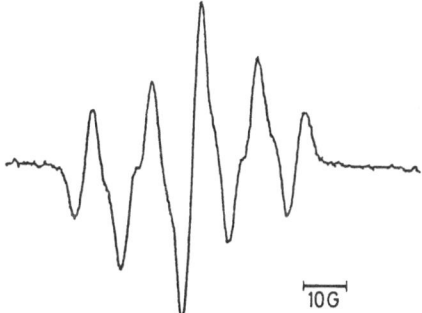

10 G

Fig. 61. ESR spectrum of propagating radical of DBI in bulk polymerization [101]. [BPO] = 0.05 M

Interestingly, itaconate esters can be polymerized at a moderate rate to yield high-molecular-weight polymers in spite of their possession of two bulky substituents. Kamachi et al. [101] observed a 5-line ESR spectrum with small shoulders as shown in Fig. 61 when di-butyl itaconate (DBI) was photopolymerized in bulk with BP at low temperature. They computer-simulated a similar spectrum by assuming the hyperfine splitting constant to be 14.1 G for two β-protons and 10.1 G for another two β-protons, and concluded that the observed spectrum can be attributed to

$$\begin{array}{c}\text{CH}_2\text{---COO-}n\text{-Bu}\\|\\\text{------CH}_2\text{---C}\cdot\\|\\\text{COO-}n\text{-Bu}\end{array}\qquad\qquad\text{7-III}$$

The propagating radical of dialkyl itaconate in the homogeneous polymerization was found by Sato et al. [102] to be stable enough to be observed by ESR even above 60 °C. Fig. 62a displays the 5-line spectra obtained when dimethyl 2,2′-azobis-isobutyrate, benzoylperoxide, and di-tert-butylperoxide were used as initiators. On the other hand, the polymerization of DBI initiated with azonitriles such as 2,2′-azobisisobutyronitrile, 2,2′-azobis(2,4-dimethylvaleronitrile), and cyclohexanecarbo-

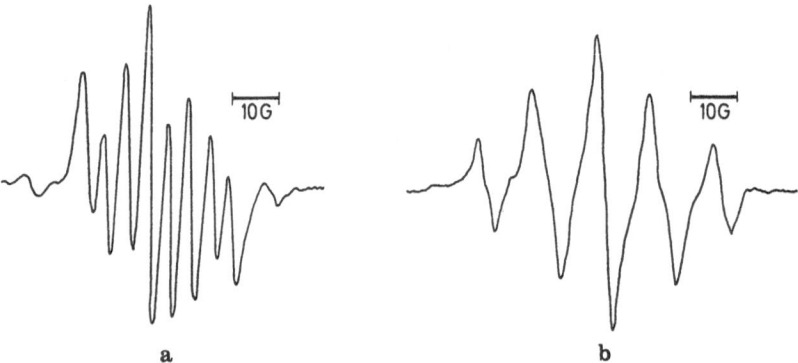

Fig. 62a and b. ESR spectra for (**a**) DBI/MAIB and (**b**) DBI/AVN systems at 61 °C [102]; (**a**) [DBI] = 1.19 M, [MAIB] = 2.90×10^{-2} M, (**b**) [DBI] = 1.14 M, [AVN] = 5.75×10^{-2} M

nitrile, yielded a 8-line spectrum overlapping over the 5-line spectrum, as shown in Fig. 62b. The intensity of the 8-line spectrum relative to the 5-line one increased at lower monomer concentration and higher temperature, as can be seen from Fig. 63. The temperature dependence of the relative intensity of the two spectra was different for different azonitrile initiator: the bulkier the initiating radical, the stronger the relative intensity of the 8-lines became when compared at the same temperature. The 8-line spectrum was explained as due to the hyperfine interaction of 4.5 G of the unpaired electron with one proton and that of 9.5 G with three protons. Sato et al. [102)]

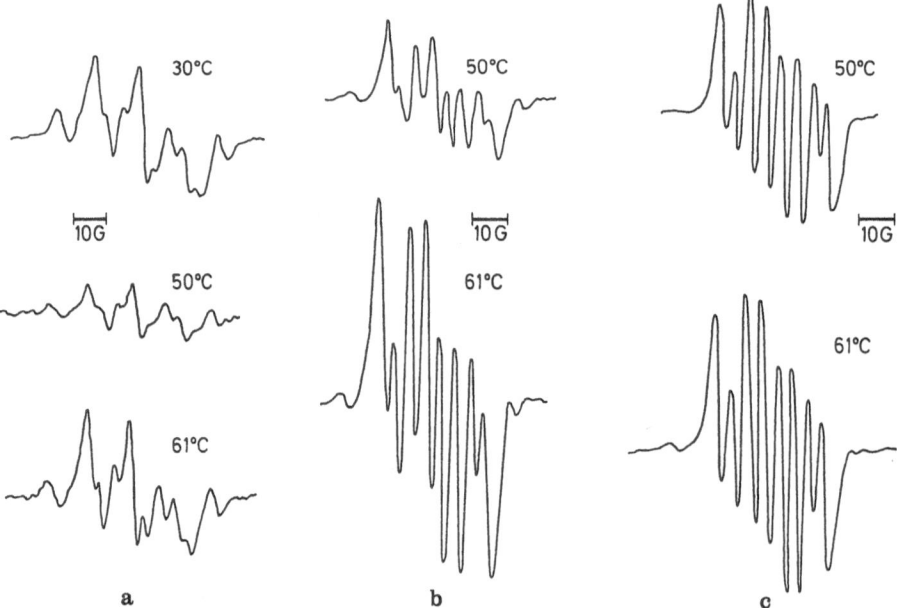

Fig. 63a–c. ESR spectrum change in polymerization of DBI with AIBN in benzene [102]; (**a**) [DBI] = 3.97 M (in bulk) (the spectrum at 30 °C, was observed under irradiation by a Xe-lamp (1 kw)), (**b**) [DBI] = 1.99 M, (**c**) [DBI] = 0.397 M; [AIBN] = 5.00×10^{-2} M

assumed two kinds of radicals: one is a growing chain radical 7-III which gives 5 lines, and the other a primary propagating radical as

$$CH_3-\underset{\underset{CN}{|}}{\overset{\overset{CH_3}{|}}{C}}-CH_2-\underset{\underset{COO-n\text{-}Bu}{|}}{\overset{\overset{CH_2-COO-n\text{-}Bu}{|}}{C}}\cdot \qquad\qquad 7\text{-IV}$$

$$CH_3-\underset{\underset{CH_3}{|}}{CH}-CH_2-\underset{\underset{CN}{|}}{\overset{\overset{CH_3}{|}}{C}}-CH_2-\underset{\underset{COO-n\text{-}Bu}{|}}{\overset{\overset{CH_2-COO-n\text{-}Bu}{|}}{C}}\cdot \qquad\qquad 7\text{-V}$$

$$\text{(cyclohexyl)}-CH_2-\underset{\underset{COO-n\text{-}Bu}{|}}{\overset{\overset{CH_2-COO-n\text{-}Bu}{|}}{\underset{CN}{C}}}\cdot \qquad\qquad 7\text{-VI}$$

The reason why the primary propagating radicals produced by azonitrile initiators have different stability is not yet clear, but it is of great interest in regard to the pronounced penultimate effect observed in the copolymerization of AN.

When the initiator concentration was about 100 times as large as those used for radical polymerization of vinyl compounds, the well resolved ESR spectra shown in Fig. 64 were observed at $-50\,°C$ in photopolymerization of vinyl ethers in bulk with TBPO [90]. Although a large amount of initiator (0.87–1.31 M) was used, no spectrum of the tert-butoxyl radical was found in the presence of vinyl ethers. Since the ESR signal to the tert-butoxyl radical appeared in the polymerization of monomers such as VAc, MMA, and BIA even at lower initiator concentrations, this finding indicates that the addition reactions of the tert-butoxyl radical to vinyl ethers are more likely to occur than predicted by the fact that high polymers are difficult to be obtained by radical polymerization of vinyl ethers. The olefinic bonds of vinyl ethers readily react with the tert-butoxyl radical to yield propagating radicals. Therefore, a polymer could be produced from vinyl ethers by radical polymerization if an optimum polymerization condition were used. Matsumoto et al. [92] found that polymerization behavior of butyl vinyl ether is similar to that of allylacetate, which has a monomer transfer constant 10^2–10^3 times larger than VAc, styrene, and MMA. This similarity suggests that the chain transfer reaction is more likely to occur in vinyl ethers than in other polymerizable vinyl monomers such as MMA and VAc. According to Kamachi et al. [90], the hyperfine splitting constants for α-protons of vinyl ethers are smaller than those for ethyl acrylate and VAc (22.2 G and 20.5 G, respectively). The following two possibilities may be invoked to account for this difference: (i) a decrease in the spin density of the unpaired electron and (ii) an increase in the s character of the propagating radical. It has been shown by an Ab initio molecular orbital calculation that the spin densities in vinyl ether radicals are higher than in the radical of MA [103]. This rules out the possibility (i). The possibility (ii) is consistent with the Ab-initio calculation which indicated the conformation of the propagating radical deviated from the planar sp^2 conformation [104]. Probably, no high polymer can be produced from vinyl ethers because of chain transfer reactions such as hydrogen abstraction of the propagating radical from monomer and solvent; note that a carbon-centered

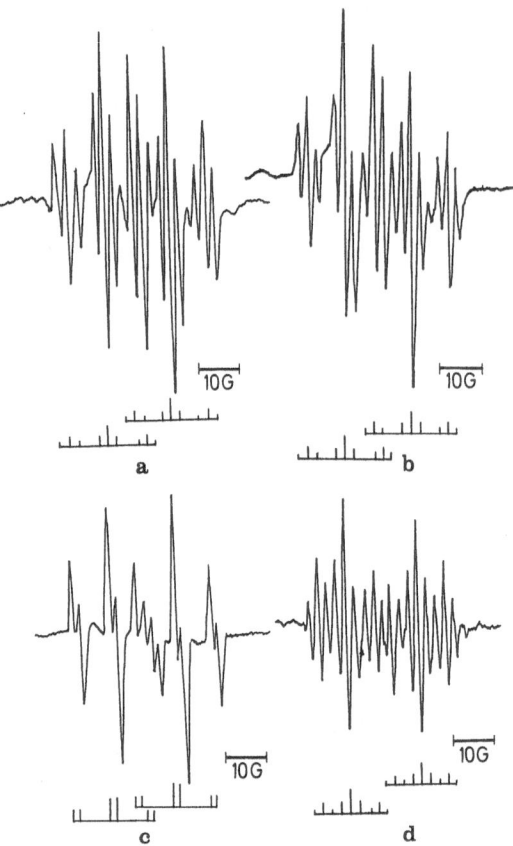

Fig. 64a–d. ESR spectra for propagating radicals of various vinyl ethers at −50 °C [90]. [DBPO] = 0.87 M. (**a**) ethyl vinyl ether; (**b**) butyl vinyl ether; (**c**) isopropyl vinyl ether; (**d**) isobutyl vinyl ether

radical deviated from the sp^2 conformation should have a reactivity higher than one having the sp^2 conformation.

8 Concluding Remarks

In this article, I have summarized applications of ESR spectroscopy to the radical polymerization of vinyl and diene compounds, with particular emphasis put on the advantages of the special cavities designed for enhancing the sensitivity of ESR spectrometers. These cavities allow determinations of the conformations of propagating radicals as well as the propagation rate constants for several monomers under conditions similar to usual radical polymerizations. However, I am afraid that the reliability of the kinetic data obtained with them is not yet high enough for precise determination of the rate constants because of experimental errors involved in the measurement of radical concentrations and polymerization rates. It is desirable that the ESR spectrometer be made about 10 times more sensitive for really reliable data to be obtained. I have added this remarks because some misunderstanding prevails about the current ESR spectrometers which are still far from sufficient in sensitivity.

9 Acknowledgement

I wish to express my sincere appreciation to Professor Shun-ichi Nozakura for his frequent discussions and encouragement. I also wish to acknowledge with many thanks the cooperation of a number of associates who took part in ESR studies of propagating radicals in the radical polymerization of vinyl and diene monomers. Finally, I am deeply indebted to Professor Hiroshi Fujita for his giving me a chance of writing this review and his valuable comments.

10 References

1. Zavoiskii, E. K.: J. Phys. USSR, 9, 211 (1945).
2. Schneider, E. E., Day, M. J., and Stein, G.: Nature 168, 645 (1951).
3. Ranby, B. and Rabek, J. F.: ESR Spectroscopy in Polymer Research, Springer Verlag, Berlin, 1977, pp. 100–172.
4. Carrington, A., and Mclachlan, A. D.: Introduction to Magnetic Resonance, Harper and Row, New York, 1972.
5. Wertz, J. E. and Bolton, J. R.: Electron Spin Resonance, McGraw-Hill, New York, 1972.
6. Atherton, N. M.: Electron Spin Resonance, Ellise-Horwood, London, 1973.
7. Poole, Jr., C. P.: Electron Spin Resonance, 2nd Ed., Interscience, 1982.
8. Ref. 3. p. 34.
9. Livingston, R. and Zeldes, H.: J. Chem. Phys., 44, 1245 (1966).
10. Fischer, H.: Z. Naturforschg., 19a, 866 (1964).
11. Fessenden, R. W. and Schuler, R. H.: J. Chem. Phys., 39, 2147 (1963)
12. McConnell, H. M.: J. Chem. Phys., 24, 764 (1956).
13. Heller, C. and McConnell, H. M.: J. Chem. Phys., 32, 1535 (1960).
14. Sullivan, P. D. and Bolton, J. R.: Avance in Magnetic Resonance, Vol. 4., Ed. J. S. Waugh, New York, Academic press, 1965, pp. 39–85.
15. Bresler, S. E., Kazbekov, E. N., and Saminskii, E. M.: Vysokomol. Soedin., 1, 132 (1959).
16. Bresler, S. E., Kazbekov, E. N., and Saminskii, E. M.: Vysokomol. Soedin., 1, 1374 (1959).
17. Bresler, S. E., Kazbekov, E. N., and Saminskii, E. M.: J. Polym. Sci., 52, 119 (1961).
18. Abraham, R. J., Melville, H. W., Ovenall, D. W., and Whiffen, D. H.: Trans. Faraday Soc., 54, 1133 (1958).
19. Ingram, D. J. E., Symons, M. C. R., and Townsend, M. G.: Trans. Faraday Soc., 54, 409 (1958).
20. Bowden, M. J. and Sutcliffe, L. H.: Trans. Faraday Soc., 60, 625 (1964).
21. Kourim, P. and Vacek, K.: Trans. Faraday Soc., 61, 415 (1965).
22. Symons, M. C. R.: J. Chem. Soc., 1963, 1186.
23. Piette, L. H.: NMR and EPR Spectroscopy. Chemical Application of EPR, Oxford: Pergamon Press, 1960, 218.
24. Sohma, J., Komatsu, T., and Kashiwabara, H.: J. Polym. Sci., B3, 287 (1965).
25. Harris, J. A., Hinojosa, O., and Author, Jr., J. C.: J. Polym. Sci., Polym. Chem. Ed., 11, 3215 (1973)
26. O'Donnell, J. H., McGarvey, B., and Morawetz, H.: J. Am. Chem. Soc., 86, 2322 (1964)
27. Bowden, M. J. and O'Donnell, J. H.: J. Phys. Chem., 72, 1577 (1968).
28. Bamford, C. H., Bibby, A., Eastmond, G. C.: Polymer 9, 626, 645, 653 (1968).
29. Sakai, Y. and Iwasaki, M.: J. Polym. Sci., part A-1, 7, 1749 (1969).
30. Iwasaki, M. and Sakai, Y.: J. Polym. Sci., part A-1, 7, 1537 (1969).
31. Kamachi, M., Kohno, M., Liaw, D. J., and Katsuki, Y.: Polym. J., 10, 69 (1978).
32. Kamachi, M., Kuwae, Y., and Nozakura, S.: Polym. J., 13, 919 (1981).

33. Kamachi, M., Liaw, D. J., and Nozakura, S.: Polym. J., *13*, 41 (1981).
34. Kamachi, M.: Adv. Polym. Sci., *38*, 55 (1981).
35. Shioji, Y., Ohnishi, S., and Nitta, I.: J. Polym. Sci., A *1*, 3373 (1963).
36. Morawetz, R. and Fadner, T. A.: Makromol. Chem., *162*.
37. Fadner, T. A. and Morawetz, R.: J. Polym. Sci., *45*, 475 (1960).
38. Adler, G. and Petropoulos, J. H.: J. Phys. Chem., *69*, 371 (1965).
39. Harris, J. A., Hinojosa, O., and Author, Jr., J. C.: J. Polym. Sci., *12*, 679 (1974).
40. Hirai, H. and Fujiwara, M.: Nippon Kagaku-kaishi, 1972, 968.
41. Kamachi, M., Umetani, H., Kuwae, Y., and Nozakura, S.: Polym. J., *15*, 753 (1983).
42. Shiga, T., Lund, A., and Kinell, P. O.: Int. Radiat. Chem., *3*, 145 (1971).
43. Kamachi, M., Umetani, H., and Nozakura, S.: Polym. J., *18*, 211 (1986).
44. Gibbs, W. E., Barton, J. M.: Vinyl Polymerization, Vol. 1, Part 1, Ham, G. E., Ed., Dekker, New York, 1967, Ch. 2.
44. Butler, G. B. and Myers, G. R.: J. Makromol. Sci., A*5*, 135 (1970).
45. Kodaira, T. and Morishita, K.: J. Polym. Sci., Polym. Lett. Ed., *11*, 347 (1973).
46. Kodaira, T., Taniguchi, M., and Sakai, M.: ACS Monograph 197, p. 107 (1982).
47. Kodaira, T. and Sumiya, Y.: Makromol. Chem., *187*, (1986).
48. Kodaira, T. and Butler, G. B.: J. Macromol. Sci., A*22*, 213 (1985).
50. Dixon, W. T. and Norman, R. O. C.: J. Chem. Soc., 1963, 3119.
51. Fischer, H.: J. Polym. Sci., B *2*, 529 (1964).
52. Fischer, H.: Z. Naturforschg. *19*a, 267, 866 (1964).
53. Fischer, H.: Kolloid-Z. u. Z. Polym., *206*, 131 (1965).
54. Fischer, H.: Makromol. Chem., *98*, 179 (1966).
55. Fischer, H.: Ad. Polym. Sci., *5*, 531 (1968).
56. Fischer, H. and Giacommetti, G.: J. Polym. Sci., C*16*, 2763 (1967).
57. Fischer, H.: Polymer Spectroscopy, Hummel, D. O. Ed., Verlag Chemie, 1974, p. 289.
58. Yoshida, H. and Ranby, B.: J. Polym. Sci., C*16*, 1333 (1967).
59. Takakura, K. and Ranby, B.: J. Polym. Sci., B*5*, 83 (1967).
60. Takakura, K. and Ranby, B.: J. Polym. Sci., C*22*, 939 (1969).
61. Takakura, K. and Ranby, B.: J. Polym. Sci., A-1, *8*, 77 (1970).
62. Smith, P. and Stevens, R. D.: J. Phys. Chem., *76*, 3141 (1972).
63. Smith, P., Gilman, L. B., Stevens, R. D., and Vignola de Hargrave, C.: J. Mag. Resonance *29*, 545 (1978).
64. Smith, P., House, D. W., and Gilman, L. B.: J. Phys. Chem., *77*, 2249 (1973).
65. Smith, P., Stevens, R. D., and Gilman, L. B., J. Phys. Chem., *79*, 2688 (1975).
66. Shiraishi, H. and Ranby, B.: Chemica Scripta *12*, 118 (1977).
67. Beckwith, A. L. J.: J. Macromol. Sci., -Chem., A*9*, 115 (1975).
68. Beckwith, A. L. J., Ong, A. K., and Solomon, D. H.: J. Macromol. Sci., A*9*, 125 (1975).
69. Howthrone, D. G. and Solomon, D. H.: J. Macromol. Sci., -Chem., A*9*, 149 (1973).
70. Johns, S. R. and Willing, R. L.: J. Macromol. Sci., -Chem., A*9*, 169 (1973).
71. Korus, O. and O'Driscoll, K. F.: Polymer Handbook, 2nd Ed., Brandrup, J., and Immergut, E. H. Eds., John Wiley, New York, N.Y., 1975, p. II-45.
72. Janzen, E. G.: Accounts Chem. Res.. *4*, 31 (1971).
73. Chalfont, G. R., Perkins, M. J., and Horsefield, A.: J. Am. Chem. Soc., *90*, 714 (1968).
74. Kunitake, T. and Murakami, S.: Polym. J., *3*, 249 (1972).
75. Kunitake, T. and Murakami, S.: J. Polym. Sci., Polym. Chem. Ed., *12*, 67 (1974).
76. Kamachi, M., Kuwae, Y., and Nozakura, S.: Polym. Bull., *6*, 143 (1981).
77. Lane, J. and Tabner, J. B.: J. Chem. Soc., Perkin Trans. 2, 1985, 1665.
78. Sato, T. and Otsu, T.: Makromol. Chem., *178*, 1941 (1977).
79. Sato, T. and Otsu, T.: Polymer *11*, 389 (1975).
80. Bevington, J. C.: Radical Polymerization, Academic Press, London, 1961.
81. Sato, T., Hibino, K., and Otsu, T.: Chem. Ind., 1973, 745.
82. Sato, T., Kita, S., and Otsu, T.: Makromol. Chem., *176*, 561 (1975).
83. Ko, M., Sato, T., and Otsu, T.: Makromol. Chem., *176*, 643 (1975).
84. Bresler, S. E., Kazbekov, E. N., Fomichev, V. N., and Shadrin, V. N.: Makromol. Chem., *157*, 167 (1972).
85. Bresler, S. E., Kazbekov, E. N., and Shadrin, V. N.: Makromol. Chem., *175*, 2875 (1974).

86. Bresler, S. E., Kazbekov, E. N., and Shadrin, V. N.: Vysokomol. Soyed., A*17*, 507 (1975).
87. Kamachi, M., Kuwae, Y., Kohno, M., and Nozakura, S.: Polym. J., *14*, 749 (1982).
88. Kamachi, M., Kuwae, Y., and Nozakura, S.: Polym. J., *17*, 541 (1985).
89. Kuwae, Y., Kamachi, M., and Nozakura, S.: Macromolecules, in press.
90. Kamachi, M. and Tanaka, K.: J. Polym. Sci., Polym. Chem. Ed., in press.
91. Odian, G.: Principles of Polymerization, 2nd Ed., Wiley-Interscience, New York, 1981, p. 181.
92. Matsumoto, A., Iwanami, K., and Oiwa, M.: Makromol. Chem., Rapid Comun., *4*, 227 (1983).
93. Kamachi, M., Liaw, D. J., and Nozakura, S.: Polym. J., *11*, 921 (1979).
94. Kamachi, M., Liaw, D. J., and Nozakura, S.: Polym. J., *13*, 41 (1981).
95. Yokota, K., Kani, M., and Isahi, Y.: J. Polym. Sci., A-1, *6*, (1968)
96. Mayer, G. and Schulz, G. V.: Makromol. Chem., *173*, 101 (1973).
97. Bywater, S. and Whalley, E.: ACS Abstracts of Papers, 13th meeting 10S, 22 (1960).
98. Otsu, T. and Toyoda, N.: Makromol. Chem., Rapid Comm., *2*, 79 (1981).
99. Otsu, T., Yamada, B., Yoshikawa, E., and Miura, H.: Polym. Preprint Japan, *35*, 125 (1986).
100. Yamada, B., Matsumoto, A., and Otsu, T.: J. Polym. Chem., Polym. Chem. Ed., *21*, 2241 (1983).
101. Kamachi, M., Tanaka, K., and Nozakura, S.: Unpublished data.
102. Sato, T., Inui, S., Tanaka, H., Ota, T., and Kamachi, M.: J. Polym. Sci., Polym. Chem. Ed., to be published.
103. Imoto, M., Sakai, S., and Ouchi, T.: J. Chem. Soc. Jpn., 1985, 9.
104. Fueno, T. and Kamachi, M.: J. Am. Chem. Soc., to be published.

Editor: H. Fujita
Received June 10, 1986

Author Index Volumes 1–82

Subject Index